MONONCHIDA

The Predaceous Nematodes

MONONCHIDA
The Predaceous Nematodes

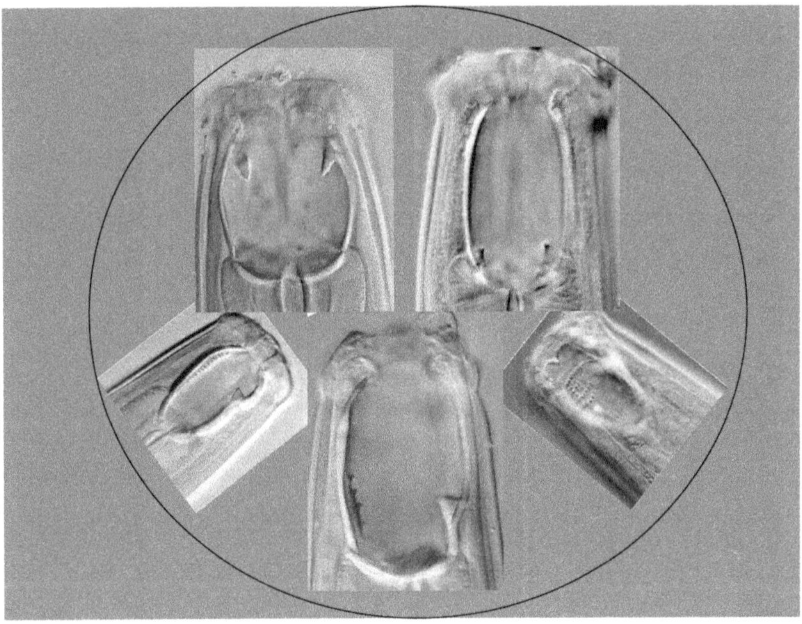

Wasim Ahmad and M. Shamim Jairajpuri

David J. Hunt and Roland N. Perry (Series Editors)

NEMATOLOGY MONOGRAPHS AND PERSPECTIVES
VOLUME 7

BRILL
LEIDEN-BOSTON
2010

This book is printed on acid-free paper.

Library of Congress Cataloging-in-Publication Data

Ahmad, Wasim.

Mononchida: the predatory soil Nematodes / Wasim Ahmad and M. Shamim Jairajpuri.

p. cm. – (Nematology monographs and perspectives; v. 7)

Includes bibliographical references and index.

ISBN 978-90-04-17464-1 (hardback: alk. paper) 1. Mononchida. I. Jairajpuri, M. Shamim (Mohammad Shamim), 1942- II. Title. III. Series.

QL391.N4A354 2010

592'.57–dc22

2010004741

ISBN: 978 90 04 17464 1

Dedicated to Prof. N.A. Cobb

Photograph by courtesy of the Nematology Laboratory, U.S. Department of Agriculture, Agricultural Research Service.

Contents

Nematology Monographs & Perspectives, 2010, Vol. 7, xv-xviii

Foreword

The Mononchida, commonly referred to as mononchs, are very well known to everyone who works on nematodes or who is interested in these small, yet very significant, organisms. They know very well that the mononchs are predatory animals, veritable "mini-tigers", as they have appropriately been called by the authors of the present book. They are found in a wide variety of terrestrial and aquatic habitats. Being comparatively large nematodes with a very distinct and well-armed mouth cavity, they are recognisable at first glance, although a great number of genera and their widely diversified species often cause difficulties in identification and classification.

Since the classical and detailed study of the group by Cobb (1917), the 'Father' of Nematology, more and more scientists have studied the mononchs and described many new taxa, particular new genera and species. However, a comprehensive review of the whole group has long been awaited. Well, Prof. W. Ahmad and Prof. M.S. Jairajpuri have presented us with a grandiose book on Mononchida that includes the history, morphology, ecology, significance and, as far as possible, the phylogeny of this very characteristic group of nematodes. Both the authors are worldwide renowned experts of the science of nematology. They have been studying nematodes, including the predatory groups, for several decades, and have published a very large number of extremely valuable papers on the theme. Moreover, Prof. Jairajpuri, together with the (late) Dr Wajih U. Khan, had already presented to our science a fine book on the Mononchida of India way back in 1982.

The present book, written with the greatest possible care by the best specialists of the group, constitutes a very helpful foundation, a true *vademecum* for any nematologist intending to work on mononchs. The work produced by the authors is an updated treasure-chest of all that the science can present at this point of time, and we, the soldiers of this science, would like to know about Mononchida. Perhaps it would have been much easier for the authors to make a compilation based only on the literature, but the true value of this book lies in the fact that it is based on several years of painstaking observations and studies by the two authors from the great seat of learning, the Aligarh Muslim University, so famous for the research in the field of nematology. It is, indeed, the result of the personal expertise of the two authors on this rather large group of nematodes that are widely found all over the world and are rightly considered great friends of man in agriculture.

The book is likely to be of great value to all nematologists, but for those who intend working on Mononchida, it will be indispensable. I am sure it will stimulate further research on mononchs for academic or maybe for practical reasons due to these being efficient predators on a wide variety of soil-inhabiting organisms, including plant-parasitic species of nematodes. Profs Ahmad and Jairajpuri have to be complimented for having produced such an excellent, invaluable and standard contribution which is a milestone indeed to the science of nematology.

István Andrássy
Department of Systematic Zoology & Ecology
Eötvös Lorand University
Budapest, Hungary

Foreword

Nematologists interested in the role of nematodes in nutrient cycling in soil, and understanding the sustainable management of nematode populations, are well aware of the importance of a taxonomic understanding of the organisms involved. Professors Ahmad and Jairajpuri bring their specialist understanding to produce an authoritative, up-to-date account of the phylogeny, morphology and taxonomy of the Mononchida, and the importance of members of the Order as predators of a range of organisms, especially nematodes in soil. The monograph brings together diverse and new information and continues the high standards of this series of publications.

As with most other animals, nematode populations are regulated by the availability of food resources (bottom-up processes) and by parasitism, competition and predation from other organisms in soil (top-down processes). The authors note that in a number of countries research on the use of members of the Mononchida as potential biological control agents has provided variable results, and there is a dearth of information on the ecology and biology of this important nematode Order. In general, predatory nematodes are more abundant in natural, undisturbed soils than in cultivated soils, but the importance of the Mononchida in the regulation of nematode pest populations is not proven and awaits more detailed research. At its simplest level, research on biological control is littered with examples where the lack of consistent responses has been because different researchers in different parts of the world have been mistakenly working on different natural enemies. For example, predators considered to be generalists may exist as subspecies which have differences in their host preference and/or behaviour that profoundly affect their impact on nematode populations. Understanding the role of the Mononchida in the biodiversity and resilience of nematode communities will require a sound taxonomic understanding of the members of the Order, based on morphological and molecular approaches. The authors are to be applauded for producing an excellent, timely monograph, which makes a major contribution to this understanding during a period of renewed interest in the ecology and management of soils. It will provide an important reference source for nematologists and soil ecologists.

Brian Kerry MBE
Nematode Interactions Unit
Rothamsted Research
Harpenden, UK

Preface

Nematodes are one of the most abundant, diverse, dynamic, ecologically and biologically significant groups of organisms. Unlike insects, they by and large always remain hidden in the niche they occupy. Many biologists thereby call them "cryptic animals", whilst plant nematologists refer to them as the "unseen enemies". They may be free-living in marine and freshwater, soil-inhabiting and, as parasites, feed on all kinds of higher and lower plants or live as parasites in all animal groups above the level of Metazoa. Due to their varied habits and habitats, it was natural for them to have different kinds of feeding habits ranging from microbivorous, saprophytic, phyto- or zooparasitic and predatory. As a result of their varied habits and habitats, these animals have undergone tremendous changes in their structures and life styles.

Predatory nematodes occur in many groups of nematodes, in fact, in all groups of nematodes, but members of one of these groups, namely those belonging to the Order Mononchida, are exclusively predaceous. The predaceous nematodes, although microscopic in size, possess varied kinds of feeding apparatus and, consequently, feed on very different and diverse kinds of prey. Literally, they are the miniature tigers, lions, cheetahs, panthers, wolves, jackals, foxes, *etc*., of the world they inhabit. Should they have been large, say of the same size as the beasts dwelling on the land, they would indeed have looked more ferocious than any of the predators mentioned above. Many of them possess a tooth or teeth and can tear or bite just as ferociously and can pounce upon their prey with similar ferocity to the tigers, lions or cheetahs. Their food/diet varies vastly and, in addition to other nematodes, they can also relish many other types of organisms. Not much attention was paid to the mononchs until the beginning of the 20th century. Whether these could be of much biological significance in controlling other soil-inhabiting groups of microorganisms, including plant-parasitic nematodes, is still highly debatable. However, there is no doubt that they reduce populations of other kinds of nematodes, irrespective of whether these are plant-parasitic, free-living or other types of parasite. Their stoma or mouth cavity is the most significant part of their body and much of their feeding ability depends on the structure of this organ. A lot still remains to be done in order to determine exactly their feeding habits and behaviour, as well as their overall significance as a group of organisms. We hope that our monographic study of this group will help to stimulate further work on various aspects of mononchs by other nematologists of the world.

Chapter 1 presents a detailed account of the morphology of mononchs, with particular emphasis on the structures that are of taxonomic importance, *i.e.*, buccal cavity (stoma), pharynx, pharyngo-intestinal junction, reproductive organs, tail, *etc.* This information should enable students, as well as researchers, to understand and follow with ease the systematics of the group. Each family forms a chapter in the book, and the historical background of the group, as well as diagnoses of the family, subfamilies and various genera, is given. Keys to the various subfamilies and genera have been incorporated. A list of species, together with their synonyms, description of the type species and key to species under each genus, is provided. Line drawings giving important taxonomic characters of the type species of each genus and, as far as possible, photomicrographs of either type species or a closely related species have been provided. The illustrations have been borrowed from many sources/authors, all of which are mentioned in the appropriate places in the text. Most of the photomicrographs are original, although some have been loaned by courtesy of Dr Jiménez-Guirado, a generosity for which we are highly grateful.

We are grateful to Prof. István Andrássy and Prof. Brian Kerry for agreeing to write the forewords for the book and to Dr David Hunt for suggestions during the compilation of this monograph. The help rendered by the members of the Nematology Research Laboratory of Aligarh Muslim University is also greatly acknowledged.

Wasim Ahmad and M. Shamim Jairajpuri
Aligarh Muslim University
Aligarh, India

Nematology Monographs & Perspectives, 2010, Vol. 7, 1-19

Chapter 1

Introduction

Mononchs: microscopic 'tigers' of the soil

The biological diversity that sustains all living organisms on this planet occurs on land, in the air and in water. Mankind has explored, as well as exploited, a large number of species of animals, plants and microbes that are found on land, although our knowledge of those inhabiting water bodies, both fresh and marine, including the vast oceans, is far from satisfactory. The interactions between plants and plants, plants and animals and between animals and animals are complex, and the same is also true when we look at the interactions between microorganisms and plants on the one hand, and with animals on the other. These clearly indicate the complexity of the various kinds of interactions, and the dependency of one or more kinds of organisms on others is largely for feeding purposes in order to fulfil energy requirements. This has given rise to the many varied feeding types of which parasitism and predation are of paramount importance to us as biologists in the wider sense and as nematologists in the restricted sense. However, the organisms that are referred to as pathogens, parasitoids and pests also enter into associations with other organisms that they may destroy or kill.

The two very important associations between organisms as mentioned above are parasitism and predation. The parasites living in or on a host, which itself is a living organism, are physiologically dependent on it for nourishment. This is a rather complex association in which the two are biochemically adapted to each other. However, in predation, predators simply kill their prey before feeding. Parasitism and predation are not restricted to any particular group of animals but have evolved independently in several groups of organism at different levels of organisation. Such relationships are extremely useful for keeping the population levels of hosts and prey species in sustainable proportions as required by the ecosystem. Consequently, the presence of both parasite and predator is ecologically essential for all kinds of habitats. Parasitism and predation are much more conspicuous in animals than in plants. The insects, nematodes and helminths serve as the best examples of parasitism, but predation is prevalent in practically every group of organisms, from the simplest *Amoeba* to the highly evolved *Homo sapiens*.

The nematodes, like insects, constitute a highly diversified group of animals occupying all kinds of habitat. They occur in unimaginable numbers and in a wide variety of shapes, sizes and forms. Every inch of this globe, including the oceans, the rivers, the pools, the ponds and the puddles, teems with millions, in fact billions, of these tiny creatures. It would perhaps not be wrong to say that nearly 90% of all metazoans in the world are nematodes. Some nematodes are 'free-living' in marine habitats and freshwater, others are soil-inhabiting and still others are parasitic on different kinds of plant and animal. The parasitic species are of considerable agricultural, clinical and veterinary importance as parasites/pests of plants and parasites of man and his livestock. The nematodes, in general, are small multicellular animals with rather simple organisation. They are considered second only to insects in their kinds and characteristics and possibly in the total number of species as well. Our knowledge of nematode parasites of man and animals goes back to antiquity due to the fact that such nematodes are usually large in body size and the diseases they cause to man and animals usually attract a lot of attention. The existence of those living in water, soils, or parasitising plants, remained largely unknown until the 18th century, perhaps because of their exceedingly small size, absence of any coloration, mostly underground and cryptic habitats and the difficulties encountered in their isolation and processing.

Among the soil-inhabiting organisms the nematodes dominate over all others, both in numbers and species. These nematodes mainly fall into four categories: *i*) microphagous, feeding upon microorganisms; *ii*) saprophagous, living on decaying organic matter; *iii*) phytophagous, parasitising plants; and *iv*) predaceous, living exclusively by predation. The phytophagous nematodes have received considerable attention, mainly because of their agricultural importance, whilst the other three groups have been comparatively neglected. The predatory nematodes feed either by piercing or sucking or by ingesting intact a wide variety of soil organisms such as protozoa, rotifers, small oligochaetes, enchytraeids, nematodes, *etc*.

The predatory species occur in many groups of nematodes. While only a few species of Aphelenchida Siddiqi, 1980, Rhabditida Chitwood, 1933, Enoplida Chitwood, 1933, Diplogastrida Maggenti, 1981 and Dorylaimida Pearse, 1942 are predaceous, all the species of Mononchida Jairajpuri, 1969 (commonly known as mononchs) feed exclusively by predation and hence this group may be of importance in agriculture. There are definite indications that in some instances they are reported to have brought about decline in the population levels of some species of phytophagous nematodes, a possibility which clearly stimulated research in various parts of the world, including India, to assess whether they can be used as agents in the biological control of plant-parasitic nematodes. However, to date, not much success has been achieved in this direction.

As early as 1917, Cobb, while studying the biology and the predatory ability of some species of mononchs had, for the first time, postulated the use of predaceous nematodes for the biological control of plant-parasitic nematodes (Fig. 1). Since then a substantial amount of research has been done in this field. Although many nematodes have shown predatory abilities *in vitro*, studies on their practical use in the field are still very much a matter for speculation. Most predaceous nematodes belong to

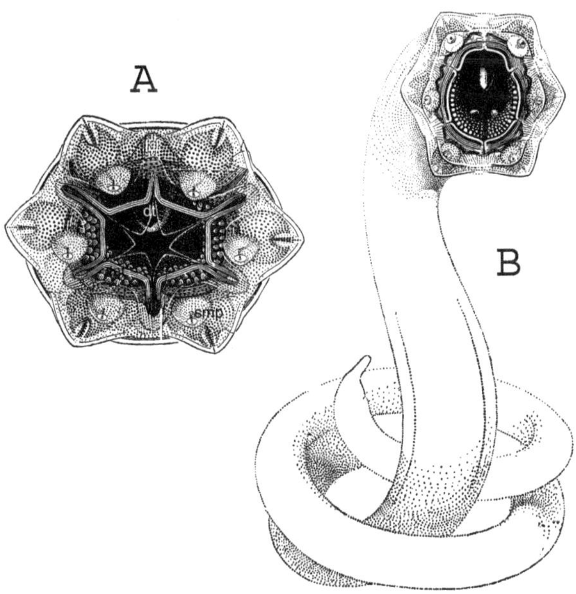

Fig. 1. *A: Face view of head showing dorsal tooth and denticles; B: Mononch with open buccal cavity about to seize its prey. After Cobb (1917).*

the Orders Mononchida (*e.g.*, *Clarkus papillatus* (Bastian, 1865) Jairajpuri, 1970a, *Iotonchus brachylaimus* Cobb, 1917, *Mononchus aquaticus* Coetzee, 1968, *etc.*); Dorylaimida (*e.g.*, *Actinolaimus* sp., *Discolaimus* sp., *Dorylaimus* sp., *Thornia* sp., *etc.*) or to Diplogastrida (*e.g.*, *Diplenteron* sp., *Diplogaster* sp., *Mononchoides potohikus* Yeates, 1969, *etc.*), although *Seinura* Fuchs, 1931 belongs to the Aphelenchida. Members of the Dorylaimida are predominantly omnivorous and feed upon fungi, algae, nematodes, enchytraeids and some other type of microfauna, whereas mononchs feed on nematodes, rotifers, protozoans, algae, fungal spores, *etc.* Species of Diplogastrida feed on both nematodes and microbes.

The role of mononchs in reducing or managing populations of plant-parasitic nematodes has been commented upon by various nematologists. Cobb (1920) and Steiner and Heinley (1922) suggested the use of *C. papillatus* for controlling populations of nematodes in sugar beet fields, although Thorne (1927) found that the species of mononchs present in sugar beet fields were usually insufficient to bring about encouraging results as far as the control of plant-parasitic nematodes was concerned. Cassidy (1931) was of the opinion that, under suitable conditions, *I. brachylaimus* could partially control populations of pest nematodes. Christie (1960) suggested that predatory nematodes should be assessed for determining the practicality of their use in the control of plant-parasitic nematodes. Further studies in this direction were done by Mulvey (1961a), Esser (1963), Esser and Sobers (1964) and Ritter and Laumond (1975) among others. The nature and source of food of the majority of mononchs has not yet been determined adequately. Of those known in some detail, a variety of soil microorganisms including, of course, species of nematodes, are fed upon (Banage, 1961; Arpin & Kilbertus, 1981). According to Webster (1972) and Jones (1974), non-specific predators, such as the mononchs, exert only partial control, and the possibility of these being successful agents of biological control is quite remote. Nelmes (1974) compared the predation abilities of *Prionchulus punctatus* Cobb, 1917 with those of predatory arthropods (Solomon, 1949; Holling, 1966). According to Cohn and Mordechai (1974), *Mylonchulus sigmaturus* Cobb, 1917 predates upon the citrus nematode, *Tylenchulus semipenetrans* Cobb, 1913, and when the predator density is high, populations of *T. semipenetrans* were usually found to be low. Consequent upon these observations, research on predatory mononchs gained the desired impetus and investigations were made on their life cycles (Maertens, 1976), predatory behaviour (Grootaert & Maertens, 1976; Jairajpuri & Azmi, 1978; Bilgrami *et al.*, 1984), ultrastructure and function of feeding apparatus (Grootaert & Wyss, 1979), predatory abilities (Small, 1979; Small & Grootaert, 1983), effect of mechanical stimulation and crowding on predation (Azmi & Jairajpuri, 1979), factors influencing predation (Nelmes, 1974; Bilgrami *et al.*, 1983, 1984), cannibalistic tendency (Azmi & Jairajpuri, 1978; Bilgrami & Jairajpuri, 1985); and intra-specific interactions (Bilgrami *et al.*, 1988).

Like other soil-inhabiting nematodes, the mononchs are cosmopolitan and are found in all kinds of soils; however, they occur in much greater numbers in undisturbed soils compared with arable land. Some species are known to adapt themselves to surprisingly varied conditions. Mononchs occasionally occur in freshwater, even at great depths in lakes and at high altitudes on mountains. For example, *Mononchus longicaudatus* Cobb, 1893, a species which occurs mainly in the tropics, is also found in temperate regions of the world, sometimes being recorded from very cold areas. It can survive both in soil and in freshwater (Cobb, 1917). In addition to freshwater, mononchs have even been recorded from salt lakes and also, in a few instances, from marine waters. Some nematologists have observed mononchs floating on the surface of the water, possibly due to repulsion between the cuticle and water. Such flotation may facilitate the rapid and wide distribution of the species concerned.

Like other kinds of predators, particularly the land-dwelling species of bird and mammal predators, the mononchs are, by and large, much larger and also stronger than their nematode prey. They can easily be recognised in any soil collection on the basis of their prominent feeding apparatus or buccal cavity, the cylindroid pharynx, strongly cuticularised pharyngeal lumen, and also by the usual presence of caudal glands in their tail and the absence of a prerectum.

Historical background and literature review

The history of literature on mononchs goes back to 1845 when Dujardin described nematode species under the names *Oncholaimus muscorum*, *O. fovearum* and *Enoplus crassiculus*. However, Bastian (1865; Fig. 2) proposed the type genus *Mononchus*, described five new species,

Fig. 2. *Some leading nematologists who have contributed to the taxonomy of Mononchida. They are from left to right: H.C. Bastian, I.N. Filipjev and G. Thorne (1st row), W.C. Clark, R.H. Mulvey and H.J. Jensen (2nd row), A. Coomans, P.A.A. Loof and M.R. Siddiqi (3rd row), M.S. Jairajpuri, W.U. Khan and W. Ahmad (4th row).*

viz., *M. truncatus*, *M. papillatus*, *M. macrostoma*, *M. tunbridgensis* and *M. cristatus*, and transferred the three species of Dujardin to this genus. During the next 50 years (1865-1915) this group of nematodes did not attract much attention and only some 20 additional species were described. However, immediately afterwards, over the following 2 years (1916 and 1917), the group received tremendous attention from one of the greatest nematologists of all time, Dr N.A. Cobb. Cobb not only worked on the systematics and morphology of the group but also provided useful information on biology and ecology. He (1916) split the genus *Mononchus* Bastian, 1865 into five subgenera, *viz.*, *Mononchus*, *Prionchulus*, *Mylonchulus*, *Iotonchus* and *Anatonchus*, and prophesied that all of these subgenera would ultimately be elevated in rank to genus level. In 1917 he published an authoritative paper on the mononchs, a paper that remains one of the most valuable contributions on this group of nematodes. Much of what he said at that time is still relevant today, despite the passage of almost a century. In the paper, he also proposed yet another subgenus, *Sporonchulus*, and described 32 new species of mononch, which brought the tally of known species to *ca* 60.

Micoletzky (1922) also did some work on this group of animals, although he recognised only 41 species. Thorne (1924; Fig. 2), a student of Cobb, was himself a world authority on nematodes. He was fascinated by mononchs and wrote an excellent paper on the species occurring in Utah, USA.

The prophecy of Cobb came true when subsequent workers started raising the subgenera that he had proposed to full generic rank. Wu and Hoeppli (1929) were the first to do this when they raised *Prionchulus* Cobb, 1916 to generic status. In 1931, Cassidy studied the mononchs from Hawaii. Filipjev (1934; Fig. 2) proposed the subfamily Mononchinae, which was raised to the family Mononchidae Filipjev, 1934 under the superfamily Tripyloidea de Man, 1876, suborder Enoplina Chitwood, 1933, Order Enoplida by Chitwood (1937) for the genera *Mononchus* and *Prionchulus* and the subgenera *Mylonchulus*, *Iotonchus*, *Anatonchus* and *Sporonchulus*. Of these four subgenera, *Anatonchus* was raised to genus status by De Coninck (1939), *Iotonchus* and *Mylonchulus* by Altherr (1950, 1953) and *Sporonchulus* by Pennak (1953).

Andrássy (1958) provided a valuable revision of the group and thus made a very useful contribution to our knowledge of mononchs. He added five more genera to the family Mononchidae, *viz.*, *Brachonchulus*, *Cobbonchus*, *Granonchulus*, *Judonchulus* and *Miconchus*.

During the latter half of the 20th century, the mononchs received much attention, possibly because of their potential role in nematode biocontrol. The taxonomy of the group has been greatly expanded by the addition of a large number of species and genera. Clark (1960a, b, c, 1961a, b, c, d, 1962, 1963), in a series of papers dealing exclusively with mononchs from New Zealand, provided valuable information on this group. He (1961d) removed Mononchidae from the suborder Enoplina, brought it under the suborder Dorylaimina Pearse, 1936 together with the new family Bathyodontidae, and gave it superfamily status. He was of the opinion that this family represented a transitional kinship between Mononchoidea Filipjev, 1934 and Dorylaimoidea de Man, 1876. He also traced the evolution and inter-relationship of mononchs. Clark (1960b) did some useful work on the pharyngo-intestinal junction of mononchs and emphasised the diagnostic importance of 'tubercles' correlating it with the shape of the buccal cavity.

The work of Clark (Fig. 2) was soon followed by another good series of publications entitled "The Mononchidae: A family of predaceous nematodes" by Mulvey (1961a, b, 1962, 1963a, b, 1967a, b; Fig. 2) from Canada. In these papers not only were a number of new species belonging to various genera of mononchs described, but also descriptions of a large number of known species were thoroughly revised, along with a key to their identification. In 1967, Mulvey and Jensen (Fig. 2) published an account of the mononchs collected from Nigeria in which they described 16 new species and proposed the following four new genera, *viz.*, *Crassibucca*, *Hadronchus*, *Polyonchulus* and *Prionchulellus*. Another notable serial publication appeared on the mononchs of South Africa authored by Coetzee (1965, 1966, 1967a, b, 1968), in which she described several known and new species of *Cobbonchus* Andrássy, 1958, *Granonchulus* Andrássy, 1958, *Mylonchulus* Cobb, 1916, *Iotonchus* Cobb, 1916, *Mononchus* and *Prionchulus*. Jensen and Mulvey (1968) gave a brief account of the morphology of mononchs while describing the "Predaceous nematodes (Mononchidae) of Oregon" in which they reported some 24 known and five new species of the genera *Mylonchulus*, *Prionchulus*, *Anatonchus* Cobb, 1916, *Miconchus* Andrássy, 1958, *Mononchus* and *Iotonchus*. Other contributions during this period included those by Williams (1958), Andrássy (1959), Buangsuwon and Jensen (1966), Yeates (1967), Altherr (1968), Lordello (1970) and Thong (1971).

Jairajpuri (1969, 1970a, b, 1971) not only reported for the first time many genera of mononchs from India but also described a number of known and new species of *Hadronchus* Mulvey & Jensen, 1967, *Miconchus*, *Iotonchus*, *Mononchus*, *Clarkus* Jairajpuri, 1970a, *Mylonchulus* and *Sporonchulus* Cobb, 1917. One of the most important contributions of Jairajpuri (Fig. 2) is on the systematics of mononchs (Jairajpuri, 1969). Previously, mononchs were regarded as belonging to a superfamily or, at the most, a suborder of Dorylaimida. He removed the mononchs from the Dorylaimida and gave them an independent Ordinal rank. In addition, the classification of mononchs at family level was considerably enhanced to bring it up to par with other nematode groups like the Dorylaimida and Tylenchida Thorne, 1949. Jairajpuri (1971) recognised two suborders: the Mononchina Kirjanova & Krall, 1969 with the two superfamilies Mononchoidea and Anatonchoidea Jairajpuri, 1969, and the Bathyodontina Coomans & Loof, 1970 with the two superfamilies Bathyodontoidea Clark, 1961d and Mononchuloidea De Coninck, 1965. Under Mononchina, five families – three under Mononchoidea and two under Anatonchoidea – were recognised. Ivanova and Dzhuraeva (1971) proposed a new family, Tigronchidae, to accommodate the new genus *Tigronchoides*, which they described from Tadzhikistan. Andrássy (1972) added a new genus, *Margaronchulus*, from the Congo, while Baqri and Jairajpuri (1974) added two new genera, *Actus* and *Paracrassibucca*, from El Salvador. Eroshenko (1975) described ten new species representing the genera *Clarkus*, *Iotonchus*, *Miconchus* and *Prionchulus* from the USSR. Jairajpuri and Khan (1977) proposed the genus *Coomansus* for those species of *Clarkus* not possessing a ventral ridge on the vertical plates of the buccal cavity, and 11 species previously placed under *Clarkus* were transferred to *Coomansus* Jairajpuri & Khan, 1977. A new species each of *Clarkus* and *Coomansus* was also described. Baqri *et al.* (1978) described two new species of *Iotonchus* and a new species of *Cobbonchus* and also recorded *Anatonchus ginglymodontus* Mulvey, 1961b from India, while Khan *et al.* (1978) added two new species of *Miconchus* from India.

Eliava (1978), in his revised classification of the Order Dorylaimida, accepted the two suborders Mononchina and Bathyodontina under Mononchida as earlier proposed by Jairajpuri (1969), but these were recognised as suborders under Dorylaimida by Coomans and Loof (1970; Fig. 2). Hunt (1978) provided a list of 14 species from seven genera of Mononchida from St Lucia and Dominica. Khan and Jairajpuri (1979) described several species of *Mylonchulus* from India and provided a key to the species of the genus, while in 1980 they provided a key to species of *Iotonchus*, in addition to describing several species of this genus from India. They also revised the classification of the family Iotonchidae and proposed a new subfamily, Hadronchinae Khan & Jairajpuri, 1979, thereby recognising two subfamilies, Iotonchinae Jairajpuri, 1969 and Hadronchinae under Iotonchidae Jairajpuri, 1969. Mohandas (1979), Mohandas and Prabhoo (1979), Sharma and Saxena (1981) and Coomans and Khan (1981) described and redescribed several species from India and Kenya.

Jairajpuri and Khan (1982), in an excellent monographic study, dealt with the whole group, providing a detailed account of the morphology and systematics of the Order Mononchida, diagnoses and identification keys to the suborders, family groups and all genera described to that date. They proposed six new genera, *viz.*, *Paramylonchulus*, *Megaonchulus*, *Comiconchus*, *Miconchoides*, *Promiconchus* and *Jensenonchus*, and synonymised the genera *Tigronchus* and *Tigronchoides*. They also described all the species of mononchs recorded thus far from India. Ray and Das (1983) added a new genus, *Hadronchulus*, from India. Patil and Khan (1982), Ahmad and Jairajpuri (1983) and Rahman and Jairajpuri (1984) described several more new species.

Siddiqi (1983; Fig. 2) discussed the phylogenetic relationship among the soil nematode Orders Dorylaimida, Mononchida, Triplonchida Cobb, 1920 and Alaimida Siddiqi, 1983 and proposed several new Orders. The bathyodontids were taken out of Mononchida and a new Order, the Bathyodontida, was proposed for them. Baqri and Baqri (1983) discussed the importance of the location of the pharyngeal gland nuclei in mononchid identification. Siddiqi (1984a) erected four new genera, *viz.*, *Caputonchus*, *Mulveyellus*, *Nigronchus* and *Truxonchus*, described many new species and proposed several new combinations. Siddiqi (1984b, c) described two new species of *Iotonchus* and a new species of *Anatonchus* from the Fiji Islands and also described *Nullonchus*, an interesting new genus of toothless mononch, and proposed a new subfamily, Nullonchinae, under Anatonchidae (Siddiqi, 1984d). He also discussed its phylogenetic relationship.

Andrássy (1985) described a new species each of *Clarkus*, *Cobbonchus* and the rare genus *Margaronchulus* from the southern hemisphere. Zell (1985) considered *Prionchulus muscorum* to be a collective species of morphologically similar forms and described six species close to *P. muscorum* from Germany. Khan and Saeed (1987) proposed a new genus, *Pakmylonchulus*, close to *Mylonchulus*. Lal and Khan (1988), Gambhir and Dhanachand (1990) and Azmi (1991) from India, Choi and Choi (1987) from Korea, Yeates (1987) from New Zealand and in (1992) from New Caledonia, Tsalolikhin (1988) from USSR, Volcy (1988) from Colombia, Popovici (1990) from Romania, Chaves (1990) from Argentina, Vinciguerra and La Rosa (1990) from subantarctic America, Loof *et al.* (1990) from Iran, and de Bruin and Heyns (1992a, b) from South Africa, all described or redescribed several species of mononch.

Andrássy (1992, 1993a, b), in a series of important publications on the taxonomy of the families Mylonchulidae, Mononchidae and Anatonchidae, recognised only one subfamily, the Mylonchulinae, with seven genera, *viz.*, *Mylonchulus*, *Brachonchulus*, *Granonchulus*, *Margaronchulus* Andrássy, 1972, *Megaonchulus* Jairajpuri & Khan, 1982, *Oligonchulus* Andrássy, 1976 and *Polyonchulus* Mulvey & Jensen, 1967 under Mylonchulidae; two subfamilies, Mononchinae and Cobbonchinae, under Mononchidae with nine genera, *viz.*, *Actus*, *Clarkus*, *Coomansus*, *Judonchulus* Andrássy, 1958, *Mononchus*, *Nigronchus* Siddiqi, 1984a, *Paramononchus* Mulvey, 1978, *Prionchulus* and *Sporonchulus* under Mononchinae; two genera, *viz.*, *Cobbonchus* and *Comiconchus* Jairajpuri & Khan, 1982 under Cobbonchinae; three subfamilies, Anatonchinae, Iotonchinae and Miconchinae under Anatonchidae with 12 genera, *viz.*, *Iotonchus*, *Caputonchus* Siddiqi, 1984a, *Hadronchoides* Jairajpuri & Rahman, 1984, *Hadronchulus* Rey & Das, 1983, *Hadronchus*, *Iotonchulus*, *Jensenonchus* Jairajpuri & Khan, 1982, *Mulveyellus* Siddiqi, 1984a, *Nullonchus* Siddiqi, 1984d, *Parahadronchus* Mulvey, 1978, *Prionchulellus* Mulvey & Jensen, 1967 and *Prionchuloides* Mulvey, 1963b under Iotonchinae; five genera *viz.*, *Crassibucca* Mulvey & Jensen, 1967, *Doronchus* Andrássy, 1993b, *Miconchus*, *Paracrassibucca* Baqri & Jairajpuri, 1974 and *Promiconchus* Jairajpuri & Khan, 1982 under Miconchinae; and three genera, *viz.*, *Anatonchus*, *Tigronchoides* Ivanova & Dzhuraeva, 1971 and *Truxonchus* Siddiqi, 1984a under Anatonchinae. Andrássy (1992, 1993a, b) also provided diagnoses of family and generic taxa and a key to species. The genera *Paramylonchulus* Jairajpuri & Khan, 1982, *Pakmylonchulus* Khan & Sayeed, 1987, *Morenchus*

Dhanachand, Renubala & Mohilal, 1991, *Sporonchuloides* Mohandas & Prabhoo, 1982 and *Miconchoides* Jairajpuri & Khan, 1982 were considered as junior synonyms. Loof (1993) also considered the genera *Paramylonchulus* and *Pakmylonchulus* as junior synonyms of *Mylonchulus*.

Loof and Winiszewska-Slipińska (1993) made systematic observations on some species of *Coomansus* and *Iotonchus* and transferred several species from *Iotonchus* to *Coomansus*. They also designated a neotype for *Coomansus zschokkei* (Menzel, 1913) Loof & Winiszewska-Slipińska, 1993 (= *Iotonchus zschokkei*) and *Iotonchus sphagni* (Brzeski, 1960) Loof & Winiszewska-Slipińska, 1993 (= *Coomansus sphagni*). Ahmad and Jairajpuri (1993) described a new species of the rare nematode genus *Promiconchus* from Dominica, whilst Jairajpuri *et al.* (1998) added a new species each of the rather rare genera *Crassibucca* and *Judonchulus* from Nicaragua and New Zealand, respectively. They also described an atypical species of the genus *Cobbonchus* having a long filiform tail from New Zealand. Anandi *et al.* (1997) and Mohilal and Dhanachand (1997) described new species of the genera *Paramylonchulus*, *Actus* and *Coomansus* from Manipur, India. Peneva *et al.* (1999) made detailed observations on the nematodes of the subfamily Anatonchinae from Bulgaria and described a new species of *Anatonchus* and *Tigronchoides*.

In recent years, several new genera and species of mononchs have been described from unexplored areas of the world. Jairajpuri *et al.* (2000, 2001a, b) described two new genera, *Parkellus* and *Micatonchus*, and several new species from Korea, whilst Khan *et al.* (2003) added two further species to *Micatonchus* from Korea. Khan *et al.* (2000) and Khan and Araki (2002) described a few species from Japan. Siddiqi (2001) recorded 16 species of *Iotonchus* from West Africa, of which 11 were new to science. Ahmad (2000) added species from Cameroon and Siddiqi and Jairajpuri (2002) described an interesting new genus and species, *Crestonchus cristatus*, from Cameroon. Zullini *et al.* (2002) described several species of mononch from Costa Rica while Andrássy (2003) described some mononchs from Alaska. Winiszewska (2002) and Susulovsky and Winiszewska (2002, 2006) and Susulovsky *et al.* (2003), in a series of papers on the genus *Prionchulus*, described several new species mainly from Poland, Russia and Ukraine. Ahmad *et al.* (2005) described mononchs from Singapore and Loof (2006) described several new and known species from Malaysia. Vinciguerra and Orselli (2006) and Orselli and Vinciguerra (2007) described several species from Ecuador. Recently, Andrássy (2008) described a new species in each of the genera *Mononchus*, *Cobbonchus*, *Iotonchus* and *Miconchus* from Ecuador and Papua New Guinea. Elbadri *et al.* (2008) described two species of *Mylonchulus* from Sudan and Jiménez-Guirado and Murillo-Navarro (2008) described a new and known species of *Mylonchulus* from Spain. Andrássy (2009) recently proposed a new genus, *Cobbonchulus*, for *Cobbonchus longicaudatus* Jairajpuri, Ahmad & Sturhan, 1998.

Phylogenetic relationships

Many different concepts regarding the placement of Dorylaimida and Mononchida and their relatives in the subclass Dorylaimia have been proposed. As per the scheme of Clark (1961d), Mononchoidea Filipjev, 1934 (now Mononchida) shares a more recent common ancestry with Dorylaimoidea de Man, 1876 (= Dorylaimida) and Diphtherophoroidea Micoletzky, 1922 than with other enoplid taxa. In the scheme of Andrássy (1976), Mononchina was considered a suborder of the Order Dorylaimida with three other suborders, *viz.*, Dorylaimina, Diphtherophorina Coomans & Loof, 1970 and Mermithina Andrássy, 1974. He considered the latter three suborders closer to one another than to Mononchina and they must therefore share a common origin. Lorenzen (1981) included Dorylaimina, Mononchina and Bathyodontina under Dorylaimida. In Siddiqi's (1983) scheme, Mononchida Jairajpuri, 1969 and Bathyodontida Siddiqi, 1983 were considered as separate Orders. Maggenti (1983, 1991) placed Dorylaimina, Nygolaimina and Diphtherophorina together in Dorylaimida, maintaining Mononchida as a separate Order containing the superfamilies Bathyodontoidea Clark, 1961d and Mononchoidea.

Molecular studies employing SSU rDNA have indicated a basal position for the Dorylaimia (Clade I in Blaxter *et al.*, 1998) on the overall tree for the phylum. Blaxter *et al.* (1998) recovered a sister taxon relationship between Dorylaimida and a clade containing Mononchida and Mermithida, but without strong statistical support. De Ley and Blaxter (2002) left the relationship among the Dorylaimida, Trichinellida and mononchs/mermithids clade unresolved. Russin *et al.* (2003) found the same relationship for Dorylaimida, Mermithida and Mononchida as did Blaxter *et al.* (1998). Mullin *et al.* (2005), using

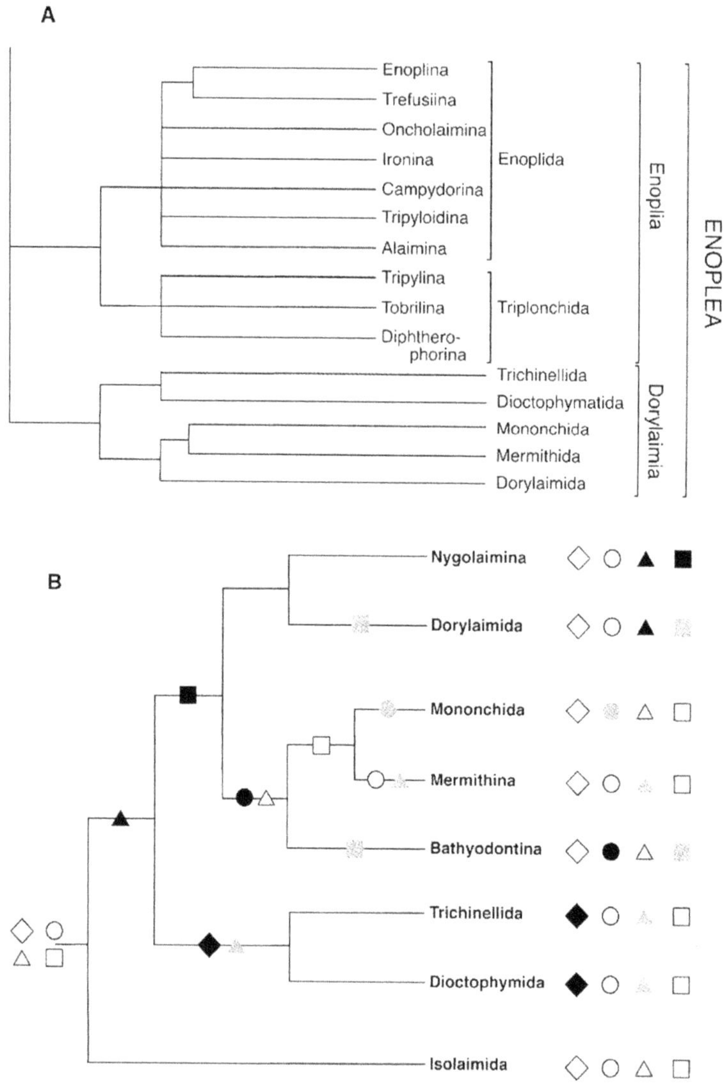

Fig. 3. *Phylogenetic relationship among nematodes of the Class Enoplia; A: Based on SSU rDNA. After De Ley and Blaxter (2004); B: Based on selected morphological characters. After Mullin* et al. *(2005).*

SSU rDNA sequences, found that the *Bathyodontus* group formed a well supported clade with the mermithid and mononchid taxa (Fig. 3). Earlier molecular studies (Blaxter *et al.*, 1998; De Ley & Blaxter, 2002; Russin *et al.*, 2003) have consistently placed Mononchida and Mermithida in a sister taxon relationship. Analysis consistently found a well supported clade containing *Mermis* Dujardin, 1842 and the sampled members of Mononchina, with *Bathyodontus* Fielding, 1950 as a sister taxon.

Traditional classification systems have generally regarded bathyodonts and mononchs as peer subtaxa within some higher level taxon, exclusive of mermithids (Clark, 1961d; Jairajpuri, 1969; Coomans & Loof, 1970; Andrássy, 1976). Coomans and van der Heiden (1978), however, indicated that bathyodonts and dorylaims shared a more recent common ancestry with each other than either did with mononchs, a view that Siddiqi (1983) also proposed. The system of classification proposed by De Ley and Blaxter (2002), now generally accepted, places the Order Mononchida, containing Bathyodontina and Mononchina, in a sister taxon relationship with Mermithida Hyman, 1951. A common feature of all these systems is the implied, if not explicitly stated, assumption of more recent common ancestry for bathyodonts and mononchs than for either of these groups with the mermithids. The results of Mullin *et al.* (2005), however, indicate that the Order Mermithida shares a more recent common ancestry with Mononchina than does Bathyodontina (Fig. 3). Their results confirm the concept of Mononchida (as an Order comprising Bathyodontina, Mermithina Andrássy, 1974 and Mononchina) as a sister taxon of the Dorylaimida. This relationship reflects the internal topology within these groups.

Yeates (1967) considered the relationships of taxa within Dorylaimina (comprising, at that time, the families Dorylaimidae, Nygolaimidae, Bathyodontidae and Mononchidae, among others), and concluded that the nygolaims, bathyodonts and mononchs descended independently from a common (pre-bathyodontid) ancestor with the dorylaims branching off from the nygolaims at some point after the divergence of the mononchid and bathyodontid lineage. Coomans and van der Heiden (1978) envisioned a similar relationship but with the Mononchina diverging first, followed by a split between the two superfamilies within the Bathyodontina, namely the Bathyodontoidea and Mononchuloidea. Recent molecular studies (Holterman *et al.*, 2006; Meldal *et al.*, 2007; van Megen *et al.*, 2009) using SSU rDNA sequences have shown Mermithida and Mononchida as sister taxa with 100% support in Bayesian analysis (Fig. 4).

The phylogenetic relationship within the Order Mononchida using SSU rDNA sequence has recently been studied (Holterman *et al.*, 2007; van Megen *et al.*, 2009) using 18S rDNA sequence data (Fig. 4). Their

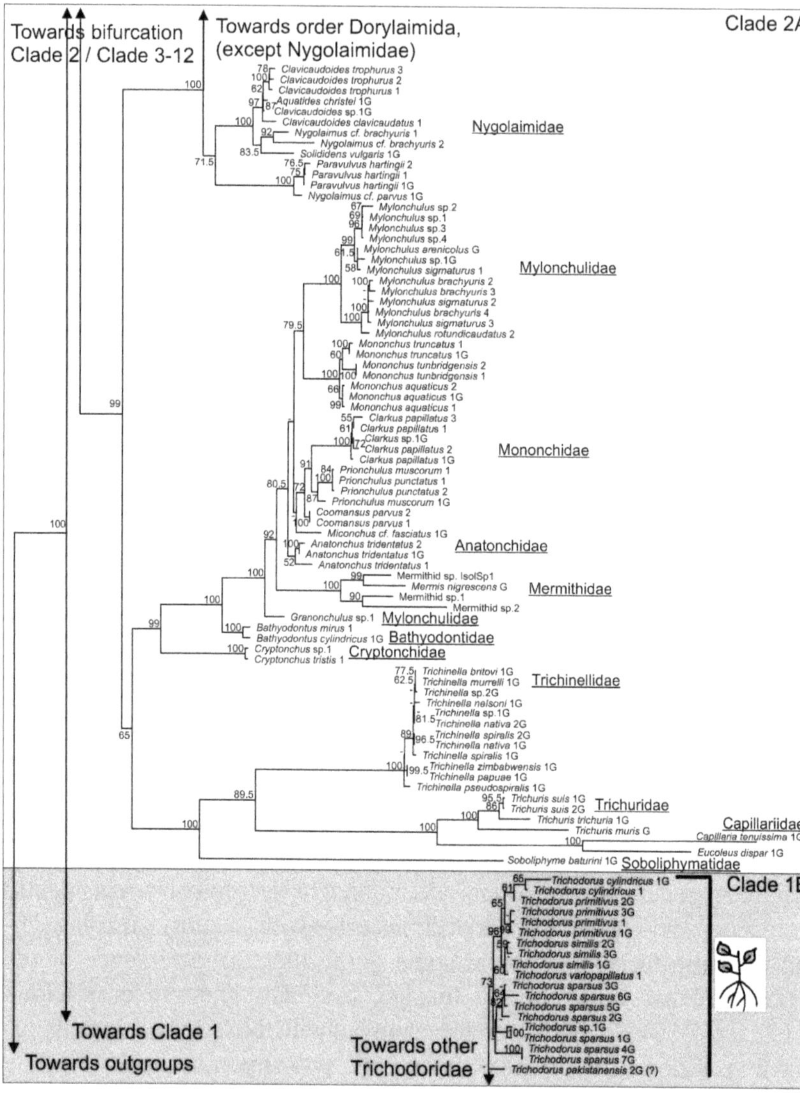

Fig. 4. *Phylogenetic relationship among Mononchida based on SSU rDNA sequence. After van Megen* et al. *(2009).*

results are, to some extent, similar to the current classification of the group. The family Mylonchulidae was shown to be polyphyletic as *Granonchulus* did not cluster with representatives of the genus *Mylonchulus*. Mononchidae also turned out to be paraphyletic, with *Mononchus* being a sister group of *Mylonchulus*. There is morphological support for this separation as both *Mononchus* and *Mylonchulus* have well developed caudal glands and spinneret. By contrast, *Clarkus*, *Coomansus* and *Prionchulus* were separated from *Mononchus*, having only rudimentary caudal glands and no spinneret. Olia *et al.* (2008) also found similar results in their analysis (Fig. 5). They recognised four major clades: *i*) clade I consisted of species of the genus *Mononchus*; *ii*) clade II represented the genera *Mylonchulus* and *Actus* (both genera belonging to the family Mylonchulidae); *iii*) clade III was represented by the genera *Prionchulus*, *Clarkus*, *Coomansus*; and *iv*) clade IV represented the two genera, *Anatonchus* and *Miconchus*, of the Anatonchoidea. The family Anatonchidae also appears to be paraphyletic (Holterman *et al.*, 2007). The position of *Granonchulus* in both analyses is questionable. Most striking is the positioning of Bathyodontidae and Cryptonchidae at the base of the Mononchida subclade. Both these families are exclusively bacterial feeding.

Nematodes of the Order Mononchida represent a fairly stable group of predatory forms much like that found in other animal groups with a predatory habit, including those of mammals. Although according to the available classification it is fairly easy to identify the genera and species of the Mononchida, their origin and evolutionary trends have not been commented upon by many workers except for Clark (1962) and, over 20 years later, by Siddiqi (1984d). Clark was somewhat disadvantaged because many genera of this group were not known at that time, whereas Siddiqi had a more informed opportunity as, by then, much was known about the group. Incidentally, both hypotheses are, in essence, geometrically opposed to one another.

Clark (1962) considered *Cobbonchus* as closer to the ancestors of the Mononchoidea (non-tuberculate), whereas *Miconchus* represented the ancestral form of the tuberculate Anatonchoidea. Siddiqi (1984d), while describing *Nullonchus*, an unusual genus which lacks dorsal as well as subventral teeth, considered it the ancestral form which gave rise to *Iotonchus*, *Miconchus* and *Anatonchus*, *etc.* In our opinion, *Nullonchus* is dangerously close to certain species of the genus *Iotonchus* which are characterised by the presence of a very small dorsal tooth. *Nullonchus* may have evolved from such species by further regression and, ultimately, loss of this small dorsal tooth.

We concur with Clark and believe that the primitive mononchs had three teeth; one on the dorsal and one on each of the subventral vertical

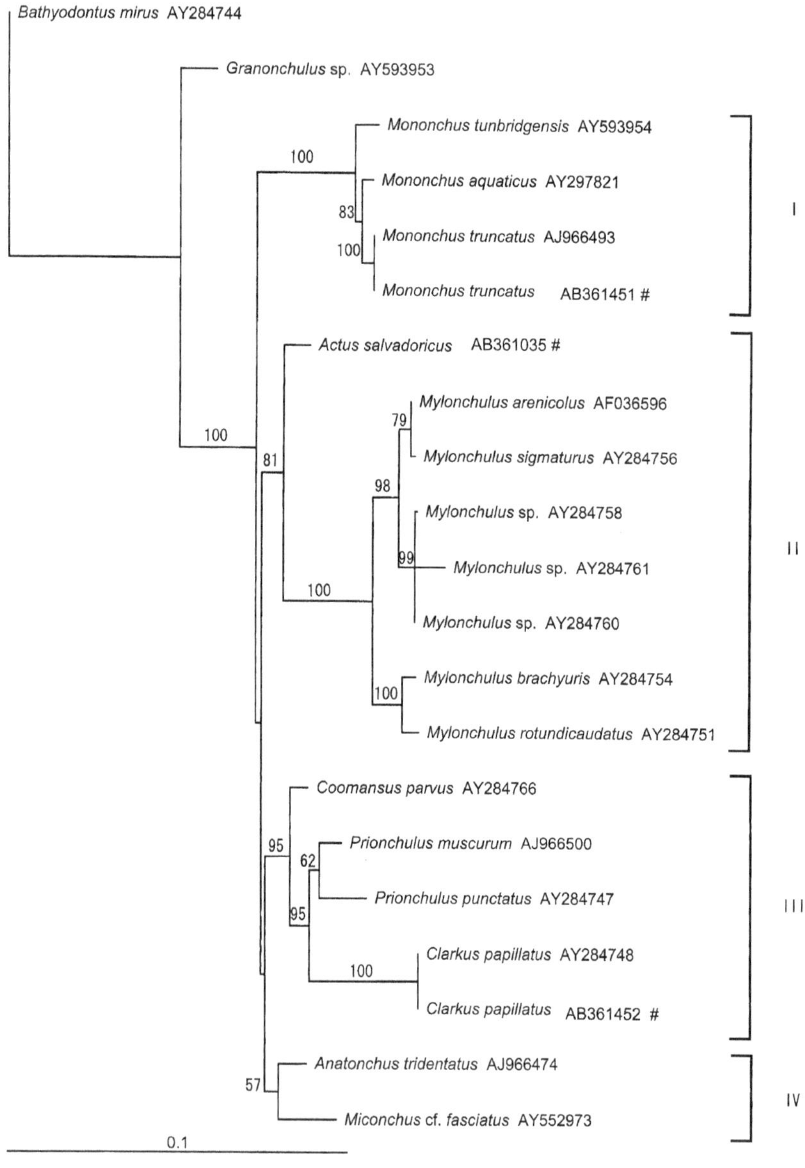

Fig. 5. *Phylogenetic relationship among Mononchida based on SSU rDNA sequence. After Olia* et al. *(2008).*

buccal plates. The ancestors of mononchs originally had a tooth on the dorsal plate and did not evolve into the present-day mononchs until additional teeth, one on each subventral plate, made their appearance, either concurrently or a little later. Together with these characters, changes in the pharynx and the rest of the body also occurred. We should perhaps look at the genus *Thalassogenus*, a relict taxon of the marine environment now inhabiting soil, which possesses three teeth, one on the dorsal and two on the subventral area of the buccal cavity, although this is probably only a case of parallel evolution.

Nematology Monographs & Perspectives, 2010, Vol. 7, 21-43

Chapter 2

Morphology

The members of the Order Mononchida Jairajpuri, 1969 can be easily recognised and differentiated from other related groups of nematodes by their somewhat large and robust bodies, in having a strongly sclerotised and well developed feeding apparatus, a long and highly muscular cylindroid pharynx, in the absence of a prerectum and in the presence of a gubernaculum, caudal glands and spinneret (Fig. 6). The body size varies from 0.55 mm (*Mylonchulus contractus*) to 7.00 mm (*Miconchus rex*). Upon death the body generally assumes a ventrally arcuate posture, sometimes becoming C-shaped, as in *Prionchulus*, *Clarkus*, *Coomansus*, *etc.*, but only rarely remaining almost straight, as in *Mononchus*. In general, the pre-vulval region is straight or only very slightly curved as compared to the post-vulval region, which shows a pronounced ventral curvature. The latter may be more evident in males due to the contraction of the copulatory muscles. Sometimes different individuals of a single species may assume slightly varying body postures but this feature is not of any taxonomic significance.

Body wall

The outermost layer of the body wall, the cuticle, is generally smooth, usually without any apparent or conspicuous markings or striations, although sometimes the inner layer(s) may possess fine transverse striations. The thickness of the cuticle (both outer and inner) varies considerably and ranges from 1 to 12 μm on different regions of body, being generally thickest near the vulva and on the tail. The body pores (dorsal, ventral and lateral) may or may not be visible or even present and, if visible, their arrangement may vary in different species or sometimes within the same species. The hypodermis (epidermis) is thickened in the dorsal, ventral and lateral positions to form four hypodermal chords which protrude into the pseudocoelomic cavity in between the somatic muscles, dividing them into four sectors. The lateral

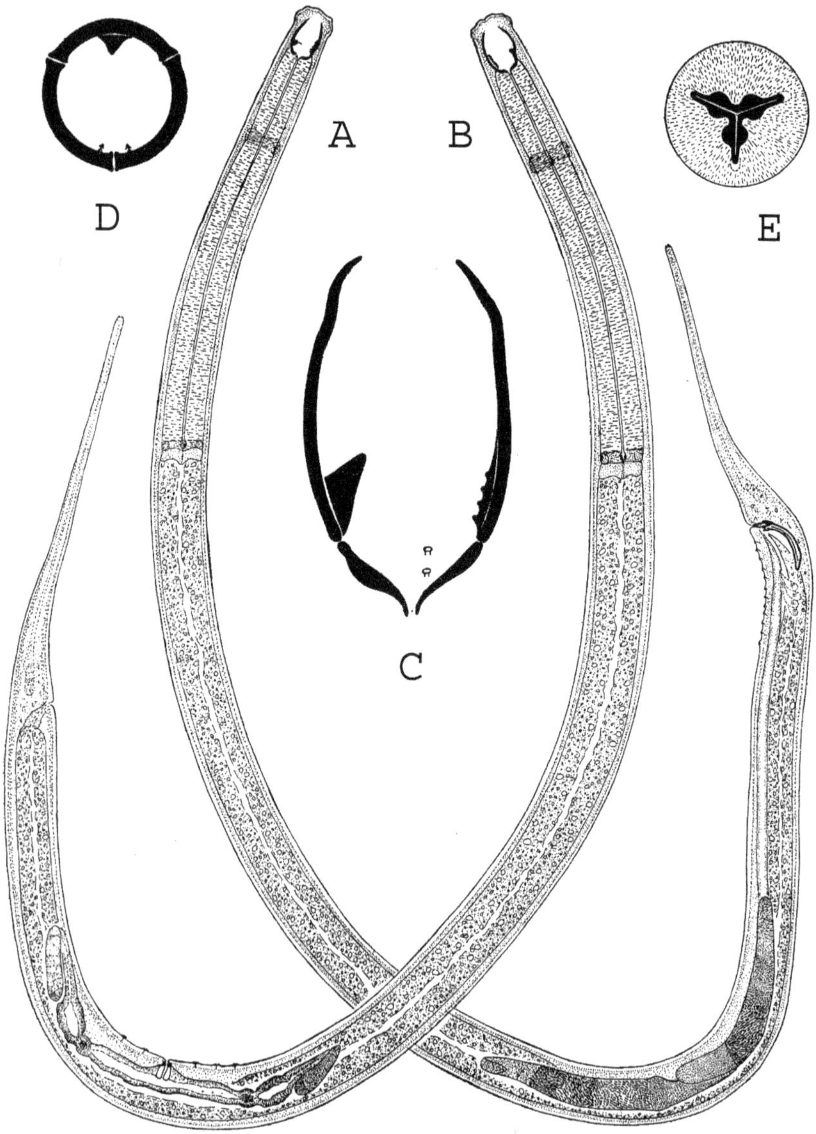

Fig. 6. *Typical mononch (*Parahadronchus shakili*). A: Entire male; B: Entire female; C: Buccal cavity; D: Transverse section of buccal cavity at level of dorsal tooth; E: Transverse section at base of pharynx. After Jairajpuri and Khan (1982).*

chords are much more prominent than the dorsal or ventral chords, occupying nearly one-sixth to one-third of the mid-body diameter. The somatic muscles are attached to the hypodermis by their flat surface, the other end projecting into the body cavity. The muscle cells are platymyarian and their arrangement is usually meromyarian. Specialised musculature associated with various body organs is very well developed (see Coomans & Lima, 1965; Baqri & Jairajpuri, 1974).

En face view

The *en face* view is usually hexagonal, and the lips are of equal size and shape, either angular or rounded, rarely amalgamated (Figs 7A-G; 8). The lips may bear a variable number of papillae along their outer margins. Each lateral lip bears only one outer papilla, while the four submedian lips possess 2-3 papillae on their outer margins. There are six papillae in the outer circlet, one on each lip. These papillae may possess dot-like structures on either side. The amphids are usually very small with slit-like apertures and are situated posterior to the lateral lips. Although the *en face* view reveals important structures, so far it has not been exploited as a taxonomic character.

Lip region

The lip region may only be marked off from the adjoining body by a slight depression or it may be clearly offset. However, in *Mononchus* and *Paramononchus* the lip region is almost continuous with the body and has almost the same diameter, but in other genera it is fairly widely expanded and is much broader than the adjoining body.

Amphids

The size and shape of amphids are variable but usually these are goblet or cup-shaped (Fig. 9A-H). The width and the position of the amphidial apertures may also be quite variable. The apertures are usually situated anterior to the apex of dorsal tooth. The amphidial chambers are connected to amphidial canals which terminate in wide sensillar pouches that are richly supplied with nerve fibres. The shape, size of amphids and the position and width of amphidial apertures may be of some diagnostic

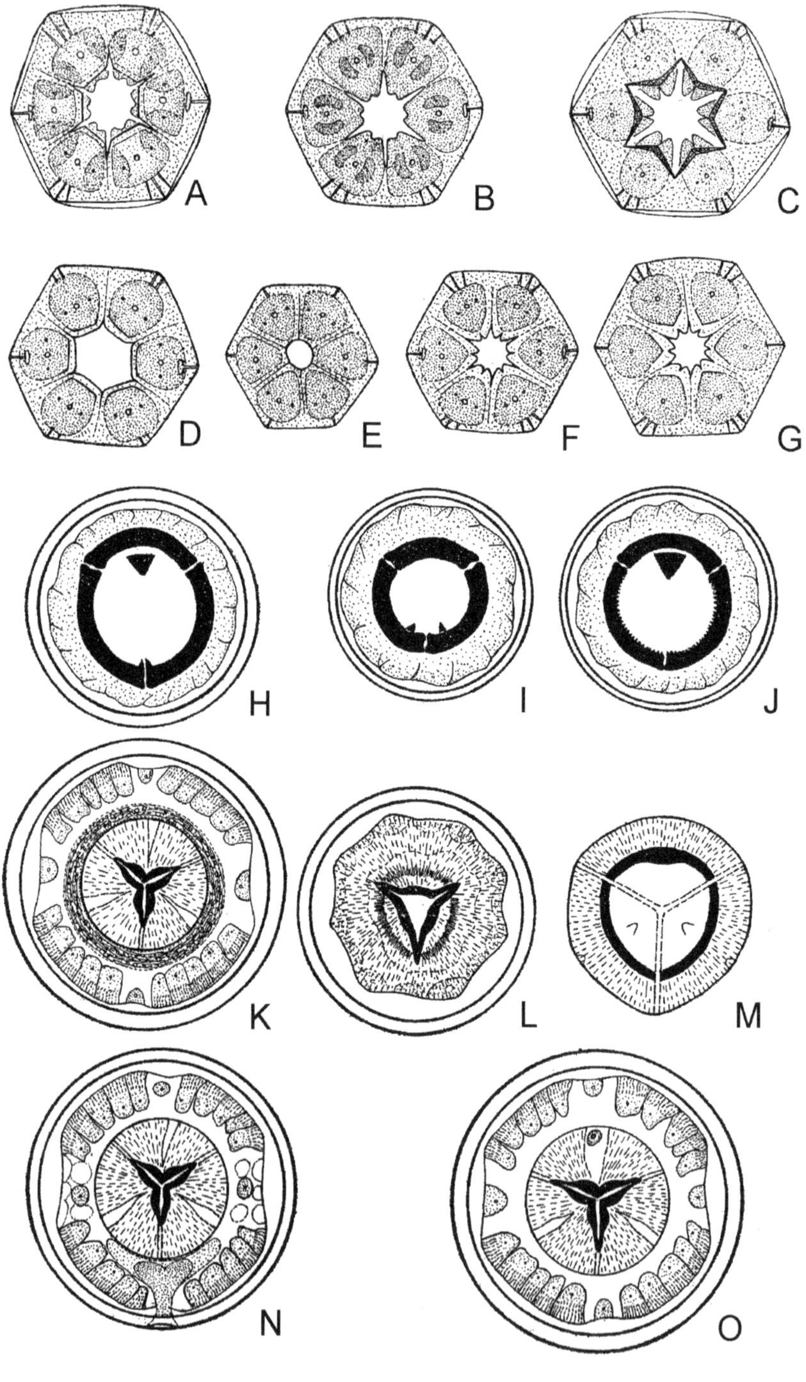
A
B
C
D
E
F
G
H
I
J
K
L
M
N
O

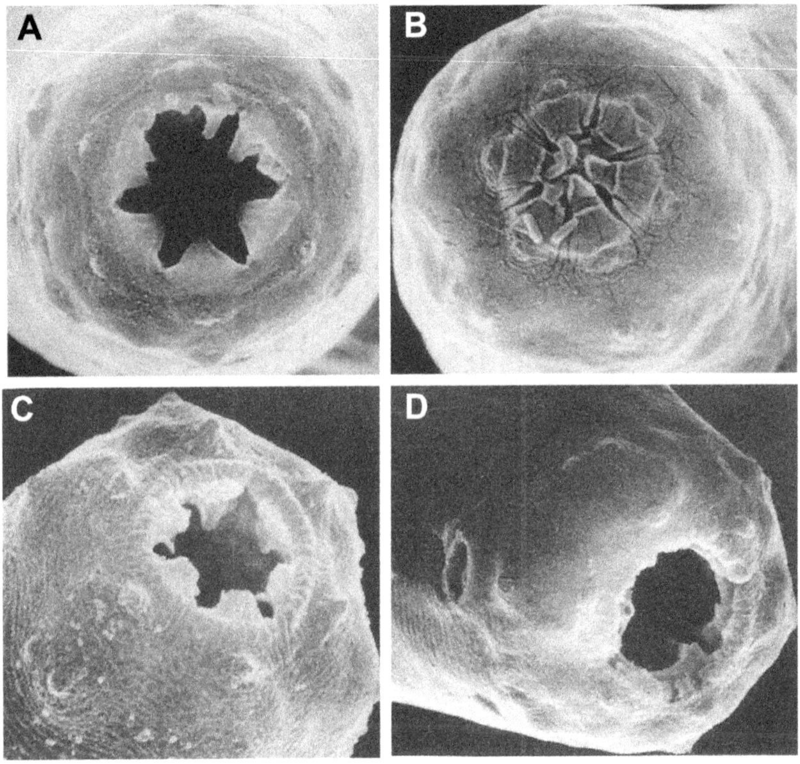

Fig. 8. *SEM of lip region. A:* Mononchus*; B-D:* Mylonchulus. *After Sauer (1985).*

value. The width of the amphidial aperture varies from one-seventh to three-quarters of the diameter of the buccal cavity.

Digestive organs

The digestive organs comprise a large, strongly developed, sclerotised stoma, usually called the buccal cavity, followed by a highly muscular,

Fig. 7. *A-G:* En face *views. A:* Miconchus citri*; B:* Coomansus indicus*; C:* Prionchulus muscorum*; D:* Mylonchulus hawaiiensis*; E:* Coomansus parvus*; F:* Clarkus elongatus*; G:* Clarkus papillatus. *H-O: H: Cross section of body at level of dorsal tooth; I: Subventral teeth; J: Transverse row of denticles; K: Nerve ring; L: Base of buccal cavity; M: Oblique walls of buccal cavity; N: Excretory pore and ampulla; O: Dorsal pharyngeal gland. After Jairajpuri and Khan (1982).*

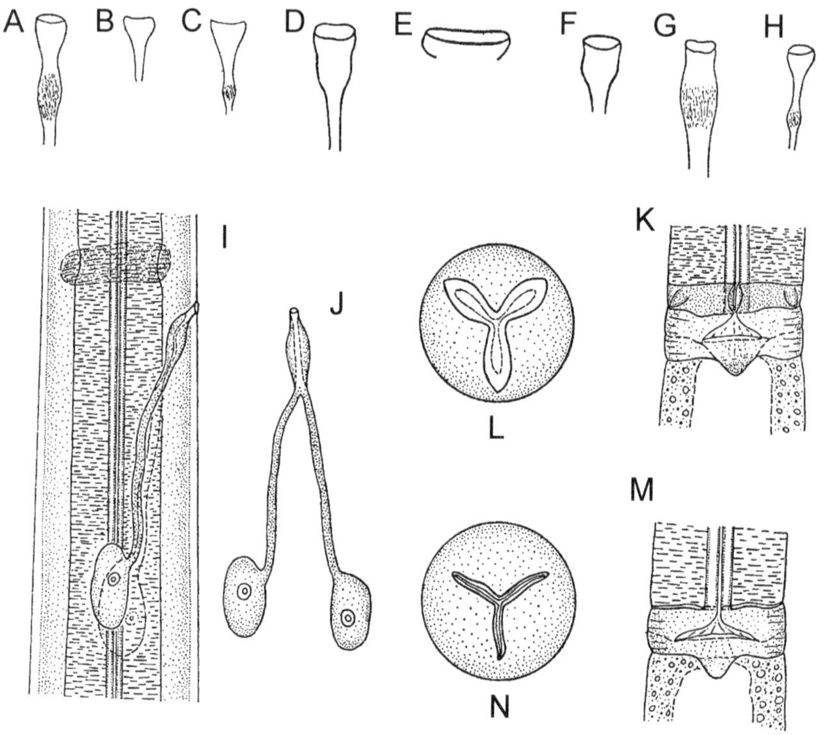

Fig. 9. *A-H: Different types of amphid; I, J: Excretory system; K-N: Pharyngo-intestinal junction. K: Tuberculate type (lateral); L: Transverse section at level of tubercles; M: Non-tuberculate type (lateral); N: Transverse section of non-tuberculate junction. After Jairajpuri and Khan (1982).*

cylindroid pharynx with very wide and sclerotised pharyngeal lumen, a well developed pharyngo-intestinal junction, tubular intestine and a short rectum opening to the exterior through an anus.

Buccal cavity

From the taxonomic as well as the structural and physiological point of view, the buccal cavity is the most important part of this animal. The shape and size of the buccal cavity and the type and position of its armature (tooth, teeth or denticles) provide virtually the main characters used in the identification of the various taxa. The buccal cavity is somewhat barrel-shaped with a tapering or flat base. The length and breadth of the buccal cavity, although variable in different species of mononch, may be useful in identification (Figs 10-12). The buccal

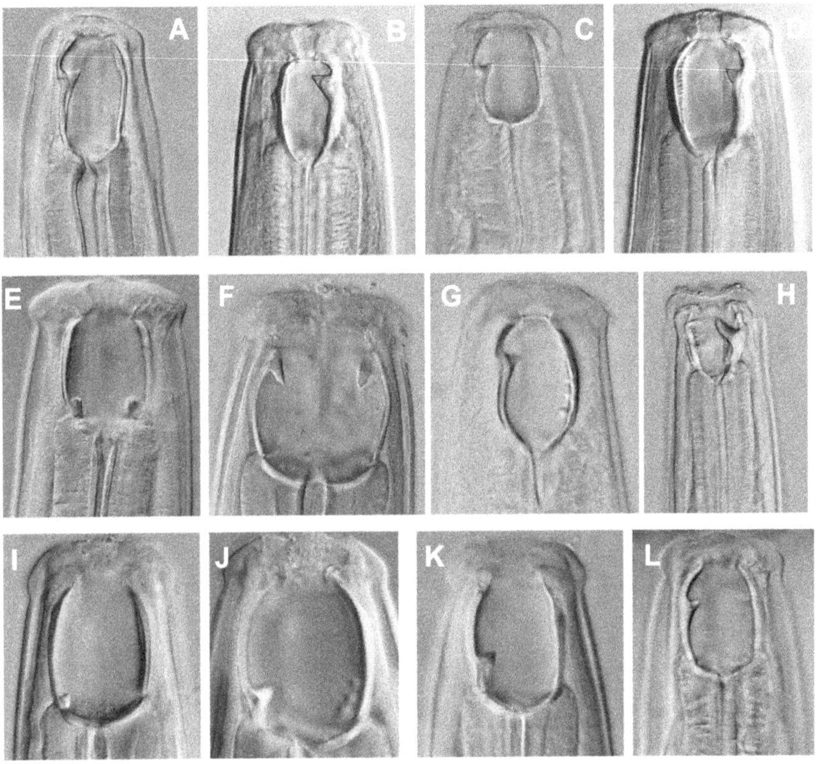

Fig. 10. *Buccal cavity. A:* Mononchus*; B:* Clarkus*; C:* Coomansus*; D:* Prionchulus*; E:* Miconchus*; F:* Anatonchus*; G:* Actus*; H:* Mylonchulus*; I:* Iotonchus*; J:* Parahadronchus*; K:* Iotonchulus*; L:* Mulveyellus.

cavity has prominent, heavily sclerotised, walls and is made up of two sets of three plates. The anterior or first set is vertical in position and is more developed than the posteriorly situated second set, which is made up of oblique basal plates. The anterior or vertical plates are formed from two plates, the proximal one being homologous to the pro- + mesorhabdions while the distal one is homologous to the metarhabdions. Andrássy (1993b) used the term 'praeparietale' for the former and 'interparietale' for the later. The oblique plates, or telorhabdions, have been termed 'postparietale' (Andrássy, 1993b). The thickness of the walls of the two sets is also variable. Each of these sets has three plates, that is, they are tripartite. One of these plates is dorsal and the other two are subventral in position. The triradiate nature of the buccal cavity becomes clearly evident in cross section. The dorsal wall of the vertical

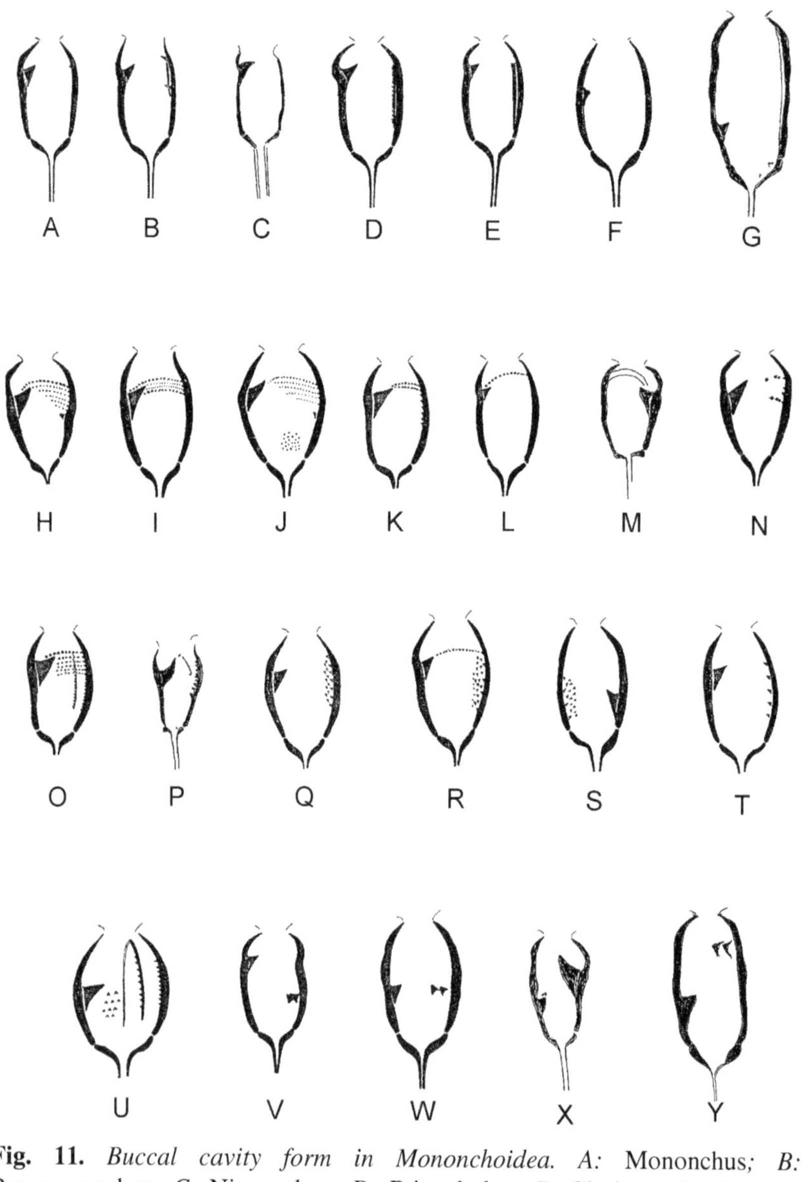

Fig. 11. *Buccal cavity form in Mononchoidea. A:* Mononchus*; B:* Paramononchus*; C:* Nigronchus*; D:* Prionchulus*; E:* Clarkus*; F:* Coomansus*; G:* Parkellus*; H:* Mylonchulus*; I:* Paramylonchulus*; J:* Brachonchulus*; K:* Polyonchulus*; L:* Margaronchulus*; M:* Margaronchuloides*; N:* Oligonchulus*; O:* Megaonchulus*; P:* Crestonchulus*; Q:* Sporonchulus*; R:* Granonchulus*; S:* Judonchulus*; T:* Actus*; U:* Prionchuloides*; V:* Cobbonchus*; W:* Comiconchus*; X:* Cobbonchulus*; Y:* Tricaenonchus.

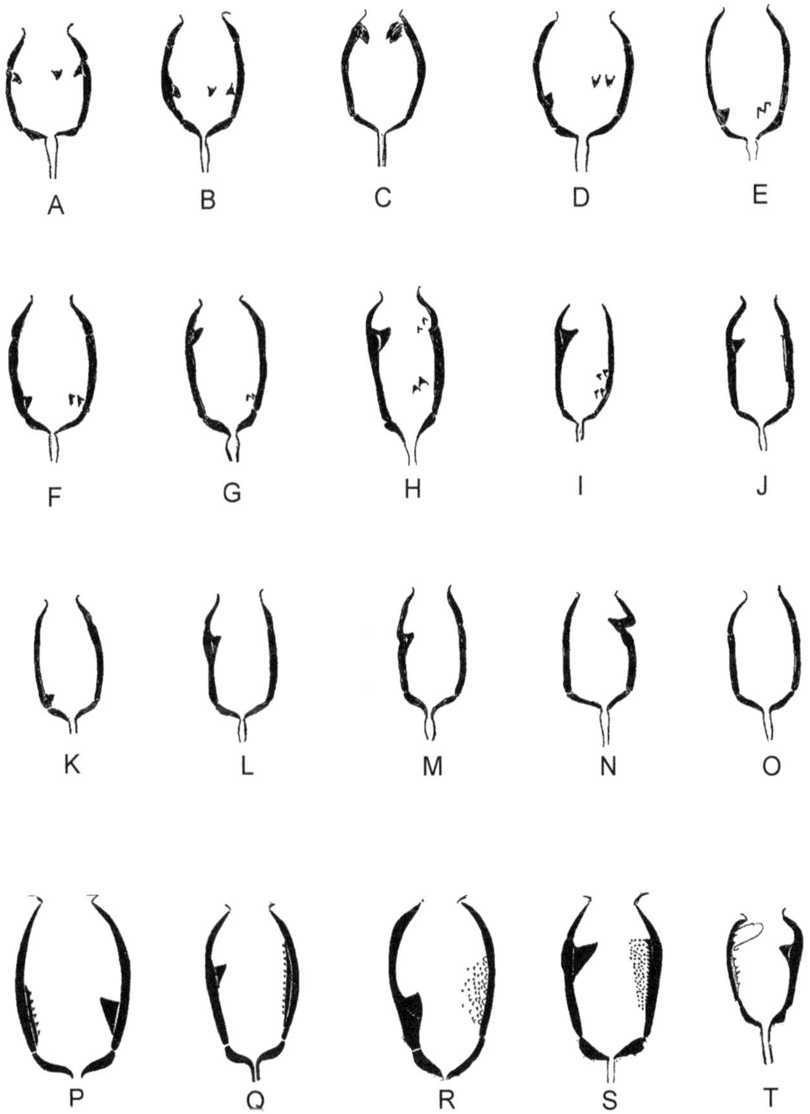

Fig. 12. *Buccal cavity form in Anatonchoidea. A:* Anatonchus*; B:* Truxonchus*; C:* Tigronchoides*; D:* Micatonchus*; E:* Doronchus*; F:* Miconchus*; G:* Promiconchus*; H:* Crassibucca*; I:* Paracrassibucca*; J:* Jensenonchus*; K:* Iotonchus*; L:* Iotonchulus*; M:* Mulveyellus*; N:* Caputonchus*; O:* Nullonchus*; P:* Parahadronchus*; Q:* Hadronchus*; R:* Hadronchulus*; S:* Hadronchoides*; T:* Prionchulellus.

set bears (except in *Nullonchus*) a large, sharply pointed, dorsal tooth, which varies in position, shape and size. It may be located at any place in the buccal cavity, ranging from the base to the anterior quarter in different genera, sometimes within the same genus. The tooth apex may be massive or minute, triangular or digitate, perpendicular or projecting anteriad or posteriad. The two subventral walls of the vertical set may or may not bear a tooth, teeth or denticles. When present, these are known as the subventral or submedian tooth, teeth or denticles. The subventral armature is absent in genera such as *Mononchus*, *Clarkus*, *Coomansus* and *Iotonchus*. The size and position of subventral teeth are also variable. They may be exactly identical to that of the dorsal tooth, as in *Miconchus* or, although similar in size, shape and position, pointing posteriad, as in *Anatonchus*. In *Micatonchus*, the dorsal tooth apex is directed anteriad whereas the subventral teeth are retrorse. The subventrals are of same size and shape as the dorsal in *Cobbonchus* but are located slightly posteriorly. In *Actus*, there are several subventral teeth arranged in a single longitudinal row on each wall. In *Prionchulus*, they are somewhat similar but are provided with a rib-like structure along the row of teeth. The subventral teeth are numerous and arranged in almost regular transverse rows in *Mylonchulus*, but are irregularly scattered in *Sporonchulus* and in some other genera. Some unusual structures are present in other genera, including spatula- or comb- or crest-like structures. Very small peg-like structures are present on the oblique subventral walls of all the type of mononchs and are called 'foramina'. The exact nature and function of these structures are not fully understood. Siddiqi (1983) attributed a chemoreceptory function to foramina (called geusids) but without supportive evidence. Grootaert and Wyss (1979), using TEM, mentioned that in *M. aquaticus* nerves were observed "in the region", although no direct link between these and the foramina were evident. Borgonie *et al.* (1995), using SEM, also failed to provide additional information about this structure and its function. They are, however, of no taxonomic value as they are present in all genera of the group.

Borgonie *et al.* (1995) made a detailed SEM study of the inner and outer surface of the buccal cavity of some mononchs. They observed 12 small liplets at the frontal region of the buccal cavity beneath the lips, two above each of the six lobes. The interior surface of these lobes is marked with fairly deep longitudinal grooves. The dorsal tooth and the subventral teeth were arranged as in Figure 13.

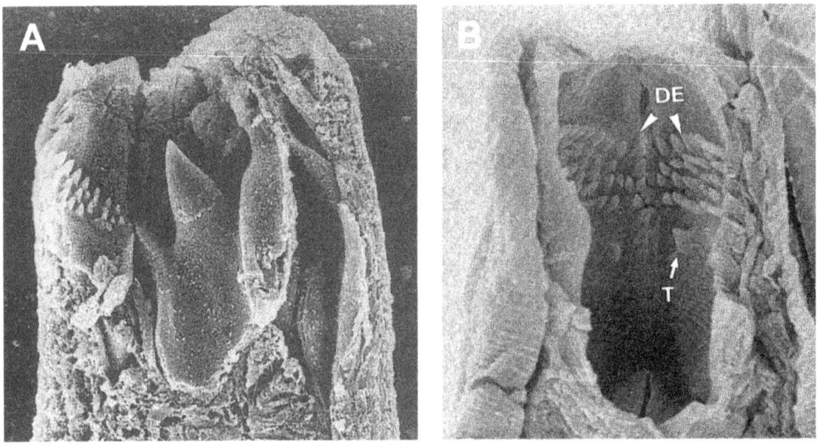

Fig. 13. *SEM of* Mylonchulus minor. *A: Dorsal plate of buccal cavity and associated tissues. The dorsal tooth sits atop a smooth, massive, base; B: Subventral teeth and denticles on subventral plate of buccal cavity (DE = denticles; T = subventral tooth). After Borgonie* et al. *(1995).*

PHARYNX

The pharynx is strongly muscular, long and cylindroid without any narrowing or swellings. The anterior portion is usually slightly expanded to enclose the base of the buccal cavity. It appears slightly narrower at the point where it is enveloped by the nerve ring; thereafter it expands gradually, but only slightly, towards its base. The pharyngeal lumen is very well developed. It is strongly cuticularised and greatly thickened as is evident in cross sections of the pharynx. The lumen is triradiate with one ventral and two subdorsal arms dividing the pharynx into three sectors, one dorsal and two subventral (Fig. 7K-O). The lumen is attached to the walls of the pharynx and is held in position by means of well developed muscles that also assist in its dilation during food intake. The pharyngeal lumen is interrupted at intervals by the orifices of the pharyngeal glands. There is one dorsal and two pairs of subventral pharyngeal glands, which open into the lumen posterior to the nerve ring. All five glands are uninucleate and are similar to those of the Dorylaimida, although their shape, size and location may be somewhat different. The glands themselves may often be obscure due to the strong musculature of the pharynx, but their openings to the lumen are clearly visible. The dorsal gland is anteriormost and its opening (DO) is located at nearly 50-60% of the pharyngeal length from the anterior extremity while the nucleus of the dorsal gland (DN) is placed at 55-66%, *i.e.*, the orifice (DO) is placed anterior to the gland nucleus. The first pair of subventral glands (S_1N) are smaller in size and are situated at 74-88%, their orifices (S_1O) being located at 65-75%. The two glands (S_1N) are placed almost at the same level. The second pair of subventrals (S_2N) is situated near the base of the pharynx at 90-95% and the glands' orifices (S_2O) are very close, lying at 92-98% of the pharyngeal length from anterior end. The subventral glands of the second pair are usually bigger than those of the first pair. Although the position of the pharyngeal gland nuclei and their orifices are relatively important taxonomic characters in the closely related Order Dorylaimida, a fact pointed out by Loof and Coomans (1970), in Mononchina this character seems to be of limited value. Perhaps the paucity of information on different species may be a reason, but it is also possible that they may not be, in reality, of much diagnostic value in this group of nematodes.

PHARYNGO-INTESTINAL JUNCTION

The pharyngo-intestinal junction is made up of transparent tissues and has a funnel-shaped valve (with or without tubercles), sphincter muscles and a conical organ. The detailed structure of this organ differs in size and shape in various species of mononchs. The lumen is triradiate and is lined with cuticle. The conical organ appears rather structureless and may be obscured by the dense granules of intestine. It also varies in size and shape and may project into the lumen of intestine. The sphincter muscles may or may not be demarcated. The structure of the pharyngo-intestinal junction offers a good taxonomic character in Mononchida. Due to the presence or absence of the tubercles, the junction is differentiated into two types: tuberculate and non-tuberculate (see Clark, 1960a).

Tuberculate type

The members of the superfamily Anatonchoidea have the tuberculate type of pharyngo-intestinal junction (Fig. 9K, L). The tubercles were first reported by Ditlevsen (1911) in *Anatonchus dolichurus* and later by Steiner (1914) in *Miconchus studeri*. There is no marked change in the apparent diameter of the pharyngeal lumen as it passes into the transparent zone. The peripheral ends of the three arms of the sclerotised lining become inflated and greatly thickened, forming conspicuous hollow 'tubercles'. From this point a narrow duct extends posteriorly and forms the anterior part of the tripartite funnel-shaped valve.

Non-tuberculate type

The members of the superfamily Mononchoidea have the non-tuberculate type of pharyngo-intestinal junction (Fig. 9M, N). There is an abrupt change in the shape of the pharyngeal lumen – it becomes narrow at the base of pharynx, opens into the transparent zone through a simple funnel and does not form tubercles.

There appears to be a more or less definite correlation between the type of pharyngo-intestinal junction and the shape of the buccal cavity. The tuberculate mononchs have buccal cavities that are wide and flat at the base, whilst the non-tuberculate mononchs have a narrow and somewhat tapering base. The tubercles are usually more clearly visible in living than in preserved material. According to Clark (1960a), the tuberculate type of valve is generally found in those genera in which the tooth or teeth lie in the posterior half of the buccal cavity. However, this generalisation no longer holds good.

INTESTINE

As in other nematode groups, the intestine in mononchs is a simple straight tube made up of a single layer of polygonal, mostly hexagonal, cells. The granules present inside the cells frequently give it a tessellate appearance. In a cross section, the inner surface of the intestine may appear to be lined by a bacillary layer composed of short rod-like structures, which are often referred to as 'cilia', although these are not always present. The cilia, if present, are more prominent at the anterior and the posterior ends of the intestine. The diameter of the intestine is more or less uniform except in the region of gonads where it may be pressed dorsally. In a species of *Mylonchulus*, *M. contractus*, it is very strongly contracted and pushed dorsally in the region of gonads, hence the specific epithet. The intestinal lumen of mononchs is exceptionally wide giving it a saccate appearance and, as they feed on soil microorganisms such as protozoans, rotifers, small oligochaetes, nematodes, *etc.*, it may contain remnants of the bodies, with heavily cuticularised portions like the spear, spicules, buccal cavity, setae, *etc.*, of the ingested animals often being visible within the lumen. Sometimes the entire bodies of the nematode(s) engulfed by a mononch can be observed, apparently intact, within the intestine. If a mononch is killed immediately after it has taken a meal consisting of nematodes, the latter, if engulfed intact, can easily be identified up to genus or even species level. When the prey is in the anterior part of the intestine it appears fairly intact, but as it passes down the lumen of the intestine it gradually gets digested and only the harder cuticularised structures may remain intact, these ultimately being passed out with other faecal matter through the anus.

RECTUM AND ANUS

The posterior part of the intestine is connected to a dorsoventrally compressed tube, the rectum. This junction is guarded by a sphincter made up of muscles of the intestino-rectal valve. The rectum is lined internally by the cuticle and is externally covered by a layer of large epithelial cells. Sometimes, unicellular rectal glands open into the rectum through a common duct. These glands may be clearly visible, *e.g.*, as in *Mononchus aquaticus*. In females, the rectum opens externally through the anus, but in males, the ejaculatory duct and the rectum join to form a cloaca. The cloacal aperture may be situated on a raised area. The rectal glands, if present, are 3-5 in number and empty into the rectum. The rectum in males is usually longer than that in females of the same species.

Nerve ring

The nerve ring is a conspicuous, broad, collar-like structure located at 22-40% of the pharyngeal length from the anterior end of body (Fig. 9I). It may show a slight ventro-dorsal slant. Its position in different species may be variable but it is doubtful if it can be of taxonomic value. A hemizonid is usually absent, but in *Mononchus* it has been found to be present posterior to the nerve ring.

Excretory system

The excretory system is usually obscure in mononchs as it is in other members of the Adenophorean group of nematodes. However, in a few species a complete excretory system has been traced, although in others only the excretory pore and the ampulla have been observed. Cobb (1917) had doubtfully reported the presence of an excretory pore a little posterior to the nerve ring in *Clarkus papillatus* and *Iotonchus trichurus*. Chitwood and Chitwood (1937) and Coomans and Lima (1965) detected the excretory pore in *Prionchulus muscorum* and *Anatonchus amiciae*, respectively. Jairajpuri and Khan (1975) reported for the first time the existence of a complete excretory system in *Prionchulus muscorum* and subsequently it was also observed in some other species. The excretory system in mononchs comprises two large uninucleate renette cells, each connected to an ampulla through separate excretory ducts that run almost parallel to the pharynx (Fig. 9I, J). The excretory pore, which is situated on the ventral surface of the body a little posterior to the nerve ring at 27-35% of the pharyngeal length from the anterior end of the body, is marked by a prominent depression in the cuticle. The shape and size of the ampulla may be variable. After its emergence from the ampulla, the excretory duct immediately bifurcates, the two ducts proceeding ventrolaterally with each terminating in a renette suspended in the lateral sector of the pseudocoelom. Sometimes the excretory duct may be obscure, but the excretory pore, ampulla and renettes are clearly visible. The excretory system in mononchs, in having two renettes, one on each side of the body, is reminiscent of the advanced Secernentean type with paired renettes and their outgrowths in the form of longitudinal canals.

Female genital organs

The female reproductive system (Figs 14, 15) may comprise one set (mono-prodelphic) or two sets (didelphic, amphidelphic). Each set possesses an ovary, oviduct and uterus leading to a common vagina and vulva. The sexual branches are usually large and well developed. The gonads are telogonic and proliferation of the germ cells takes place only at the apical end of the ovary. The terminal and growth zones are clearly demarcated.

Ovary

The ovary is reflexed with its terminal end near the ventral surface of the body. The oocytes have large nuclei and are usually arranged in a single row except at the tip of ovary where proliferation of the germ cells takes place. The germinal zone is smaller than the growth zone.

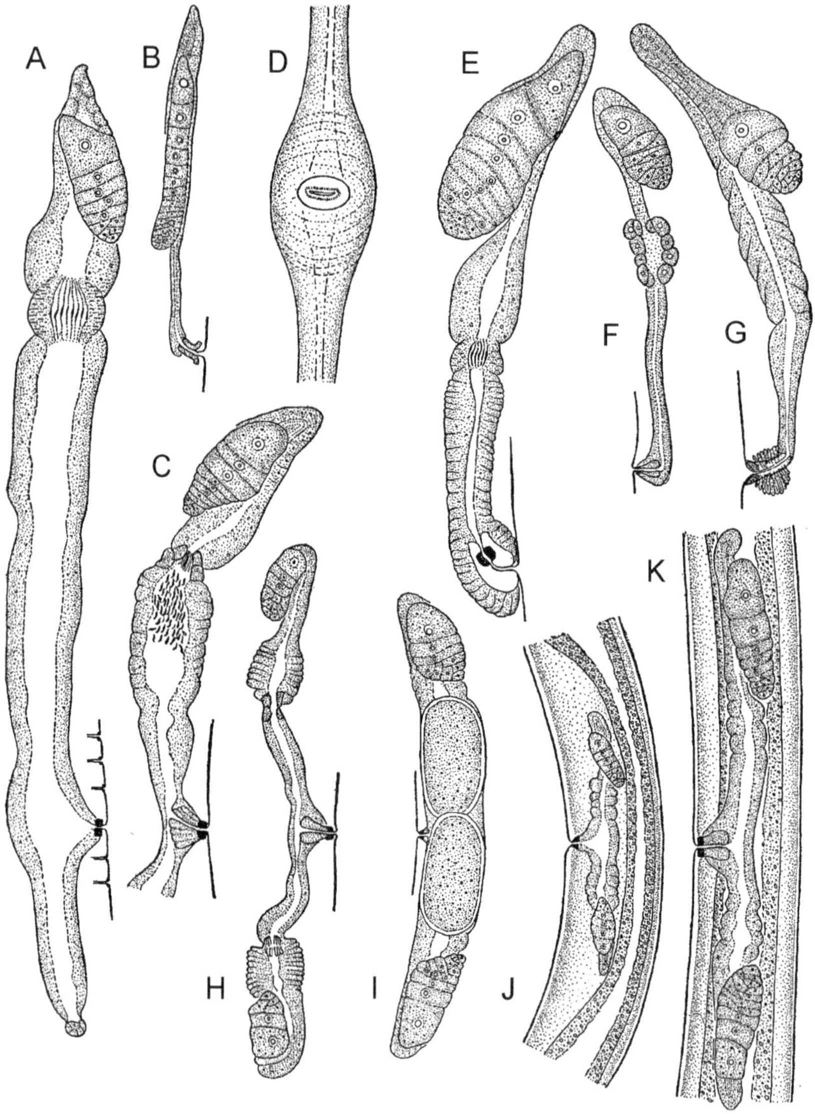

Fig. 14. *Female genital system. A:* Parahadronchus andamanicus*; B:* Mononchulus nodicaudatus*; C:* Iotonchus narari*; D: Vulval slit (dorsoventral view); E:* Iotonchus baqrii*; F:* Paramylonchulus mulveyi*; G:* Mulveyellus jairi*; H:* Prionchulus muscorum*; I:* Clarkus papillatus*; J:* Mylonchulus contractus*; K:* Mononchus tunbridgensis. *After Jairajpuri and Khan (1982).*

Oviduct

The oviduct is made up of high, columnar, epithelial cells and consists of a narrow distal and an enlarged proximal part. The latter may also function as a spermatheca. The junction of the uterus and the oviduct is usually guarded by a well developed muscular valve, the sphincter (Fig. 15J). The latter may be poorly developed, obscure or absent. In *Prionchulus muscorum* the sphincter is absent, but in *Parahadronchus shakili* and *Miconchus aquaticus* it is very prominent.

Uterus

The uterus is hollow and sac-like, its walls being made up of circular and oblique muscles. The inner lining is composed of squamous epithelial cells. The uterus may function as an organ for the storage of sperm. Fertilisation occurs in this region and it may also contain egg(s) (Fig. 15G, H). Usually a single egg is present in the uterus at a time, although rarely there may be more. The mononchs are oviparous and a single female produces several eggs during its life span. The eggs do not have any characteristic shape, but are usually one and a half to two times longer than broad. The outer surface of the egg shell is smooth, but in species like *Prionchulus punctatus* it may be echinulate.

Vagina

The vagina is a short, thick-walled tube rarely exceeding half of the corresponding body diameter and is encircled by very strong muscles. In some species, uninucleate gland cells may also be associated with it (*e.g.*, *Mononchus aquaticus*). These cells may be clearly visible in live specimens but become obscure upon fixation. The lining of the vagina is not highly cuticularised, although prominent cuticularised pieces may occur at the vulva-vagina junction. The shape, size, and often the position, of these cuticularised structures are variable. They may be crescent-shaped, oval, rectangular, or tear-drop shaped. These pieces are usually situated at the vulva-vagina junction, but in *Iotonchus baqrii* they are farther away (9-12 μm) from the vulva (Fig. 14E).

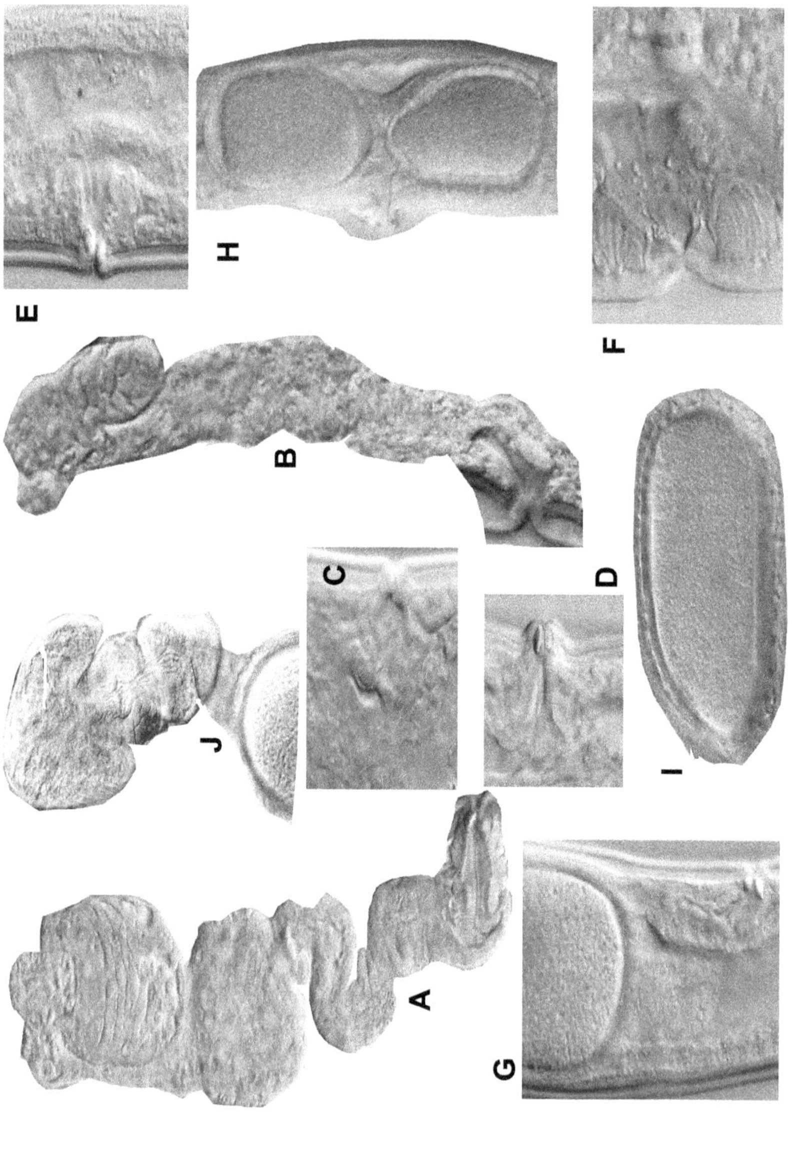
E
H
F
B
D
C
J
I
A
G

Male genital organs

The male reproductive system (Fig. 16) comprises two testes, *vas deferens* and an ejaculatory duct. The copulatory apparatus is made up of spicules, gubernaculum, accessory pieces and the ventromedian supplements. The male gonad is telogonic.

TESTES

The testes are paired, opposed and outstretched as in the Dorylaimida. The distal part of each testis is occupied by spermatogonia, which are arranged in many rows, the proximal part being filled with sperm. The sperm of different species of mononchs vary in shape and size. The testes are connected to the seminal vesicles, which are dilated portions of the male gonoduct and function as organs for the storage of developing sperm. The seminal vesicles join a common *vas deferens*, which is composed of tubular and glandular regions. This is followed by an ejaculatory duct, which gradually narrows posteriorly and opens into the rectum to form the cloaca.

SPICULES

The spicules are paired, ventrally arcuate, structures with a knobbed head and bidentate terminus (Fig. 16). Some species have broad spicules with a bluntly rounded terminus (*Mylonchulus orbitus*). The spicules may be long and slender (*Mononchus*) or short and thick (*Clarkus*, *Prionchulus*), and their shape and length are relatively important diagnostic features.

GUBERNACULUM

The gubernaculum is variable in length and thickness (Fig. 16). It may be keel-like (*Crassibucca penicula*), or the anterior part may be thickened while the posterior part is long and thin (*Hadronchus*, *Parahadronchus*). Some species have a gubernaculum with bifurcated

Fig. 15. *Female genital system. A, B: Mono-prodelphic gonad; C: Vulval region of prodelphic species; D-F: Vulval region of amphidelphic species; G: Prodelphic gonad showing part of egg; H: Amphidelphic gonad showing an egg in each uterus; I: Egg with punctations.*

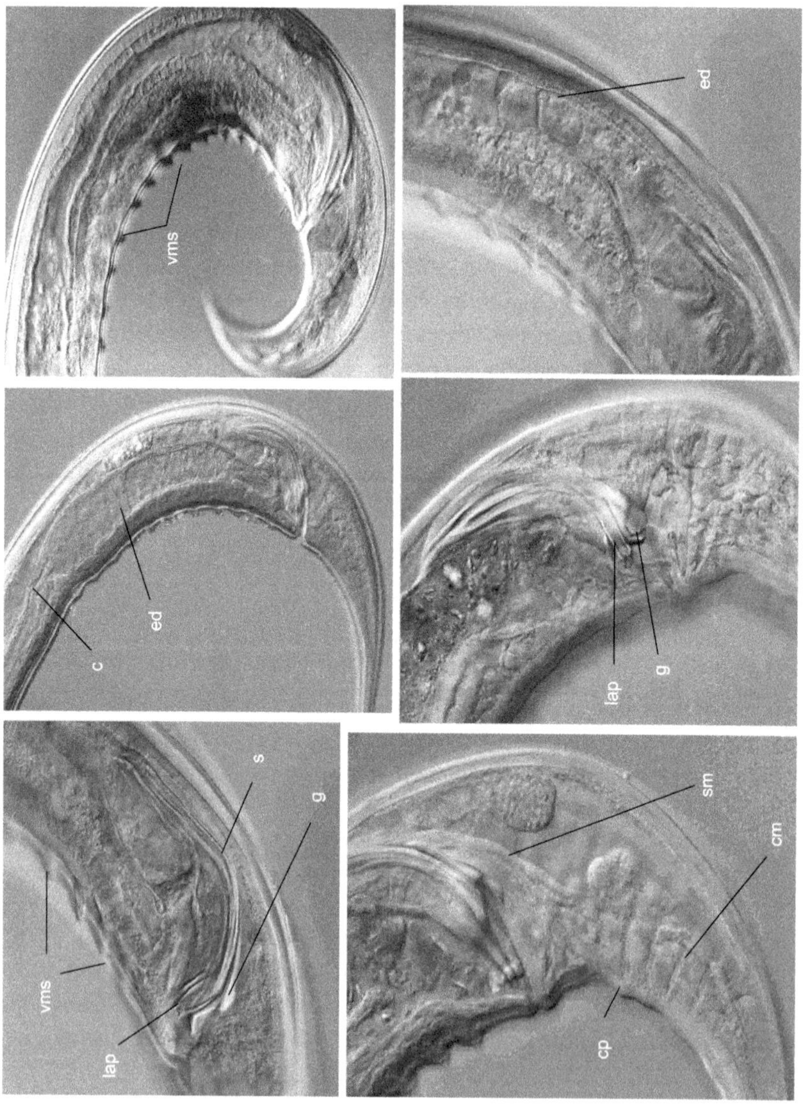
vms
ed
c
ed
lap
g
s
g
vms
lap
sm
cm
cp

tips enveloping the spicules (*Iotonchus vulvapapillatus*, *Polyonchulus megadontus*). The shape and size of the gubernaculum is useful in species identification.

LATERAL ACCESSORY PIECES

The lateral accessory pieces, also known as lateral guiding pieces, are present on either side of the spicules near their terminus (Fig. 16). Lateral accessory pieces have more or less the same shape and structure in all species. They are elongate and rectangular and end in a bidentate terminus. Rarely, they may be absent, *e.g.*, *Clarkus papillatus*, *Iotonchus basidontus*, *Polyonchulus cobbi*, *etc*. The presence or absence and size of the lateral accessory pieces are good taxonomic characters.

VENTROMEDIAN SUPPLEMENTS

The ventromedian supplements are prominently raised structures located near the cloacal opening or within the range of the spicules (Fig. 16). However, some may be placed more anteriorly. The number and arrangement of these supplements, although slightly variable within a species, are helpful in species identification. They may be spaced regularly or irregularly. Sometimes one or more partially developed supplements may also be present anterior to the series. The number of the supplements varies from nine (*Mylonchulus contractus*) to 50 (*Mononchus superbus*). The number and disposition of the ventromedian supplements are taxonomically relatively important.

Tail

The tail in both sexes in mononchs is similar (except in *Doronchus* where sexual dimorphism occurs), but varies greatly in shape and size amongst the species (Fig. 17), *viz.*, very long filiform (*Iotonchus trichurus*), moderately long filiform (*Hadronchus*, *Parahadronchus*), elongate (*Mononchus*, *Miconchus*), elongate-conoid (*Sporonchulus*), conoid

Fig. 16. *Morphology of male posterior region. Abbreviations: s = spicule; g = gubernaculum; lap = lateral accessory piece; vms = ventromedian supplements; ed = ejaculatory duct; c = constriction between* vas deferens *and ejaculatory duct; cp = caudal papillae; cm = caudal muscles; sm = spicular muscles.*

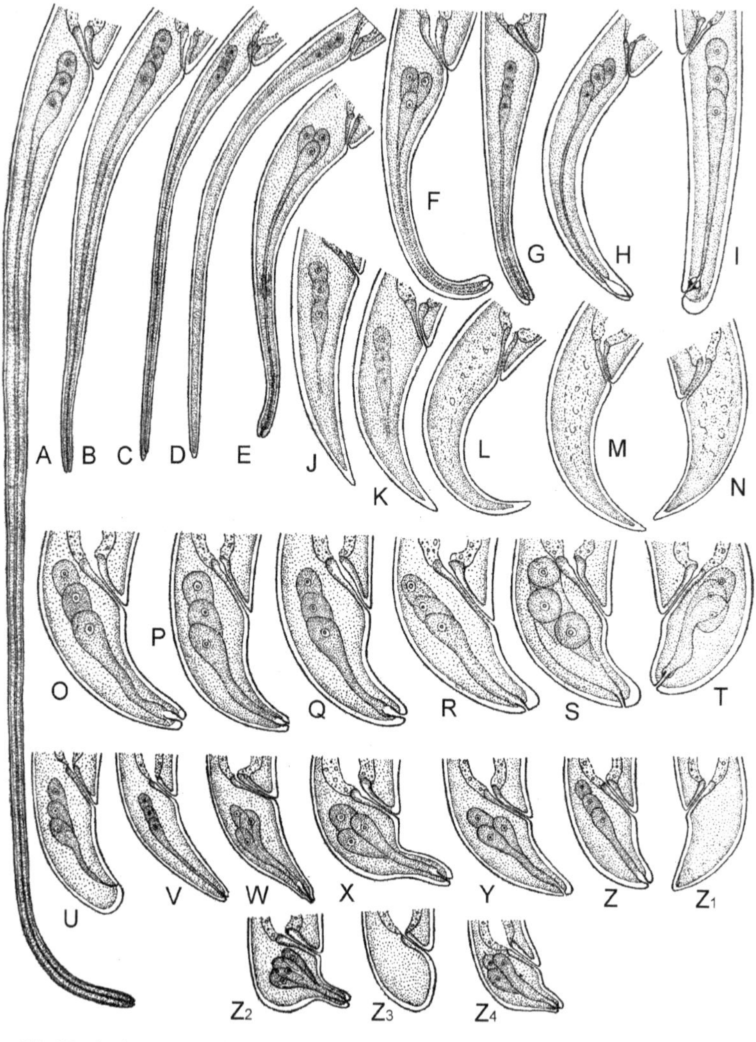

Fig. 17. *Variation in tail shape. A:* Iotonchus trichurus; *B:* Parahadronchus shakili; *C:* Iotonchus indicus; *D:* I. tarjani; *E:* I. basidontus; *F:* Mononchus tunbridgensis; *G:* M. aquaticus; *H:* Miconchus thornei; *I:* Mononchulus nodicaudatus; *J:* Miconchus citri; *K:* Coomansus pretoriensis; *L:* C. parvus; *M:* Clarkus sheri; *N:* Prionchulus muscorum; *O:* Mylonchulus hawaiiensis; *P:* M. minor; *Q:* M. lacustris; *R:* M. brachyuris; *S:* M. nainitalensis; *T:* M. muradi; *U:* M. rotundicaudatus; *V:* Sporonchulus vagabundus; *W:* Paramylonchulus mulveyi; *X:* M. sigmaturus; *Y:* M. contractus; *Z:* M. orbitus; Z_1: M. subsimilis; Z_2: Paramylonchulus index; Z_3: Iotonchus ovatus; Z_4: Cobbonchus indicus. *After Jairajpuri and Khan (1982).*

(*Prionchulus, Clarkus, Coomansus*), bluntly-conoid (*Mylonchulus orbitus*), convex-conoid to obtusely rounded (*Mylonchulus muradi*), bluntly rounded to hemispherical (*Iotonchus ovatus, Mylonchulus striatus*). Caudal pores are present but may or may not be clearly visible. The length and shape of tail are regarded as one of the most important diagnostic characters in Mononchida at the species or genus level. The number and disposition of the caudal pores may also be important but have yet to be exploited fully for taxonomic purposes.

CAUDAL GLANDS

Three unicellular glands are usually present in the anterior part of the tail (Fig. 17). In species where they are well developed, these glands join to form a common duct that opens to the exterior at or near the tail tip. The gland opening is guarded by a well developed cuticularised structure, the 'spinneret'. It is generally situated terminally (*Mononchus, Paramononchus, Actus, Sporonchulus, Mylonchulus minor, M. lacustris, M. hawaiiensis, Paramylonchulus, Anatonchus, Iotonchus basidontus, I. parabasidontus, I. shafii, Miconchus thornei, Parahadronchus, etc.*), but may be dorsally subterminal (*Mylonchulus contractus, M. brachyuris, M. nainitaliensis, Iotonchus indicus, I. acuticaudatus, etc.*) or rarely ventrally subterminal (*Mylonchulus rotundicaudatus*). The glands are either grouped (*Mylonchulus brachyuris, M. index, M. californicus, Paramylonchulus mulveyi*) or arranged in tandem (*Mononchus, Iotonchus, Anatonchus, Parahadronchus, Mylonchulus lacustris, M. minor, M. hawaiiensis*). Their arrangement differs from species to species but may also be variable within a species (*Mylonchulus hawaiiensis*). Caudal glands may be entirely absent (*Prionchulus, Clarkus, Coomansus, Mylonchulus subsimilis, M. exacutus, Hadronchus, Prionchulellus*). Sometimes, when the caudal glands are poorly developed the spinneret may be obscure or absent (*Iotonchus tarjani, Miconchus citri*). The presence or absence and the nature and disposition of the caudal glands and spinneret are considered to be important taxonomic characters.

Nematology Monographs & Perspectives, 2010, Vol. 7, 45-50

Chapter 3

Order Mononchida Jairajpuri, 1969

DIAGNOSIS

Medium to large nematodes, generally well over 1 mm long. Cuticle thick. Lip region expanded, lips and labial papillae well developed. Buccal cavity spacious, strongly sclerotised, posterior quarter or entire length embedded in pharyngeal tissue, provided with tooth or teeth, with or without longitudinal ventral ridges. Subventral walls either smooth or provided with large teeth or small denticles arranged in longitudinal ridges or scattered irregularly. Amphids small, goblet-shaped, aperture post-labial. Pharynx long, cylindrical, strongly muscular, with greatly thickened lumen. Pharyngo-intestinal junction tuberculate or non-tuberculate. Pharyngeal glands uninucleate; one dorsal and two pairs of subventrals, their orifices located posterior to nerve ring. Excretory system, when visible, with an ampulla and a pair of excretory ducts each terminating in a large uninucleate cell, the renette. Female genital system amphidelphic, mono-prodelphic or mono-opisthodelphic, with or without vulval papillae. Males lacking adanal pair of supplements but with a series of ventromedian papillae. Spicules paired and identical, gubernaculum and lateral accessory pieces present. Caudal glands three, uninucleate. Spinneret terminal or subterminal. Inhabitants of soil, rarely of freshwater. Predatory in habit.

TYPE SUBORDER

Mononchina Kirjanova & Krall, 1969

OTHER SUBORDER

Bathyodontina Coomans & Loof, 1970*

---

* See Chapter 9.

Suborder Mononchina Kirjanova & Krall, 1969

DIAGNOSIS

Buccal cavity large and wide, posterior quarter embedded in pharyngeal tissue. Dorsal tooth prominent, ventral ridges and subventral teeth may or may not be present. Ventral ridges with or without denticles; subventral teeth may be as large as dorsal tooth and/or in form of denticles. Dorsal pharyngeal gland nucleus located posterior to dorsal pharyngeal gland orifice, but anterior to subventral gland opening. First subventral gland nucleus located posterior to first subventral gland orifice. Well developed gubernaculum and prominent lateral accessory pieces present.

TYPE SUPERFAMILY

Mononchoidea Filipjev, 1934

OTHER SUPERFAMILY

Anatonchoidea Jairajpuri, 1969

Key to superfamilies of Mononchina Kirjanova & Krall, 1969

1 – Buccal cavity with a narrow base; pharyngo-intestinal junction non-tuberculate . . . . **Mononchoidea**
– Buccal cavity with a broad and flat base; pharyngo-intestinal junction tuberculate . . . . **Anatonchoidea**

Superfamily Mononchoidea Filipjev, 1934

DIAGNOSIS

Buccal cavity thick-walled, tapering at base, dorsal tooth medium to large. Subventral teeth, if present, may be as large as dorsal tooth or smaller. Ventral longitudinal ridges, if present, with or without denticles. Pharyngo-intestinal junction non-tuberculate. Tail usually short conoid or cylindroid, rarely filiform, caudal glands poorly to well developed, rarely absent.

Type family

Mononchidae Filipjev, 1934

Other families

Cobbonchidae Jairajpuri, 1969
Mylonchulidae Jairajpuri, 1969

Key to families of Mononchoidea Filipjev, 1934

1 – Subventral and dorsal teeth nearly equal in size .... **Cobbonchidae**
– Subventral teeth, if present, much smaller (= denticles) than dorsal tooth 2
2 – Denticles forming transverse rows, or scattered, or both .... **Mylonchulidae**
– Denticles either absent or arranged along a longitudinal rib .... **Mononchidae**

Superfamily Anatonchoidea Jairajpuri, 1969

Diagnosis

Buccal cavity broad and flattened at base. Dorsal tooth of varying size, position and direction. Subventral teeth, if present, directed anteriorly or posteriorly or in form of two longitudinal rows of denticles. Pharyngo-intestinal junction tuberculate. Tail conoid or long cylindroid, may be filiform, caudal glands poorly developed, may be absent.

Type family

Anatonchidae Jairajpuri, 1969

Other family

Iotonchidae Jairajpuri, 1969

Key to families of Anatonchoidea Jairajpuri, 1969

1 – Subventral and dorsal teeth nearly equal in size .... **Anatonchidae**

– Subventral teeth, if present, much smaller (denticles) than dorsal and arranged in longitudinal rows . . . . **Iotonchidae**

Outline classification of Mononchida

Order Mononchida Jairajpuri, 1969
Suborder Mononchina Kirjanova & Krall, 1969
Superfamily Mononchoidea Filipjev, 1934
Family Mononchidae Filipjev, 1934
Subfamily Mononchinae Filipjev, 1934
Mononchus Bastian, 1865
= *Mononchus* (*Mononchus* Cobb, 1916)
Paramononchus Mulvey, 1978
Nigronchus Siddiqi, 1984a
Subfamily Prionchulinae Andrássy, 1976
Prionchulus Cobb, 1916
= *Mononchus* (*Prionchulus* Cobb, 1916)
Clarkus Jairajpuri, 1970a
Coomansus Jairajpuri & Khan, 1977
Parkellus Jairajpuri, Tahseen & Choi, 2001b
Family Mylonchulidae Jairajpuri, 1969
Subfamily Mylonchulinae Jairajpuri, 1969
Mylonchulus Cobb, 1916
= *Mononchus* (*Mylonchulus* Cobb, 1916)
= *Pakmylonchulus* Khan & Saeed, 1987
Brachonchulus Andrássy, 1958
Polyonchulus Mulvey & Jensen, 1967
Margaronchulus Andrássy, 1972
Oligonchulus Andrássy, 1976
Megaonchulus Jairajpuri & Khan, 1982
Paramylonchulus Jairajpuri & Khan, 1982
Crestonchulus Siddiqi & Jairajpuri, 2002
Margaronchuloides **gen. n.**
Subfamily Sporonchulinae Jairajpuri, 1969
Sporonchulus Cobb, 1917
= *Mononchus* (*Sporonchulus* Cobb, 1917)
Granonchulus Andrássy, 1958
Judonchulus Andrássy, 1958

Prionchuloides Mulvey, 1963b
Actus Baqri & Jairajpuri, 1974
Family Cobbonchidae Jairajpuri, 1969
Subfamily Cobbonchinae Jairajpuri, 1969
Cobbonchus Andrássy, 1958
Comiconchus Jairajpuri & Khan, 1982
Tricaenonchus Andrássy, 1996
Cobbonchulus Andrássy, 2009
Superfamily Anatonchoidea Jairajpuri, 1969
Family Anatonchidae Jairajpuri, 1969
Subfamily Anatonchinae Jairajpuri, 1969
Anatonchus Cobb, 1916
= *Mononchus* (*Anatonchus* Cobb, 1916)
= *Tigronchus* Kirjanova & Krall, 1969
Tigronchoides Ivanova & Dzhuraeva, 1971
Truxonchus Siddiqi, 1984a
Micatonchus Jairajpuri, Tahseen & Choi, 2001a
Subfamily Miconchinae Andrássy, 1976
Miconchus Andrássy, 1958
= *Miconchoides* Jairajpuri & Khan, 1982
Crassibucca Mulvey & Jensen, 1967
Paracrassibucca Baqri & Jairajpuri, 1974
Promiconchus Jairajpuri & Khan, 1982
Doronchus Andrássy, 1993b
Family Iotonchidae Jairajpuri, 1969
Subfamily Iotonchinae Jairajpuri, 1969
Iotonchus Cobb, 1916
= *Mononchus* (*Iotonchus* Cobb, 1916)
Jensenonchus Jairajpuri & Khan, 1982
Caputonchus Siddiqi, 1984a
Mulveyellus Siddiqi, 1984a
Nullonchus Siddiqi, 1984d
Iotonchulus Andrássy, 1993b
Subfamily Hadronchinae Khan & Jairajpuri, 1980
Prionchulellus Mulvey & Jensen, 1967
Hadronchus Mulvey & Jensen, 1967
Parahadronchus Mulvey, 1978
Hadronchulus Ray & Das, 1983
= *Morenchus* Dhanachand, Renubala & Mohilal, 1991

Hadronchoides Jairajpuri & Rahman, 1984

Suborder Bathyodontina Coomans & Loof, 1970

Superfamily Cryptonchoidea Chitwood, 1937

Family Cryptonchidae Chitwood, 1937

Cryptonchus Cobb, 1913
= *Gymnolaimus* Cobb, 1913
= *Ditlevsenia* Micoletzky, 1925
= *Cryptonchulus* Sukul, 1969

Family Bathyodontidae Clark, 1961d

Bathyodontus Fielding, 1950
= *Mirolaimus* Andrássy, 1956

Superfamily Mononchuloidea De Coninck, 1965

Family Mononchulidae De Coninck, 1965

Mononchulus Cobb, 1918
Oionchus Cobb, 1913
= *Enoplochilus* Kreis, 1932

Nematology Monographs & Perspectives, 2010, Vol. 7, 51-88

Chapter 4

Family Mononchidae Filipjev, 1934

Chitwood (1937) proposed the family Mononchidae for the placement of all species of mononch. However, at that time only two genera, *viz.*, *Mononchus* Bastian, 1865 (with subgenera *Mononchus*, *Mylonchulus*, *Anatonchus* and *Sporonchulus*) and *Prionchulus* (Cobb, 1916) Wu & Hoeppli, 1929, were known. The family Mononchidae was later raised to the rank of a superfamily by Clark (1961d). When Jairajpuri (1969) proposed the Order Mononchida, he restricted this family to those mononchs which had their dorsal tooth in the anterior to middle region of the buccal cavity and a non-tuberculate pharyngo-intestinal junction. He recognised only two genera, *viz.*, *Mononchus* and *Prionchulus*, but in 1970 added a new genus, *Clarkus*, to this group (Fig. 18). Baqri and Jairajpuri (1974) and Jairajpuri and Khan (1977) added *Actus* and *Coomansus*, respectively, to this group. Andrássy (1976) proposed the subfamily Prionchulinae with *Prionchulus* as its type genus. Mulvey (1978) added the new genus *Paramononchus*. Jairajpuri and Khan (1982) recognised two subfamilies, Mononchinae for the genera *Mononchus* and *Paramononchus*, and Prionchulinae for the genera *Prionchulus*, *Clarkus*, *Coomansus* and *Actus*. Siddiqi (1984a) added a new genus, *Nigronchus*, to the subfamily Mononchinae. Andrássy (1993a) accepted two subfamilies, Mononchinae with nine genera, *viz.*, *Mononchus*, *Actus*, *Clarkus*, *Coomansus*, *Prionchulus*, *Judonchulus*, *Nigronchus*, *Paramononchus* and *Sporonchulus*; and Cobbonchinae with two genera, *viz.*, *Cobbonchus* and *Comiconchus*. Jairajpuri *et al.* (2001b) added the genus *Parkellus* to this family. Zullini and Peneva (2006) also followed the same grouping. We do not agree with Andrássy (1993a) in grouping *Cobbonchus* and related genera in Mononchidae, and the family Cobbonchidae is retained here as originally proposed by Jairajpuri (1969) and also accepted by Jairajpuri and Khan (1982). The various genera belonging to this family are grouped here into two subfamilies – Mononchinae with the genera *Mononchus*, *Paramononchus* and

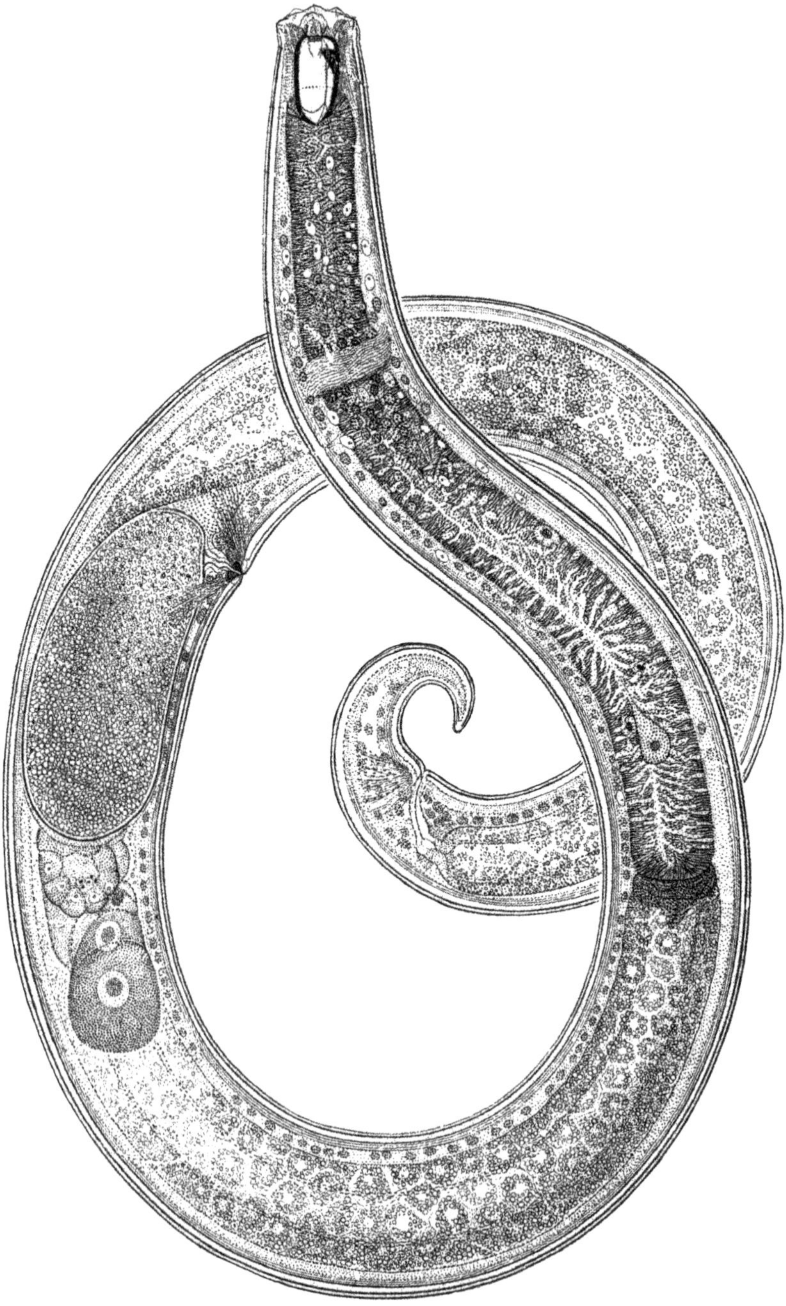

Fig. 18. Clarkus papillatus *(Bastian, 1865) Jairajpuri, 1970. After Cobb (1917).*

Nigronchus, and Prionchulinae with the genera *Prionchulus*, *Clarkus*, *Coomansus* and *Parkellus*.

DIAGNOSIS

Medium to large-sized nematodes. Buccal cavity spacious, more or less oval, tapering at base, armed at least with a large dorsal tooth, predominantly located in anterior part of buccal cavity and directed anteriad. Subventral walls either smooth or provided with large teeth or small denticles arranged in longitudinal ridges or scattered irregularly. Pharyngo-intestinal junction simple, non-tuberculate. Caudal glands and spinneret present or absent.

TYPE SUBFAMILY

Mononchinae Filipjev, 1934

OTHER SUBFAMILY

Prionchulinae Andrássy, 1976

Key to subfamilies of Mononchidae Filipjev, 1934

1 – Subventral walls of buccal cavity with fine transverse or longitudinal ribs at level of dorsal tooth; caudal glands and spinneret well developed .... **Mononchinae**
– Subventral walls of buccal cavity with weak to well developed denticulate or non-denticulate longitudinal ridges along entire vertical plate; caudal glands and spinneret absent or obscure .... **Prionchulinae**

Subfamily Mononchinae Filipjev, 1934

DIAGNOSIS

Lips and labial papillae poorly developed, lip region almost continuous with body. Dorsal tooth in anterior half or anterior third of buccal cavity. Spicules and accessory pieces extremely long and slender. Tail elongate, terminus rounded. Caudal glands with duct and spinneret well developed.

TYPE GENUS

Mononchus Bastian, 1865

OTHER GENERA

Nigronchus Siddiqi, 1984a
Paramononchus Mulvey, 1978

Key to genera of Mononchinae Filipjev, 1934

1 – Lip region sharply offset by constriction; subventral walls of buccal cavity without transverse or longitudinal ribs; female genital system mono-prodelphic . . . . *Nigronchus*
– Lip region either faintly or not offset; subventral walls of buccal cavity with a thin transverse and/or longitudinal rib; female genital system amphidelphic, very rarely mono-prodelphic 2
2 – Subventral walls of buccal cavity with two fine transverse ribs at level of dorsal tooth . . . . *Mononchus*
– Subventral walls of buccal cavity with short longitudinal ridges armed with two pairs of small toothlets at level of dorsal tooth . . . . *Paramononchus*

Genus *Mononchus* Bastian, 1865
(Figs 19; 20A-C)

DIAGNOSIS

Body 1-6 mm long. Lip region almost continuous with body; lips and labial papillae poorly developed. Buccal cavity elongate-cylindroid, thin-walled, tapering at base. Dorsal tooth well developed, lying in anterior third to sixth of buccal cavity and directed anteriad. Opposite to dorsal tooth, each subventral wall possessing thin, transverse rib. Short and weak longitudinal rib may be present in anterior third of buccal cavity. Pharyngo-intestinal junction non-tuberculate. Female genital system amphidelphic, in one species mono-opisthodelphic. Vulva median, at 46-62% of body length. Males common; spicules and accessory pieces extremely long and slender, gubernaculum crescent-shaped, supplements 10-47, regularly arranged. Tail of both sexes

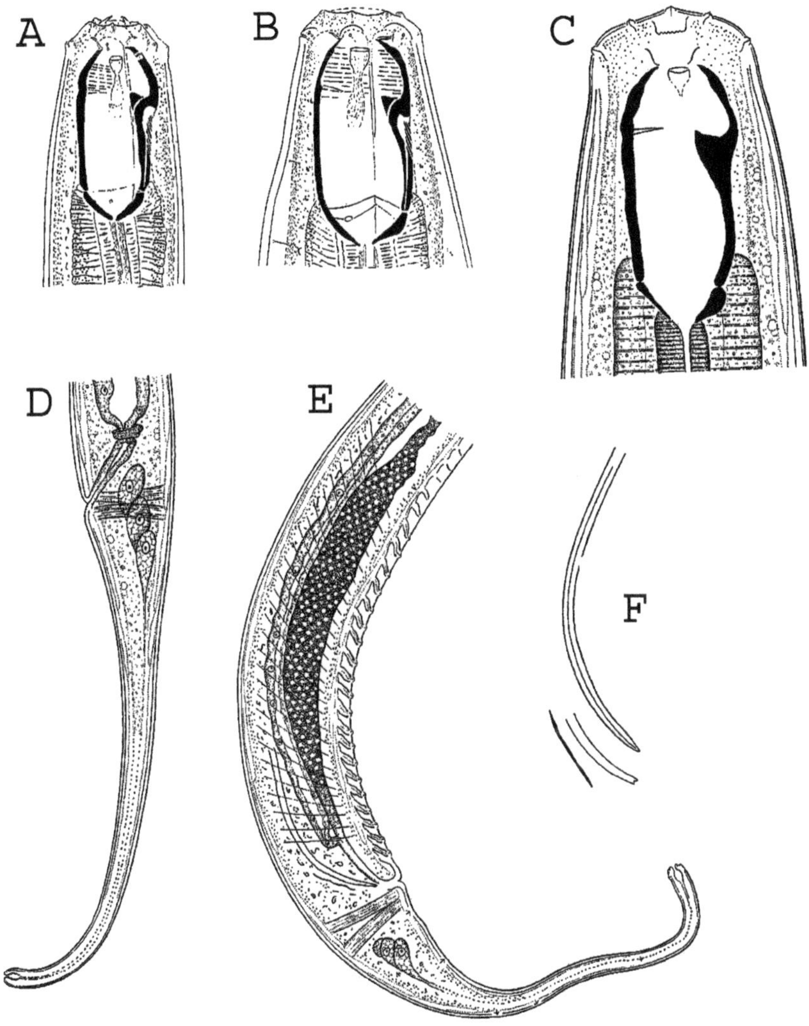

Fig. 19. Mononchus truncatus *Bastian, 1865. A-C: Anterior region; D: Female posterior region; E: Male posterior region; F: Spicule, lateral guiding piece and gubernaculum. A, B after Coomans and Khan (1981); C-F after Jiménez-Guirado* et al. *(2007).*

similar, either short and obtuse or elongate, cylindroid, 1.2-10 anal body diam. long, tip rounded to clavate. Caudal glands and spinneret well developed. Generally found in moist soils.

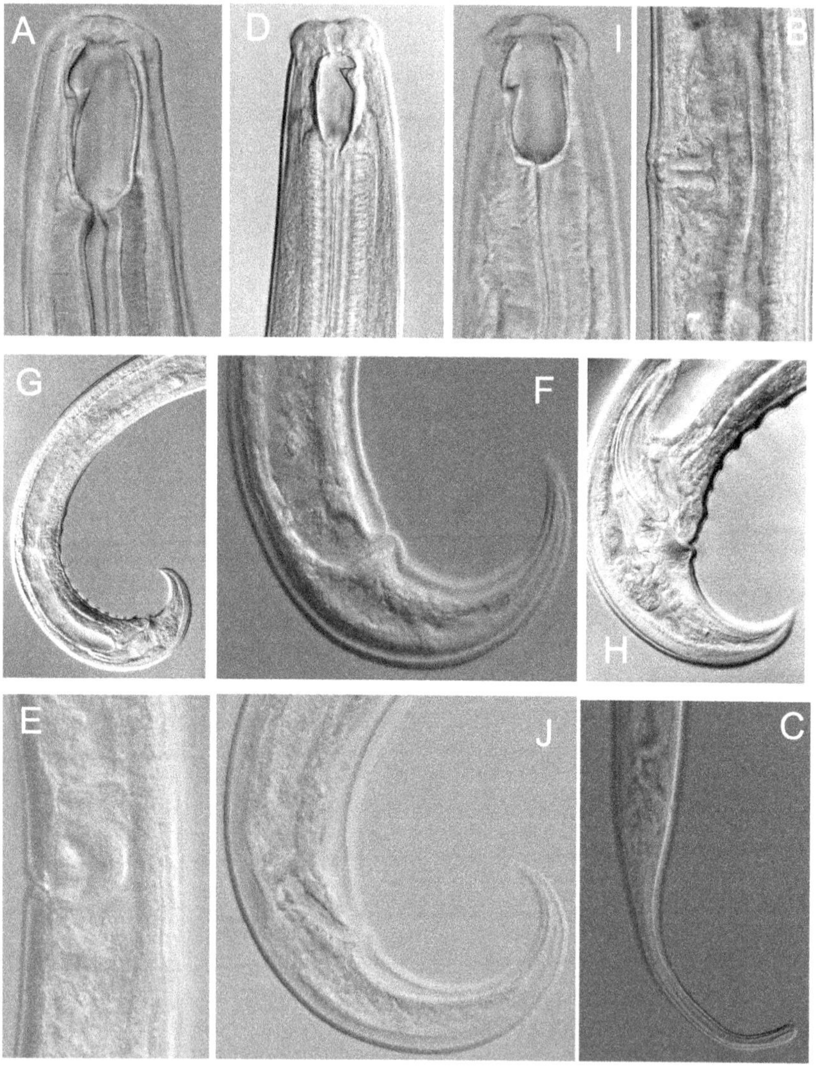

Fig. 20. *A-C:* Mononchus truncatus *Bastian, 1865. A: Anterior region; B: Vulval region; C: Tail region; D-H:* Clarkus papillatus *(Bastian, 1865) Jairajpuri, 1970. D: Anterior region; E: Vulval region; F: Female tail region; G: Male posterior region; H: Male tail region; I, J:* Coomansus parvus *(de Man, 1880) Jairajpuri & Khan, 1977. I: Anterior region; J: Female tail region.*

TYPE SPECIES

Mononchus truncatus Bastian, 1865
= *M. macrostoma* Bastian, 1865
= *M. longicaudatus* Cobb, 1893
= *M. macrostoma longicaudatus* Cobb, 1893 (Micoletzky, 1922)
= *M. macrostoma armatus* von Daday, 1894
= *M. megalaimus* Cobb, 1917
= *M. brevicavatus* Kreis, 1924
= *M. acutarius* Eroshenko, 1972
= *M. fusiformis* Eroshenko, 1972
= *M. bellus* Andrássy, 1985
= *M. oryzae* Dhanam & Jairajpuri, 1999

OTHER SPECIES

M. agilis Gagarin & Mataphonov, 2004
M. angarensis Gagarin, 1984
M. aquaticus Coetzee, 1968
M. clarki Altherr *in* Altherr & Delamare Deboutteville, 1972
M. himalayensis Rawat & Ahmad, 2000
M. italicus Andrássy, 1959
= *Clarkus italicus* (Andrássy, 1959) Jairajpuri, 1970a
M. laminatus Zullini, Loof & Bongers, 2002
M. loofi Winiszewska, 1998
M. maduei Schneider, 1925
M. mulveyi Andrássy, 1985
= *Mononchus maduei apud* Mulvey, 1978
M. niddensis Skwarra, 1921
M. nudus Gagarin, 1991
M. piperae Dhanam & Jairajpuri, 1999
= *Mononchus sandur* Eisendle, 2008 **syn. n.**
M. pulcher Andrássy, 1993a
M. sinensis Soni & Nama, 1983
M. superbus Mulvey, 1978
M. syrmatus Andrássy, 2008
M. tajmiris Gagarin, 1991
M. tunbridgensis Bastian, 1865

SPECIES INQUIRENDAE

M. allgeni Meyl, 1957a
M. macrostoma filicaudatus Schneider, 1937
M. macrostoma pseudoparvus Micoletzky, 1922
M. neotunbridgensis Kannan, 1961a
M. obtusus Cobb, 1917

Mononchus truncatus Bastian, 1865

(Figs 19; 20A-C)

MEASUREMENTS

Female: L = 1.65-2.4 mm; a = 26-40; b = 3.3-4.3; c = 5.5-8.6; c′ = 6-8; V = 48-56.
Male: L = 1.7-2.2 mm; a = 32-45; b = 3.9-4.1; c = 9-10.

DESCRIPTION

Lip region slightly offset, 22-29 × 6-8 μm. Amphidial apertures slit-like, just posterior to lateral lip. Buccal cavity barrel-shaped, 40-50 × 18-22 μm, *ca* 2.5 times as long as broad. Dorsal tooth large, its apex at 67-70% of buccal cavity length from its base, located posterior to transverse subventral rib. Female genital system amphidelphic. Vulva a transverse slit with slightly thickened lips. Vagina with small *pars refringens*. Uterus and oviduct separated by well developed sphincter. Tail elongate, tapering uniformly, curved ventrally in posterior half, 250-280 μm long. Caudal glands grouped, spinneret terminal. Male with 95-110 μm long spicules. Gubernaculum 22 μm long, accessory pieces 26 μm long. Supplements numbering 15-21.

DISTRIBUTION

Cosmopolitan.

Key to species of *Mononchus* Bastian, 1865

1 – Body large, 2.7-6 mm long 2
– Body small, 0.9-2.4 mm long 8

2 – Tail very long, 13-15 anal body diam. long; ♀: L = 1.59-1.68 mm; a = 32-35; b = 3.5-3.9; c = 4.2-5.0; c′ = 13-15; V = 47-50; ♂: unknown . . . . *syrmatus*

– Tail shorter, <10 anal body diam. long 3

3 – Tail short, *ca* 2-3 anal body diam. long 4

– Tail longer, *ca* 5-8 anal body diam. long 7

4 – Posterior third of tail in both sexes digitate, ventrally bent; ♀: L = 2.8-3.4 mm; a = 35-44; b = 2.8-3.3; c = 16-20; c′ = 2.5; V = 50-60; ♂: L = 2.7-3.0 mm; a = 30-33; b = 3.1-3.3; c = 20-23; supplements 22-26 + 6 (Canada) . . . . *mulveyi*

– Posterior third of tail not digitate, more or less straight 5

5 – Male supplements < 30 (22-27); ♀: L = 3.5-5.3 mm; a = 24-29; b = 4-5; c = 18-23; c′ = 2; V = 55-62; ♂: L = 3.4-4.3 mm; a = 28-34; b = 4.2-4.6; c = 26-39; supplements 22-27 (Germany, Russia, Lithuania) . . . . *maduei*

– Male supplements > 30 (37-47) 6

6 – Buccal cavity 100-120 μm, almost thrice as long as broad; ♀: L = 3.2-4.4 mm; a = 17-18; b = 3.2-3.5; c = 13-20; c′ = 1.9-2.5; V = 55-60; ♂: L = 3.6-4.2 mm; a = 18-21; b = 3.5-3.9; c = 16-20; supplements 41-47 (Russia) . . . . *tajmiris*

– Buccal cavity 80-90 μm, almost twice as long as wide; ♀: L = 5.1-5.6 mm; a = 37-43; b = 4.6-5.3; c = 16-19; c′ = 3; V = 48-59; ♂: L = 3.6-4.8 mm; a = 29-45; b = 3.5-4.0; c = 17-24; supplements 37-39 (Russia) . . . . *angarensis*

7 – Body 5-6 mm long; apex of dorsal tooth at 76-77% of buccal cavity length from its base; female tail *ca* 400 μm long; ♀: L = 5.0-5.6 mm; a = 44-47; b = 3.8-4.5; c = 13; c′ = 5; V = 51-53; ♂: L = 4.5-6.0 mm; a = 39-46; b = 4.0-4.6; c = 19-24; supplements 36-40 + 7-9 (Canada) . . . . *superbus*

– Body 3.5 mm long or shorter; apex of dorsal tooth at 83-88% of buccal cavity length from its base; female tail *ca* 200 μm; ♀: L = 2.7-3.5 mm; a = 32-43; b = 3.8-4.2; c = 12-15; c′ = 7-8; V = 51-54; ♂: L = 3.1 mm; a = 36; b = 3.7; c = 17; supplements 29-33 (Germany, Poland, former Czechoslovakia, Denmark, Russia, Canada) . . . . *niddensis*

8 – Female monodelphic, ovary posterior to vulva; tail short, plump; ♀: L = 1.1 mm; a = 28; b = 3.7; c = 15; c′ = 2.4; V = 54; ♂: unknown (Italy) . . . . *italicus*
– Female didelphic; tail generally long, only exceptionally short . . 9

9 – Tail very short, 1.5 anal body diam. long, bluntly rounded; spinneret of caudal glands subdorsal; ♀: L = 1.8 mm; a = 25; b = 3.5; c = 33; c′ = 1.2; V = 59-60; ♂: unknown (USA-Massachusetts) . . . . *clarki*
– Tail elongate, 3-8 anal body diam. long; spinneret of caudal glands terminal 10

10 – Buccal cavity unusually small and narrow, 18-23 μm long with almost straight walls 11
– Buccal cavity of usual shape, 25-50 μm long, with concave walls 12

11 – Walls of buccal cavity very thick in middle, dorsal tooth apex at 67-72% of buccal cavity length from its base; ♀: L = 1.12-1.31 mm; a = 26-30; b = 5.0-5.6; c = 9.4-10.9; c′ = 3.9-4.8; V = 50-53; ♂: unknown (Poland) . . . . *loofi*
– Walls of buccal cavity of normal thickness, apex of dorsal tooth quite close to beginning of buccal cavity (85-94% from its base); ♀: L = 0.9-1.4 mm; a = 20-35; b = 4.2-5.3; c = 8.3-10.3; c′ = 4-6; V = 51-55; ♂: L = 1.2-1.3 mm; a = 34; b = 4.6-4.9; c = 13; supplements 16-17 (The Netherlands, UK, Switzerland, former Czechoslovakia, Russia, Georgia, India, Japan, South Africa, Canada, USA-Alabama; Virginia, Surinam, Australia, Costa Rica) . . . . *tunbridgensis*

12 – Tail short, 3-4 anal body diam. long; ♀: L = 1.6-2.1 mm; a = 22-35; b = 3.9-4.6; c = 9-12; c′ = 2.8-4.1; V = 49-57; ♂: unknown (Russia) . . . . *nudus*
– Tail not short, 5-8 anal body diam. long 13

13 – Buccal cavity 40-50 μm long 14
– Buccal cavity 25-38 μm long 15

14 – Vulval papillae present; tail strongly ventrally curved in posterior half; ♀: L = 1.6-1.9 mm; a = 33-38; b = 3.6-3.7; c = 5.4-6.5; c′ = 8.8-10; V = 48-54; ♂: unknown (India) . . . . . . *himalayensis*
– Vulval papillae absent; tail not as strongly ventrally curved; ♀: L = 1.6-2.4 mm; a = 26-40; b = 3.3-4.3; c = 5.5-8.6; c′ = 6.8;

V = 48-56; ♂: L = 1.7-2.2 mm; a = 32-45; b = 3.9-4.1; c = 9-10; supplements 15-21 (The Netherlands, Belgium, Germany, Denmark, UK, Ireland, Finland, Poland, Switzerland, Austria, former Czechoslovakia, Hungary, former Yugoslavia, Spain, France, Italy, Russia, Belarus, Romania, Georgia, Kyrgyzstan, Uzbekistan, Tadzhikistan, Mongolia, Nepal, Japan, India, Sumatra, Ivory Coast, Ghana, Uganda, Zaire, Kenya, South Africa, Mauritius, Canada, USA-Hawaii, Mexico, Puerto Rico, Costa Rica, Colombia, Venezuela, Argentina, Peru, Australia)........... *truncatus*

15 – Vulval lips unsclerotised; ♀: L = 1.3-1.8 mm; a = 32-38; b = 4.0-4.6; c = 7.2-8.7; c′ = 5-7; V = 47-50; ♂: unknown (India)...... .... *sinensis*

– Vulval lips sclerotised.................................... 16

16 – Buccal cavity > 30 μm, *ca* twice as long as broad 17

– Buccal cavity < 30 μm, 2.5 times as long as broad.......... 18

17 – Dorsal tooth apex at 78-80% of buccal cavity length from its base, subventral walls with short longitudinal ridge in anterior third; ♀: L = 1.7-2.1 mm; a = 35-38; b = 4.4-4.8; c = 8.7-9.7; c′ = 6-7; V = 49-52; ♂: unknown (Hungary, USA-California, Chile)..... .... *pulcher*

– Dorsal tooth apex at 65-70% of buccal cavity length from its base, subventral walls without longitudinal ridge; ♀: L = 1.3-2.0 mm; a = 29-39; b = 3.8-5.1; c = 7.5-8.3; c′ = 6.1-8.1; V = 43-53; ♂: unknown.. *piperae*

18 – Subventral lamina in anterior part of buccal cavity present; amphidial aperture posterior to dorsal tooth; ♀: L = 1.50-1.86 mm; a = 21-32; b = 4.2-4.7; c = 9.7-12; c′ = 3.6-4.9; V = 50-55; ♂: unknown (Costa Rica)............................ *laminatus*

– Subventral lamina in anterior part of buccal cavity absent; amphidial aperture anterior to dorsal tooth; ♀: L = 1.2-1.8 mm; a = 19-37; b = 4-5; c = 8-12; c′ = 5-7; V = 46-50; ♂: L = 1.7-2.0 mm; a = 26-32; b = 4.3-4.9; c = 11-26; supplements 10-13 (Hungary, UK, Italy, Russia, Kyrgyzstan, India, Nigeria, South Africa, Mauritius, Mexico, St Lucia, Costa Rica)...... *aquaticus*

Note

Not included in key because of non-availability of literature: *M. agilis*.

Genus *Paramononchus* Mulvey, 1978
(Figs 21I; 22)

DIAGNOSIS

Body 2.5-4.0 mm long. Lip region almost continuous with body; lips and labial papillae poorly developed. Buccal cavity oblong, more than twice as long as broad. Dorsal tooth large, situated in anterior quarter of buccal cavity, directed anteriad. Opposite to it each subventral wall bearing a fine transverse rib connected with a short longitudinal ridge armed with two pairs of small toothlets. Pharyngo-intestinal junction non-tuberculate. Female genital system amphidelphic. Tail elongate, in posterior part cylindrical, 7-8 anal body diam. long. Caudal glands and spinneret distinct. Male not known.

TYPE SPECIES

Paramononchus arcticus Mulvey, 1978

OTHER SPECIES

P. alimovi Tsalolikhin, 1990

Paramononchus arcticus Mulvey, 1978
(Figs 21I; 22)

MEASUREMENTS

Female: L = 3.5-4.0 mm; a = 40-44; b = 3.9-4.7; c = 9-11; V = 50-54.

DESCRIPTION

Lip region almost continuous with body, lips and labial papillae poorly developed. Amphidial aperture slit-like. Buccal cavity thick-walled, 44-48 × 19-21 μm or *ca* twice as long as broad. Dorsal tooth large, its apex at 80-83% of buccal cavity length from its base, opposed by a short denticulate rib with two small subventral teeth at its base. Transverse rib extending from ventral wall of buccal cavity to level of dorsal tooth apex. Female genital system amphidelphic. Vulva a transverse slit. Vagina with small *pars refringens*. Vulval papillae absent. Tail 300-400 μm long, uniformly cylindroid, tapering to a blunt terminus. Caudal glands present. Spinneret terminal. Male not known.

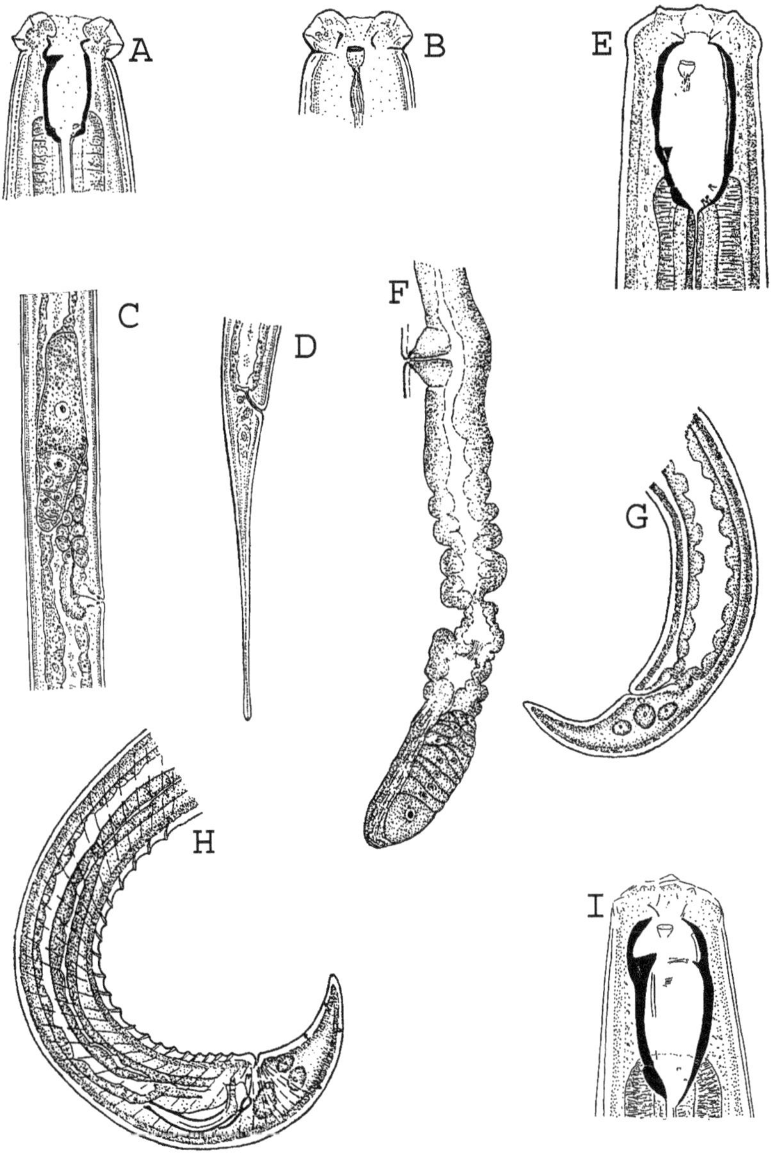

Fig. 21. *A-D:* Nigronchus parvus *Siddiqi, 1984. A: Anterior region; B: Anterior end showing amphid; C: Female genital branch; D: Female posterior region; E-H:* Parkellus parkus *Jairajpuri, Tahseen & Choi, 2001. E: Anterior region; F: Female genital branch (posterior); G: Female posterior region; H: Male posterior region; I:* Paramononchus arcticus *Mulvey, 1978. A-D after Siddiqi (1984); E-H after Jairajpuri* et al. *(2001); I after Zullini and Peneva (2006).*

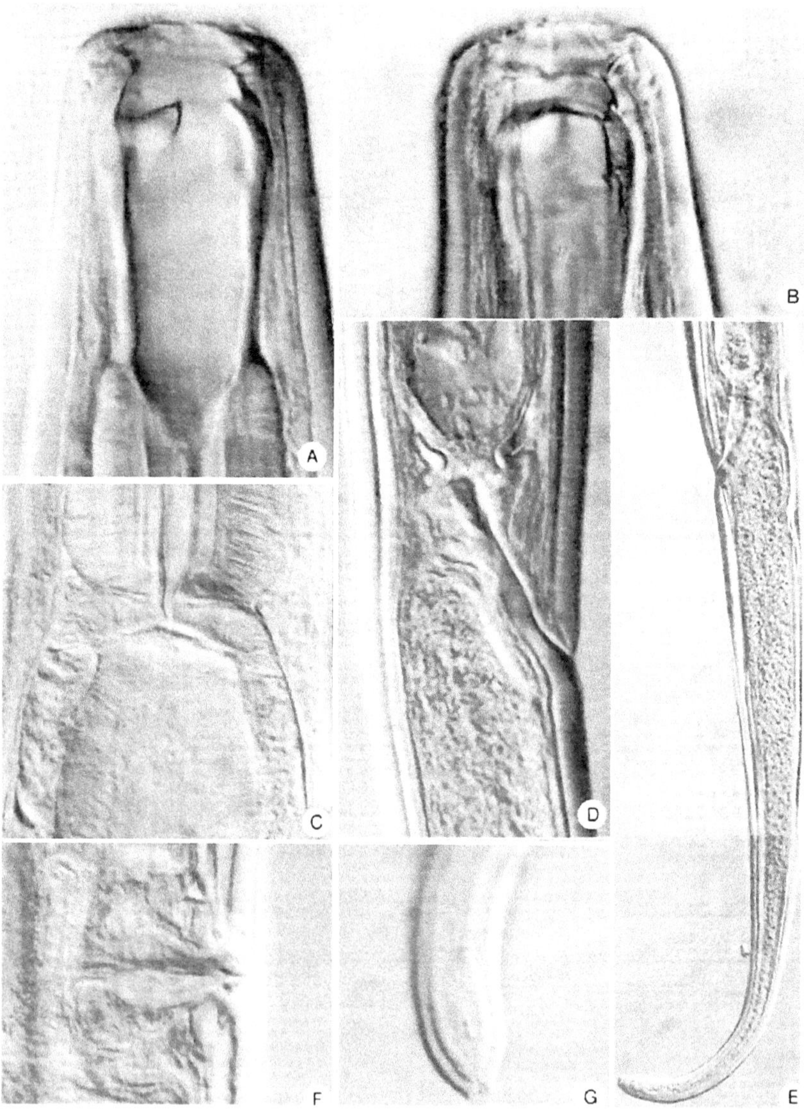

Fig. 22. Paramononchus arcticus *Mulvey, 1978. A, B: Anterior region; C: Pharyngo-intestinal junction; D: Anal region; E: Female posterior region; F: Vulval region; G: Tail tip with terminal spinneret. After Mulvey (1978).*

DISTRIBUTION

Canada.

Key to species of *Paramononchus* Mulvey, 1978

1 – Large species; ♀: L = 3.5-4.0 mm; a = 40-44; b = 3.9-4.7; c = 9-11; V = 50-54; c′ = 7-8; tail = 300-400 μm; ♂: unknown (Canada) . . . . *arcticus*

– Small species; ♀: L = 2.5-2.9 mm; a = 32-39; b = 3.8-4-3; c = 9.7-10; V = 50; c′ = 7-8; tail = 260-290 μm; ♂: unknown (Russia-Far East) . . . . *alimovi*

Genus *Nigronchus* Siddiqi, 1984a
(Fig. 21A-D)

DIAGNOSIS

Body small, 0.6-0.7 mm long. Lip region broader than adjoining body, sharply separated by a constriction. Buccal cavity oblong with nearly parallel walls, armed with a single dorsal tooth located in anterior quarter of buccal cavity and directed anteriad. No other teeth or ribs present. Pharyngo-intestinal junction non-tuberculate. Female genital system mono-prodelphic. Tail elongate, *ca* seven anal body diam. long. Caudal glands and spinneret conspicuous. Male not known.

TYPE AND ONLY SPECIES

Nigronchus parvus Siddiqi, 1984a

Nigronchus parvus Siddiqi, 1984a
(Fig. 21A-D)

MEASUREMENTS

Female: L = 0.60-0.73 mm; a = 23-27; b = 3.4-4.4; c = 6.3-6.8; c′ = 6.8-7.5; V = 52-56.

DESCRIPTION

Lip region wider than adjoining body, offset by sharp constriction. Lips partially separated, papillae raised. Amphidial aperture slit-like, slightly anterior to apex of dorsal tooth. Buccal cavity 14.5-15.0 × 6.5-

7.0 μm. Dorsal tooth at anterior end of buccal cavity, directed anteriad. Female genital system mono-prodelphic. Vulva a small pore. Vagina with small *pars refringens*. Short post-uterine sac present. Tail conoid in anterior third then gradually tapering to appear filiform. Caudal glands present. Terminal opening indistinct. Male not known.

DISTRIBUTION

Nigeria.

Subfamily Prionchulinae Andrássy, 1976

DIAGNOSIS

Lips and labial papillae well developed. Lip region set off from body. Dorsal tooth usually large, may or may not be opposed by a denticulate or non-denticulate longitudinal ridge. Tail short, conoid. Caudal glands and spinneret generally absent or obscure.

TYPE GENUS

Prionchulus Cobb, 1916

OTHER GENERA

Clarkus Jairajpuri, 1970a
Coomansus Jairajpuri & Khan, 1977
Parkellus Jairajpuri, Tahseen & Choi, 2001a

Key to genera of Prionchulinae Andrássy, 1976

1 – Subventral walls of buccal cavity with serrate or denticulate longitudinal ridges .... *Prionchulus*
– Subventral walls of buccal cavity either smooth or with non-denticulate ridge 2

2 – Ventral ridge originating abruptly by tooth-like anterior projection .... *Clarkus*
– Ventral ridge, if present, originating gradually without a tooth-like projection 3

3 – Dorsal tooth in anterior half of buccal cavity .... *Coomansus*
– Dorsal tooth in posterior half of buccal cavity .... *Parkellus*

Genus *Prionchulus* Cobb, 1916

(Figs 23; 24A-F)

DIAGNOSIS

Body 1-4 mm long. Lip region offset, lips and labial papillae prominent. Buccal cavity large, heavily sclerotised, barrel-shaped, tapering at base. Dorsal tooth stout, directed anteriad, situated in anterior half of buccal cavity, opposed by two longitudinal. denticulate ridges. Pharyngo-intestinal junction non-tuberculate. Female genital system amphidelphic. Spicules stout, arcuate; gubernaculum simple or bidentate; accessory pieces present or absent. Tail short, conoid, ventrally arcuate. Caudal glands absent or very poorly developed. Spinneret absent.

TYPE SPECIES

P. muscorum (Dujardin, 1845) Cobb, 1916
= *Oncholaimus muscorum* Dujardin, 1845
= *Mononchus muscorum* (Dujardin, 1845) Bastian, 1865
= *Mononchus* (*Prionchulus*) *muscorum* (Dujardin, 1845) Bastian, 1865 (Cobb, 1916)
= *Mononchus bastiani* de Man, 1876
= *Mononchus* (*Prionchulus*) *longicollis* Cobb, 1917
= *Mononchus papillatus macrodon* Fuchs, 1930
= *Prionchulus medius* Eroshenko, 1975
= *Prionchulus paucidentatus* Zell, 1985
= *Prionchulus ferox* Winiszewska & Susulovsky, 2003 **syn. n.**
= *Prionchulus olexandri* Winiszewska & Susulovsky, 2003 **syn. n.**
= *P. prasadi* Saha, Lal & Singh, 2006 **syn. n.**
= *P. ukhrum* Mohilal & Dhanachand, 1997 **syn. n.**
= *P. pachydermis* Khan, Choi, Lee & Choi, 2000 **syn. n.**

OTHER SPECIES

P. altherri Zell, 1985
P. auritus Andrássy, 1985
P. bogdanowiczi Susulovsky & Winiszewska, 2006
P. borealis Andrássy, 2003
P. brevicaudatus Winiszewska, 2002
P. ctenodentatus (Tysowsky, 1915) Mulvey, 1967b

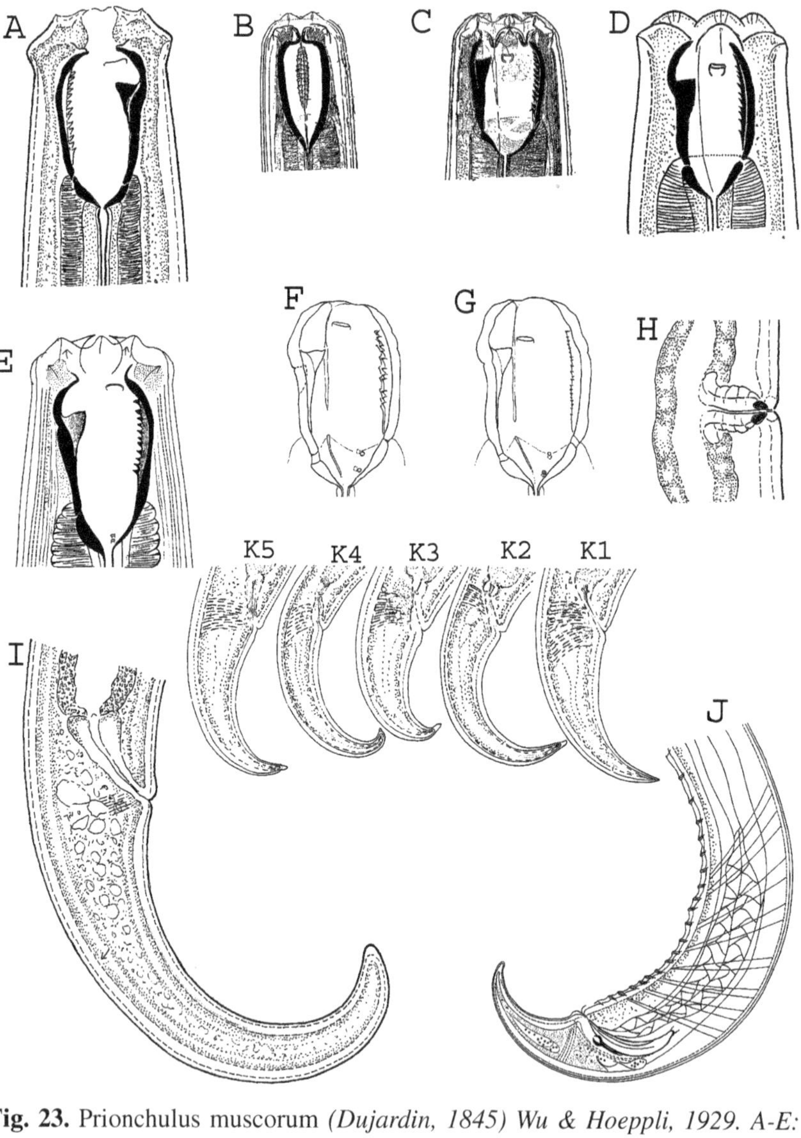

Fig. 23. Prionchulus muscorum *(Dujardin, 1845) Wu & Hoeppli, 1929. A-E: Anterior regions; F, G: Buccal cavity; H: Vulval region; I: Female posterior region; J: Male posterior region; K1-K5: Female posterior ends. B, C after Cobb (1916); D after Clark (1960); E, J after Mulvey (1967); F, G, K1-K5 after Arpin* et al. *(1984).*

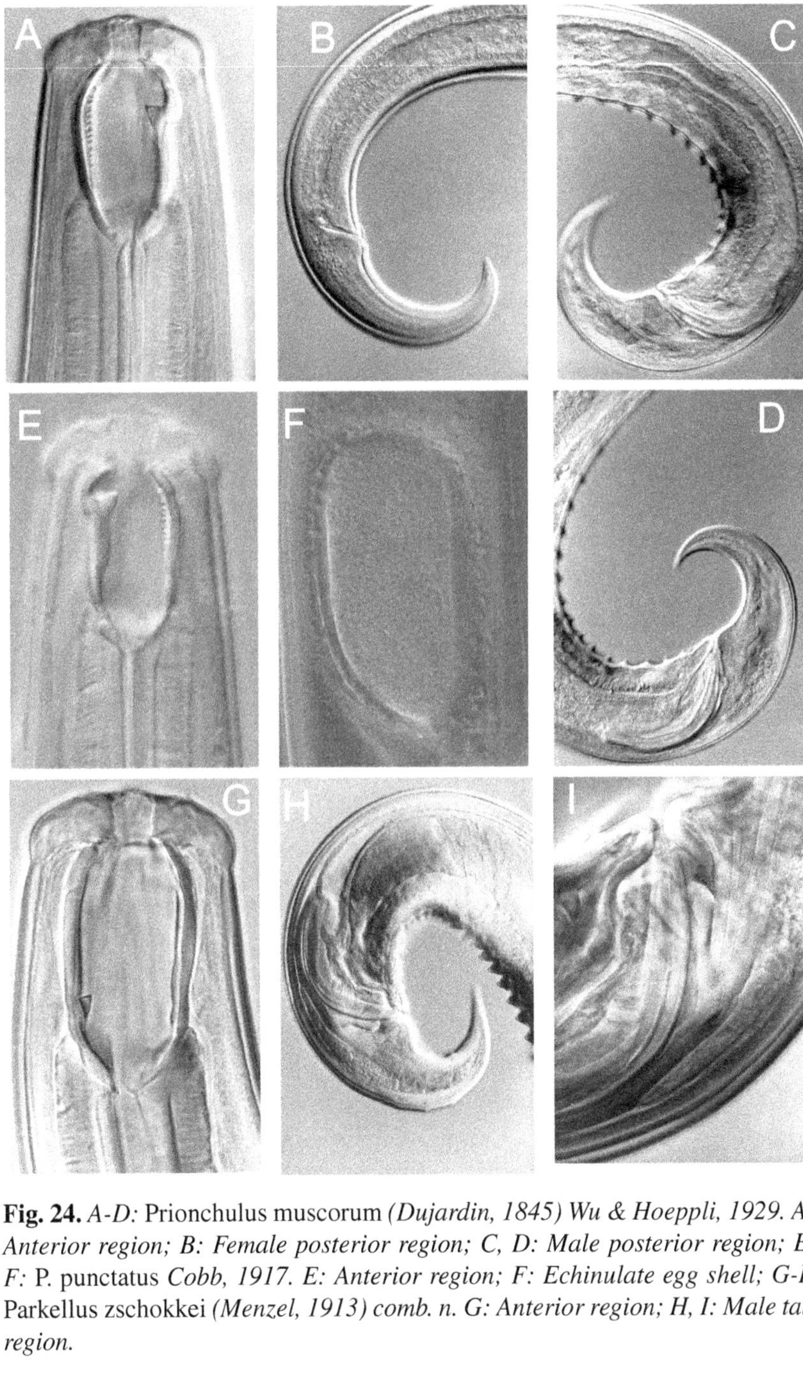

Fig. 24. *A-D:* Prionchulus muscorum *(Dujardin, 1845) Wu & Hoeppli, 1929. A: Anterior region; B: Female posterior region; C, D: Male posterior region; E, F:* P. punctatus *Cobb, 1917. E: Anterior region; F: Echinulate egg shell; G-I:* Parkellus zschokkei *(Menzel, 1913) comb. n. G: Anterior region; H, I: Male tail region.*

= *Mononchus ctenodentatus* Tysowsky, 1915
P. fagi Zell, 1985
P. fistulosus Susulovsky & Winiszewska, 2002
P. grazynae Orselli & Vinciguerra, 2007
P. hygrophilus Susulovsky & Winiszewska, 2006
P. kamchaticus Susulovsky, Winiszewska & Gagarin, 2003
P. kralli Winiszewska & Susulovsky, 2004
P. longus (Thorne, 1929) Goodey, 1951
= *Mononchus longus* Thorne, 1929
= *Mononchus* (*Prionchulus*) *longus* Thorne, 1929 (Goodey, 1951)
P. major Gagarin, 2001
P. mordax Andrássy, 1993a
= *P. koreanus* Khan, Choi, Lee & Choi, 2000 **syn. n.**
P. obtusidentatus Vinciguerra, 1986
P. pinophilus Winiszewska & Susulovsky, 2004
P. polonicus Winiszewska & Susulovsky, 2004
P. pseudolongus Winiszewska & Susulovsky, 2004
P. punctatus Cobb, 1917
= *Mononchus* (*Prionchulus*) *punctatus* Cobb, 1917
= *Mononchus muscorum macrolaimus* Cobb, 1916
P. sedatus Gagarin, 2000
P. septentrionalis Winiszewska & Susulovsky, 2004
P. spectabilis (Ditlevsen, 1911) Cobb, 1916
= *Mononchus spectabilis* Ditlevsen, 1911
= *Mononchus* (*Prionchulus*) *spectabilis* Ditlevsen, 1911 (Cobb, 1916)
P. thornei Winiszewska & Susulovsky, 2003
P. vescus Eroshenko, 1975
= *Prionchulus minus* Zell, 1985
P. zelli Winiszewska & Susulovsky, 2003
= *Prionchulus bastiani* Zell, 1985
= *Prionchulus oxycercus* Zell, 1985

Prionchulus muscorum **(Dujardin, 1845) Wu & Hoeppli, 1929**
(Figs 23; 24A-D)

MEASUREMENTS

Female: L = 1.45-3.00 mm; a = 21-33; b = 3.0-5.0; c = 11-28; V = 61-78.

Male: L = 1.85-2.81 mm; a = 23-32; b = 3.2-4.2; c = 13-29; c′ = 1.8-2.6.

DESCRIPTION

Lip region 32-56 μm diam., 10-12 μm high. Amphidial aperture 3-7 μm wide, located at 8-29 μm from anterior end of body and 31-53 μm from base of buccal cavity. Buccal cavity 39-64 μm long, 18-35 μm broad. Dorsal tooth of medium to large size, situated in anterior half of buccal cavity, its apex 28-48 μm or 68-83% of buccal cavity length from its base. Subventral walls with two longitudinal, denticulate ridges. Excretory system well developed, comprising an excretory pore, ampulla, excretory duct and renettes. Female genital system amphidelphic. Vagina with *pars refringens*. Uterine eggs measuring 77-101 × 60-80 μm, with smooth shell. Tail conoid, arcuate ventrally, 96-185 μm or 2.0-3.5 anal body diam. long. Caudal glands and spinneret absent. Male rare. Spicules 85-120 μm long. Gubernaculum 30 μm long, lateral accessory pieces poorly developed. Supplements 14-20.

DISTRIBUTION

Cosmopolitan.

Key to species of *Prionchulus* Cobb, 1916

1 – Female tail < 2 anal body diam. long, short, plump, conical; ♀: L = 2.35-2.42 mm; a = 25-29; b = 4.1-4.9; c = 31-34; c′ = 1.4-1.6; V = 64-65; ♂: unknown (Poland) . . . . *brevicaudatus*
– Female tail > 2 anal body diam. long, ventrally curved with pointed or broadly rounded tip 2

2 – Body small, usually <2 mm long 3
– Body large, usually >2 mm long 9

3 – Buccal cavity > 40 μm long; subventral denticles strong 4
– Buccal cavity < 40 μm long; subventral denticles rather weak . . 5

4 – Body strongly ventrally curved in posterior half; vagina unsclerotised; ♀: L = 1.60-1.74 mm; a = 25-32; b = 3.4-3.6; c = 11-15; c′ = 3.2-3.5; V = 61-64; ♂: unknown (USA-Alaska) . . . . *borealis*

– Body not as strongly curved in posterior half; vagina sclerotised; ♀: L = 1.43-2.0 mm; a = 23-27; b = 3.2-3.7; c = 11-15; c′ = 2.9-4.0; V = 62-69; ♂: unknown (Taiwan) . . . . *bogdanoviczi*

5 – Apex of dorsal tooth at 90% of buccal cavity length from its base; ♀: L = 1.6-1.9 mm; a = 20-30; b = 3.9-4.8; c = 12-15; V = 58-66; c′ = 2.5; ♂: unknown (Germany) . . . . *altherri*

– Apex of dorsal tooth at <85% of buccal cavity length from its base 6

6 – Lips barely protruding; amphidial apertures at level of anterior end of buccal cavity; tail > 100 μm long 7

– Lip protruding; amphidial apertures at level of dorsal tooth apex; tail 78-95 μm long 8

7 – Buccal cavity 33-34 μm; subventral denticles 9-11; ♀: L = 1.30-1.67 mm; a = 22-24; b = 4.1-4.9; c = 12-14; c′ = 2.8-3.1; V = 60-61; ♂: unknown (Korea) . . . . *mordax*

– Buccal cavity 34-40 μm; subventral denticles 14-16; ♀: L = 1.74-2.10 mm; a = 21-27; b = 3.8-4.3; c = 12.4-16.8; c′ = 2.2-3.2; V = 57-63; ♂: unknown (Ecuador) . . . . *grazynae*

8 – Buccal cavity 35-40 μm; subventral denticles 9-11, weak; tail slightly ventrally curved; ♀: L = 1.58-1.66 mm; a = 24-26; b = 3.9-4.3; c = 17-20; c′ = 2-3; V = 65-65.5; ♂: unknown (Estonia) . . . . *kralli*

– Buccal cavity 31-33 μm; subventral denticles 12-15, weak; tail strongly ventrally curved; ♀: L = 1.13-1.57 mm; a = 17-29; b = 3.1-3.9; c = 12-19; c′ = 2.1-2.5; V = 61-69; ♂: L = 1.16 mm; a = 18; b = 3.5; c = 16; spicules 63 μm; supplements 16 (Germany, Poland, Russia-Far East, USA-Alaska) . . . . *vescus*

9 – Body cuticle with numerous pronounced body pores along body length; ♀: L = 2.0-2.4 mm; a = 22-31; b = 3.6-4.3; c = 14-20; c′ = 2.4-3.4; V = 61.5-67; ♂: L = 1.94-2.7 mm; a = 25-30; b = 3.8-4.6; c = 16-18; c′ = 2.3-3.0; spicules 81-86 μm; supplements 17 (Ukraine) . . . . *fistulosus*

– Body cuticle with usual body pore arrangement 10

10 – Female gonad long, uterus elongate with a separate muscular sphincter on each end, spermatheca distinct; ovaries short 11

– Female gonad short; uterus lacking sphincters and spermatheca; ovaries comparatively long 18

11 – Amphidial apertures at level of dorsal tooth apex 12
– Amphidial apertures near anterior margin of buccal cavity or between anterior margin and dorsal tooth apex 14

12 – Apex of dorsal tooth at 75% of buccal cavity length from its base; subventral denticles 18; ♀: L = 2.1-3.1 mm; a = 24-31; b = 3.8-4.6; c = 14-22; c′ = 2.5-3.0; V = 61-69; ♂: L = 2.0-2.9 mm; a = 23-27; b = 3.5-4.5; c = 20-27; supplements 17-27 (Italy, Nepal, Canada, USA-Colorado) . . . . *longus*
– Apex of dorsal tooth at >80% of buccal cavity length from its base; subventral denticles fewer than 15 13

13 – Labial region rounded; labial papillae distinctly protruding; tail comparatively long; ♀: L = 2.1 mm; a = 28; b = 4.2; c = 15.6; c′ = 3.2; V = 61.7; ♂: L = 2.33 mm; a = 34; b = 4.4; c = 18.9; spicules 100 μm; supplements 25 (Russia) . . . . *pinophilus*
– Labial region truncate, almost flat; labial papillae not protruding; tail comparatively short; ♀: L = 2.19-2.53 mm; a = 24-34; b = 4.5-5.4; c = 21-31; c′ = 1.7-2.9; V = 65-70; ♂: L = 2.13-2.80 mm; a = 27-40; b = 4.5-5.1; c = 22-33; c′ = 1.1-1.6; spicules 82-100 μm; supplements 25-29 (USA) . . . . *thornei*

14 – Subventral denticles strong, 8-9 in number; distance between first and the last denticle 25-25 μm; vulval papillae present; ♀: L = 2.5-4.0 mm; a = 31-42; b = 4.4-5.7; c = 24-31; c′ = 2.5-3.0; V = 54-57; ♂: L = 2.2-2.4 mm; a = 35-50; b = 4.1-5.3; c = 25-45; spicules 104-124 μm; supplements 18-29 (Germany, Switzerland, Hungary, Denmark, Greenland, Sweden, Finland, Lithuania) . . . . *spectabilis*
– Subventral denticles moderate to fine, always >10 in number; distance between first and last denticle < 20 μm; vulval papillae absent 15

15 – Cytoplasmic portion in terminal part of tail rapidly narrowing, creating a depression in dorsal surface; ♀: L = 2.25 mm; a = 33; b = 4.3; c = 18; c′ = 3.0; V = 68.8; ♂: L = 2.2-2.4 mm; a = 40-45; b = 4.0-4.3; c = 22-23; c′ = 1.9-2.1; spicules 75-82 μm; supplements 13-17 (USA-Alaska) . . . . *septentrionalis*
– Cytoplasmic portion in terminal part of tail not rapidly narrowing, tail terminus smooth, no depression on dorsal surface 16

16 – Labial papillae distinctly protruding; tail strongly curved, almost hooked; ♀: L = 1.8-2.3 mm; a = 23-31; b = 3.4-3.9; c = 9.8-

12.5; c′ = 3.5-5.0; V = 58-63; ♂: L = 1.69-1.96 mm; a = 24-29; b = 3.3-3.6; c = 13.5-14.0; c′ = 2.8-3.3; spicules 53-82 μm; supplements 14-18 (Ukraine) . . . . *hygrophilus*

– Labial papillae not protruding; tail slightly curved 17

17 – Lip region broadly rounded; vagina 26-35 μm long; *pars proximalis vaginae* 26-39% of body diam.; ♀: L = 1.7-2.2 mm; a = 21.5-30.5; b = 4.2-5.1; c = 15.7-21.6; c′ = 1.8-3.1; V = 60-68; ♂: L = 1.7-2.0 mm; a = 26-33; b = 4.1-4.6; c = 18.5-26; c′ = 1.5-2.1; spicules 77-85 μm; supplements 18-22 (Poland) . . . *polonicus*

– Lip region flattened; vagina 40-45 μm; *pars proximalis vaginae* 42-51% of body diam.; ♀: L = 1.97-2.65 mm; a = 26-31; b = 4.2-5.2; c = 15-24; c′ = 2.1-3.5; V = 61-68; ♂: L = 1.7-2.3 mm; a = 27-37; b = 4.2-4.7; c = 17.4-23.6; c′ = 1.4-2.1; spicules 61-87 μm; supplements 19-22 (Poland) . . . . *pseudolongus*

18 – Large animals, *ca* 3 mm long; tail tip sharply pointed; ♀: L = 2.7-3.1 mm; a = 23-32; b = 4.3-4.6; c = 13-15; c′ = 2.5; V = 60-64; ♂: L = 2.7 mm; a = 30; b = 4.0; c = 16; supplements 16 (Germany) . . . . *fagi*

– Smaller animals, at most 2.5 mm long and generally shorter . . . 19

19 – Dorsal tooth obtuse with broadly rounded apex; ♀: L = 2.1-2.4 mm; a = 26-27; b = 3.8-4.0; c = 13-14; c′ = 2.5; V = 63-65; ♂: unknown (Italy) . . . . *obtusidentatus*

– Dorsal tooth pointed with sharp apex 20

20 – Apex of dorsal tooth at 80% or more of buccal cavity length from its base 21

– Apex of dorsal tooth at <80% of buccal cavity length from its base 22

21 – Labial papillae large, conoid, posterior ones ear-like; buccal cavity 46-48 μm; egg shell not echinulate; ♀: L = 2.2-2.4 mm; a = 30-32; b = 4.0-4.2; c = 13-14; V = 59-62; c′ = 3.3-4.2; ♂: unknown (Sri Lanka) . . . . *auritus*

– Labial papillae low, posterior ones not ear-like; buccal cavity 35-45 μm; egg shell echinulate; ♀: L = 1.5-2.2 mm; a = 25-39; b = 3.6-4.8; c = 14-22; c′ = 2-4; V = 59-67; ♂: L = 2.0-2.4 mm; a = 32-34; b = 3.7-4.5; c = 20-25; supplements 18-21 (The Netherlands, Germany, Belgium, UK, Poland, Switzerland,

Austria, Romania, former Yugoslavia, France, Nepal, Korea, Iran, Canada, USA, Mexico, Costa Rica, Colombia) .... *punctatus*

22 – Tail tip subdigitate, somewhat bent dorsally; ♀: L = 2.1-2.6 mm; a = 21-31; b = 3.9-4.5; c = 12-15; c′ = 2.5; V = 59-66; ♂: unknown (Germany) .... *zelli*

– Tail tip not subdigitate, ventrally bent 23

23 – Amphidial apertures level with tip of buccal cavity; apex of dorsal tooth at 72-76% of buccal cavity length from its base; tail strongly ventrally curved, tip rather sharp; ♀: L = 1.8-2.6 mm; a = 25-33; b = 3.3-4.4; c = 10-18; c′ = 3.5; V = 58-67; ♂: L = 2.2 mm; a = 36-37; b = 4.0-4.1; c = 18-20; supplements 16-22 (The Netherlands, Belgium, Germany, Denmark, UK, Sweden, Poland, Switzerland, Austria, Hungary, Romania, former Yugoslavia, Spain, France, Italy, Russia, Iran, Korea, Mongolia, India, China, Canary Islands, Egypt, Kenya, Zaire, Mauritius, Canada, USA, St Lucia, Dominica, Colombia, Brazil, Costa Rica, New Zealand) .... *muscorum*

– Amphidial apertures level with dorsal tooth apex; apex of dorsal tooth at 76-79% of buccal cavity length from its base; tail rather plump, ventrally bent with broadly rounded tip; ♀: L = 1.9-2.4 mm; a = 22-30; b = 4.0-4.5; c = 16-22; c′ = 2.1-2.8; V = 62-67; ♂: unknown (Poland, France) .... *ctenodentatus*

Note

Not included in key because of non-availability of literature: *P. kamchaticus*, *P. major*, *P. sedates*.

Genus *Clarkus* Jairajpuri, 1970a

(Figs 18; 20D-H; 25A-G)

Diagnosis

Buccal cavity barrel-shaped, provided with non-denticulate ventral ridge. Dorsal tooth situated in anterior half of buccal cavity, apex directed anteriad. Pharyngo-intestinal junction non-tuberculate. Female genital system amphidelphic. Spicules and lateral accessory pieces short, stout, gubernaculum simple, crescent-shaped. Tail short, conoid, ventrally arcuate. Caudal glands generally absent; if present, poorly developed; spinneret absent or inconspicuous.

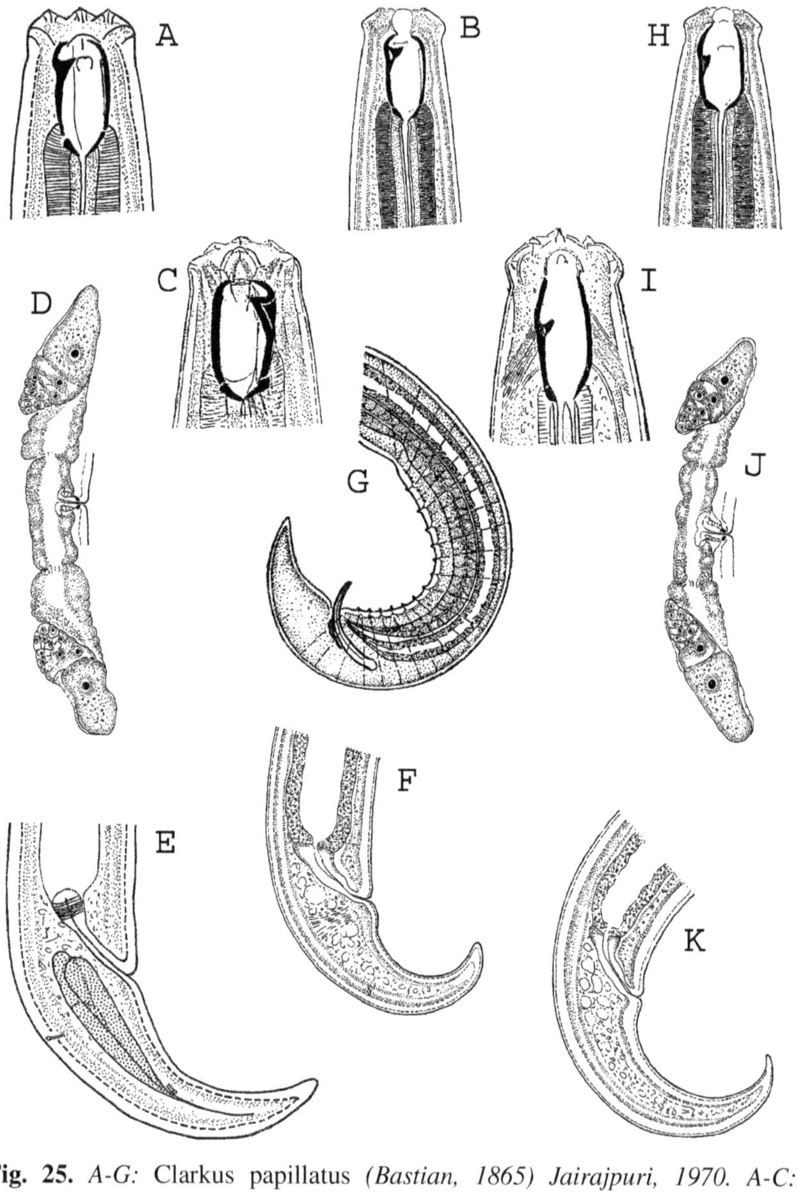

Fig. 25. *A-G:* Clarkus papillatus *(Bastian, 1865) Jairajpuri, 1970. A-C: Anterior region; D: Female genital branch; E, F: Female posterior region; G: Male posterior region; H-K:* Coomansus parvus *(de Man, 1880) Jairajpuri & Khan, 1977. H, I: Anterior region; J: Female genital branch; K: Female posterior region. A, F after Clark (1960); C, I after Cobb (1916); G after Jairajpuri and Khan (1982).*

TYPE SPECIES

C. papillatus (Bastian, 1865) Jairajpuri, 1970a
= *Mononchus papillatus* Bastian, 1865
= *Mononchus* (*Prionchulus*) *papillatus* Bastian, 1865 (Schneider, 1939)
= *Prionchulus papillatus* (Bastian, 1865) Schneider, 1939
= *Mononchus cristatus* Bastian, 1865
= *Clarkus koreanus* Choi, Khan & Lee, 1999 **syn. n.**

OTHER SPECIES

C. diversus Andrássy, 1983
C. dorsalis Yeates, 1992
C. elongatus Jairajpuri & Khan, 1977
C. longinquus Andrássy, 2003
C. macropapillatus (Mulvey, 1967a) Jairajpuri, 1970a
= *Mononchus macropapillatus* Mulvey, 1967a
C. patricius Susulovsky, 1994
C. propapillatus (Clark, 1960c) Jairajpuri, 1970a
= *Mononchus propapillatus* Clark, 1960c
C. pulcherrimus Andrássy, 1985
C. sheri (Mulvey, 1967a) Jairajpuri, 1970a
= *Mononchus sheri* Mulvey, 1967a
= *Mononchus jugalis* Coetzee, 1968
= *Clarkus jugalis* (Coetzee, 1968) Jairajpuri, 1970a
C. sveltus (Altherr, 1968) Jairajpuri & Khan, 1982
= *Mononchus sveltus* Altherr, 1968

SPECIES INQUIRENDA

Clarkus conoidus Mohilal & Dhanachand, 1997

Clarkus papillatus (Bastian, 1865) Jairajpuri, 1970a
(Figs 18; 20D-H; 25A-G)

MEASUREMENTS

Female: L = 0.7-1.4 mm; a = 17-29; b = 3.1-4.0; c = 12-19; c′ = 2.0-3.0; V = 58-68.
Male: L = 1.0-1.1 mm; a = 23-26; b = 3.2-3.4; c = 14-15.

DESCRIPTION

Lip region faintly offset, 23-26 μm diam. and 7-9 μm high. Amphidial aperture at level of dorsal tooth apex. Buccal cavity cup-shaped, 20-26 × 10-15 μm, *ca* half as wide as long and half as wide as head diam. Dorsal tooth apex at 79-85% of buccal cavity length from its base. Anterior margin of tooth straight, horizontal or nearly so, posterior edge merging smoothly into buccal cavity wall without any re-entrant angle. Subventral walls of buccal cavity with a non-denticulate ridge extending from point opposite dorsal tooth apex to posterior limit of vertical plate. Pharyngo-intestinal junction non-tuberculate. Female genital system amphidelphic. Vulva transverse; vagina with *pars refringens*. Usually only one mature egg in entire genital system, occasionally one in each uterus. Tail conoid, ventrally arcuate, 58-85 μm long, with three caudal glands but no terminal duct and opening. Male with buccal cavity 25-28 × 13-14 μm. Spicules 52-61 μm, gubernaculum 16-18 μm long. Lateral guiding pieces absent. Supplements 10-14. Tail 60-70 μm long, without caudal glands and spinneret.

DISTRIBUTION

Cosmopolitan.

Key to species of *Clarkus* Jairajpuri, 1970a

1 – Tail > 200 μm or six anal body diam. long; ♀: L = 2.0-2.2 mm; a = 31-33; b = 5.2-5.4; c = 9-10; c′ = 5.7-6; V = 45-46; ♂: unknown (Australia) . . . . *pulcherrimus*
– Tail < 200 μm or 1.5-4 anal body diam. long 2

2 – Ventral ridge originating conspicuously anterior to dorsal tooth apex 3
– Ventral ridge originating at level of dorsal tooth apex 4

3 – Tail 140-180 μm or 3-4 anal body diam. long; buccal cavity 45-50 μm; ♀: L = 2.2-2.5 mm; a = 24-30; b = 3.8-4.6; c = 15-18; c′ = 3-4; V = 60-64; ♂: unknown (India) . . . . *elongatus*
– Tail 100-110 μm or 2.5-3 anal body diam. long; buccal cavity 35-40 μm; ♀: L = 1.5-2.2 mm; a = 21-30; b = 3.9-4.6; c = 14-21; c′ = 2.5-3; V = 58-64; ♂: L = 1.8 mm; a = 33; b = 4.3; c = 22; supplements ten (India, South Africa, USA, Mexico) . . . . *sheri*

4 – Tail straight, terminus somewhat finger-like; ♀: L = 1.1-1.2 mm; a = 20-23; b = 3.3-3.5; c = 16-17; c′ = 2.4-2.5; V = 62-64; ♂: unknown (Hungary) . . . . *diversus*
– Tail ventrally curved but not finger-like 5

5 – Buccal cavity 32-40 μm long; body 2 mm or longer 6
– Buccal cavity 20-28 μm long; body 1.5 mm or shorter 8

6 – Tail strongly bent, hooked, 1.5-2 anal body diam. long; ♀: L = 1.9-2.0 mm; a = 30; b = 3.8-4.2; c = 14-21; c′ = 1.5-2; V = 59-64; ♂: L = 1.6-21 mm; a = 28-40; b = 3.6-4.0; c = 23-32; supplements 13-17 (Canada) . . . . *macropapillatus*
– Tail less bent, not hooked, 2.5-3 anal body diam. long 7

7 – Front edge of dorsal tooth curved anteriad; ♀: L = 2.0-2.7 mm; a = 41-56; b = 3.8-4.7; c = 19-23; c′ = 2.5-3; V = 65-69; ♂: L = 2.0-2.7 mm; a = 41-60; b = 3.2-4.5; c = 19-25; supplements 14-19 (Germany) . . . . *sveltus*
– Front edge of dorsal tooth straight, perpendicular to body axis; ♀: L = 1.6-2.1 mm; a = 28-33; b = 3.8-5.2; c = 16-29; c′ = 2.5-3.2; V = 57-61; ♂: unknown (New Caledonia) . . . . *dorsalis*

8 – Lateral accessory pieces in the spicular apparatus present; supplements ten; ♀: L = 0.9-1.6 mm; a = 21-36; b = 2.9-3.9; c = 10-19; c′ = 2-2.5; V = 58-69; ♂: L = 1.1 mm; a = 32; b = 3.5; c = 11.5 (India, New Zealand) . . . . *propapillatus*
– Lateral accessory pieces in the spicular apparatus absent; supplements 12-16; ♀: L = 0.7-1.4 mm; a = 17-29; b = 3.1-4.0; c = 12-19; c′ = 2-3; V = 58-60; ♂: L = 1.0-1.1 mm; a = 23-26; b = 3.2-3.4; c = 14-15 (The Netherlands, Belgium, Germany, Denmark, UK, Ireland, Norway, Greenland, Switzerland, Austria, former Czechoslovakia, Hungary, Poland, Romania, former Yugoslavia, Spain, France, Italy, Russia, Lithuania, Moldavia, Georgia, Uzbekistan, Kyrgyzstan, Tadzhikistan, Iran, Nepal, India, Ghana, Zaire, Kenya, Mauritius, Canada, USA, Mexico, Venezuela, Colombia, Brazil, Argentina, Hawaii, Australia, New Zealand, Campbell Islands) . . . . *papillatus*

Genus *Coomansus* Jairajpuri & Khan, 1977
(Figs 20I, J; 25H-K)

Diagnosis

Body 0.7-5.8 mm long. Buccal cavity barrel-shaped, mostly as long as labial diam. Dorsal tooth of medium size, situated in anterior half or middle of buccal cavity, its apex always in anterior half. Longitudinal ridge opposite to dorsal tooth generally present but rather weak, gradually merging anteriorly into ventral wall. Pharyngo-intestinal junction non-tuberculate. Female genital system amphidelphic, vulva at 45-75% of body length. Spicules comparatively short, supplements 8-28 in number. Tail conoid, ventrally arcuate, 2-6 anal body diam. long in male, somewhat shorter than female. Caudal glands poorly developed, terminal duct absent.

Type species

C. parvus (de Man, 1880) Jairajpuri & Khan, 1977
= *Mononchus parvus* de Man, 1880
= *Clarkus parvus* (de Man, 1880) Jairajpuri, 1970a
= *Mononchus cobbi* Cassidy, 1931 (Cassidy, 1933)
= *Mononchus mauiensis* Oliveira, 1940
= *Clarkus parvulus* Coetzee, 1968
= *Coomansus parvulus* (Coetzee, 1968) Jairajpuri & Khan, 1977
= *Iotonchus coomansi* Baqri, Baqri & Jairajpuri, 1978

Other species

C. arvensis (Eroshenko, 1972) Jairajpuri & Khan, 1977
= *Clarkus arvensis* Eroshenko, 1972
C. campbelli (Allgén, 1929) Jairajpuri & Khan, 1977
= *Mononchus campbelli* (Allgén, 1929) Jairajpuri, 1970a
C. composticola (Clark, 1960c) Jairajpuri & Khan, 1977
= *Mononchus composticola* Clark, 1960c
= *Clarkus composticola* (Clark, 1960c) Jairajpuri, 1970a
C. fletcherensis Mulvey, 1978
C. gerlachei (de Man, 1904) Jairajpuri & Khan, 1977
= *Mononchus gerlachei* de Man, 1904
= *Clarkus gerlachei* (de Man, 1904) Jairajpuri, 1970a

C. indicus Jairajpuri & Khan, 1982
C. intestinus (Vinciguerra & La Rosa, 1990) Andrássy, 1993a
= *Clarkus intestinus* Vinciguerra & La Rosa, 1990
C. kastrolli (Altherr, 1958) Jairajpuri & Khan, 1977
= *Mononchus kastrolli* Altherr, 1958
= *Clarkus kastrolli* (Altherr, 1958) Jairajpuri, 1970a
C. magellanicus Jiménez-Guirado, Wouts & Bell, 1998
C. major (Cobb, 1893) Loof & Winiszewska-Slipińska, 1993
= *Mononchus major* Cobb, 1893
= *Clarkus major* (Cobb, 1893) Jairajpuri, 1970a
C. meridionalis Jiménez-Guirado, Wouts & Bello, 1998
C. mesadenus (Clark, 1960c) Jairajpuri & Khan, 1977
= *Mononchus mesadenus* Clark, 1960c
= *Clarkus mesadenus* (Clark, 1960c) Jairajpuri, 1970a
C. obliquoris (Eroshenko, 1972) Jairajpuri & Khan, 1977
= *Clarkus obliquoris* Eroshenko, 1972
C. ouinnensis (Yeates, 1992) Andrássy, 1993a
= *Clarkus ouinnensis* Yeates, 1992
C. pretoriensis (Coetzee, 1968) Jairajpuri & Khan, 1977
= *Mononchus pretoriensis* Coetzee, 1968
= *Clarkus pretoriensis* (Coetzee, 1968) Jairajpuri, 1970a
C. splendidus Andrássy, 1993a
C. ulsani Choi, Khan & Lee, 1999
C. venezolanus (Loof, 1964) Jairajpuri & Khan, 1977
= *Mononchus venezolanus* (Loof, 1964) Jairajpuri, 1970a
= *Coomansus icarus* Jairajpuri & Khan, 1982

Coomansus parvus **(de Man, 1880) Jairajpuri & Khan, 1977**
(Figs 20I, J; 25H-K)

MEASUREMENTS

Female: L = 0.7-1.4 mm; a = 17-29; b = 2.9-4.5; c = 10-19; c′ = 2-3; V = 58-67.

DESCRIPTION

Lip region set off, distinctly wider than adjoining body. Amphids cup-shaped, aperture slit-like, situated anterior to apex of dorsal tooth. Buccal cavity 21-32 × 10-16 μm, tapering at base. Apex of dorsal tooth at 13-18 μm or 59-68% of buccal cavity length from its base. Tooth slightly anteriorly directed, anterior margin straight. Subventral vertical wall of buccal cavity lacking ridge. Female genital system amphidelphic. Vulva transverse, without vulval papillae; vagina with *pars refringens*. Uterine egg 70-115 × 40-54 μm. Tail conoid, ventrally arcuate, 56-105 μm long. Caudal glands and spinneret absent. Male not known.

DISTRIBUTION

Cosmopolitan.

Key to species of *Coomansus* Jairajpuri & Khan, 1977

1 – Dorsal tooth apex at 50-60% of buccal cavity length from its base 2
– Dorsal tooth apex at 65-90% of buccal cavity length from its base 5

2 – Large species (L = 2.6-3.0 mm); buccal cavity 42-44 μm; median vulva; ♀: L = 2.6-3.0 mm; a = 28-30; b = 5.0-5.6; c = 14-17; c′ = 3.0-3.5; V = 48-52; ♂: unknown (Ecuador)...... *splendidus*
– Smaller species (<2 mm); buccal cavity < 40 μm; vulva at 60-70% 3

3 – Body very small, 0.7 mm long, plump; buccal cavity 20 μm long; ♀: L = 0.7 mm; a = 14; b = 3.2-3.4; c = 11-14; c′ = 2; V = 60-66; ♂: unknown (Russia-Far East)........ *obliquoris*
– Body > 1.8 mm long, less plump; buccal cavity 30-40 μm long 4

4 – Dorsal wall of buccal cavity strongly arched near tooth; tail 85-110 μm long; body > 1.6 mm; ♀: L = 1.6-1.8 mm; a = 26-34; b = 3.8-4.0; c = 15-19; c′ = 2.5-3; V = 60-69; ♂: unknown (India, South Africa)........ *pretoriensis*
– Dorsal wall of buccal cavity with or without arch near tooth; tail 65-80 μm long; body < 1.5 mm; ♀: L = 1.1-1.4 mm; a = 24; b = 3.0-3.3; c = 15-19; c′ = 2-2.6; V = 70-72; ♂: unknown (India) .... *indicus*

5 – Species large, 2-6 mm long........ 8
– Species small, <1.5 mm long........ 6

6 – Tooth apex at anterior third of buccal cavity; ♀: L = 0.7-1.2 mm; a = 17-26; b = 2.9-4.5; c = 10-19; c′ = 2-3; V = 58-67; ♂: unknown (Germany, The Netherlands, Poland, Austria, Italy, former Czechoslovakia, Hungary, Romania, Spain, France, Russia, Belarus, Lithuania, Georgia, Uzbekistan, Korea, India, Iran, Kenya, South Africa, Canada, USA, Venezuela, Colombia, Costa Rica, Brazil, Hawaii) . . . . *parvus*
– Tooth apex at anterior fourth of buccal cavity 7

7 – Tail 70 μm or 2-2.5 anal body diam. long; ♀: L = 1.1 mm; a = 17-24; b = 3.5; c = 14-16; c′ = 2-2.5; V = 61-62; ♂: unknown (Russia-Far East) . . . . *arvensis*
– Tail 90-100 μm or 3-3.5 anal body diam. long; ♀: L = 1.0-1.1 mm; a = 22-29; b = 3.3-3.9; c = 10-12; c′ = 3.3.5; V = 62-64; ♂: unknown (New Caledonia) . . . . *ouinnensis*

8 – Tail comparatively long, 4-6 anal body diam. 9
– Tail shorter, 2-3.5 anal body diam. 10

9 – Vulval region with papillae; tail 4-4.5 anal body diam. long; body > 3 mm; ♀: L = 3.1-3.8 mm; a = 32-40; b = 4-6; c = 10-16; c′ = 4-4.5; V = 47-55; ♂: L = 3.0-3.7 mm; a = 31-41; b = 4.0-4.7; c = 15-22; supplements nine (New Zealand) . . *mesadenus*
– Vulval region without papillae; tail 5.5-6 anal body diam. long; body < 3 mm; ♀: L = 2.1-2.6 mm; a = 27-42; b = 3.7-5.2; c = 7-11; c′ = 5.5-6; V = 45-55; ♂: L = 3.2-3.4 mm; a = 35-39; b = 4.5-4.7; c = 18-22; supplements 12 (New Zealand, New Hebrides) . . . . *composticola*

10 – Buccal cavity 60-75 μm long 11
– Buccal cavity 30-50 μm long 12

11 – Buccal cavity wide, barrel-shaped; male with 18-20 supplements; ♀: L = 3.4-3.6 mm; a = 31-37; b = 3.8-4.1; c = 22-25; c′ = 2.5-3; V = 57-60; ♂: L = 3.0-3.5 mm; a = 33-40; b = 3.3-3.9; c = 23-29; supplements 18-20 (Canada) . . . . *fletcherensis*
– Buccal cavity oblong; male with 11-15 supplements; ♀: L = 3.7-5.3 mm; a = 42-68; b = 3.8-4.5; c = 33-47; c′ = 2; V = 63-71; ♂: L = 3.5-5.9 mm; a = 39-50; b = 3.5-4.4; c = 39-59; supplements 11-15 (Germany) . . . . *kastrolli*

12 – Buccal cavity subrectangular, flattened at base, 32-36 μm long; ♀: L = 1.77-2.02 mm; a = 26-30; b = 4.7-5.3; c′ = 2.6-3.5; V =

51-57; ♂: L = 1.84-1.97 mm; a = 29-30; b = 5.0; c = 18-21; c′ = 1.5-1.9; supplements 11-12; spicules 97-98 μm (Chile) .... *magellanicus*

– Buccal cavity barrel-shaped, narrow at base, >40 μm long 13

13 – Dorsal wall of buccal cavity distinctly arched near tooth, its anterior part broader than posterior; ♀: L = 1.9-3.0 mm; a = 27-34; b = 4.0-4.5; c = 10-14; c′ = 3; V = 52-57; ♂: L = 2.1-2.9 mm; a = 29-33; b = 3.7-4.3; c = 17-28; supplements 10-11 (Nigeria, South Africa, New Zealand, Campbell Islands) .... *campbelli*

– Dorsal wall of buccal cavity either not or only barely arched near tooth, both parts nearly equal in diam. 14

14 – Body 3-4 mm long. 15

– Body 2-3 mm long 16

15 – Dorsal tooth obtuse, apex at 82% of buccal cavity length from its base; buccal cavity walls thick; ♀: L = 3.0-4.3 mm; a = 27-40; b = 4.8-5.7; c = 13-20; c′ = 3-4; V = 49-55; ♂: L = 3.1-4.0 mm; a = 34-44; b = 4.7-5.5; c = 18-26; supplements 9-12 (Comoro Islands, Canada, Antarctic) .... *geerlachei*

– Dorsal tooth sharp, apex at 75% of buccal cavity length from its base; buccal cavity walls rather thin; ♀: L = 3.4 mm; a = 34; b = 5.3; c = 20; c′ = 3.3; V = 55; ♂: L = 3.4 mm; a = 34; b = 5.3; c = 20; supplements 12 (Germany, Australia-Tasmania) ... *major*

16 – Dorsal tooth apex at >80% of buccal cavity length from its base 17

– Dorsal tooth apex at <75% of buccal cavity length from its base; ♀: L = 2.2 mm; a = 21-34; b = 4.2-4.5; c = 18-19; c′ = 2.5-3; V = 60-61; ♂: unknown (India, Mexico, Venezuela) .... *venezolanus*

17 – Posterior part of intestine with some constriction; dorsal tooth apex at 80-83% of buccal cavity length from its base; ♀: L = 2.0-2.8 mm; a = 22-26; b = 4.0-5.4; c = 12-14; c′ = 3.0-3.6; V = 55-61; ♂: L = 2.1-2.9 mm; a = 21-27; b = 4.1-5.2; c′ = 15-19; supplements 8-11 (Chile, Argentina-Tierra del Fuego, Antarctic-South Shetland Islands) .... *intestinus*

– Posterior part of intestine without constriction; dorsal tooth apex at 85-95% of buccal cavity length from its base; ♀: L = 2.16-2.65 mm; a = 28-37; b = 4.2-4.8; c = 16-21; c′ = 2.8-3.5; V = 54-57; ♂: L = 2.09-2.67 mm; a = 30-35; b = 4.3-5.2; c = 23-

24; c′ = 1.8-1.9; supplements 12-13; spicules 93-110 μm (New Zealand) . . . . *meridionalis*

Genus *Parkellus* Jairajpuri, Tahseen & Choi, 2001b
(Fig. 21E-H)

DIAGNOSIS

Body large, labial papillae prominent. Buccal cavity spacious. Dorsal tooth situated in posterior half of buccal cavity, apex directed anteriad, always posterior to middle of buccal cavity. Thin longitudinal ridge present along subventral vertical walls opposite dorsal tooth. Pharyngo-intestinal junction non-tuberculate. Female genital system amphidelphic. Spicules and lateral guiding pieces prominent, gubernaculum with sleeves. Tail similar in both sexes, short conoid. Caudal glands poorly developed or absent. Spinneret absent.

TYPE SPECIES

Parkellus parkus Jairajpuri, Tahseen & Choi, 2001b

OTHER SPECIES

P. acuticaudatus (Eroshenko, 1975) **comb. n.**
= *Coomansus acuticaudatus* (Eroshenko, 1975) Loof & Winiszewska-Slipińska, 1993
= *Iotonchus acuticaudatus* Eroshenko, 1975
P. cobbi (Eroshenko, 1975) **comb. n.**
= *Coomansus cobbi* (Eroshenko, 1975) Loof & Winiszewska-Slipińska, 1993
= *Iotonchus cobbi* Eroshenko, 1975
P. menzeli (Loof & Winiszewska-Slipińska, 1993) **comb. n.**
= *Coomansus menzeli* Loof & Winiszewska-Slipińska, 1993
P. monticola (Eroshenko, 1975) **comb. n.**
= *Coomansus monticola* (Eroshenko, 1975) Loof & Winiszewska-Slipińska, 1993
P. mucronatus (Eroshenko, 1975) **comb. n.**
= *Coomansus mucronatus* (Eroshenko, 1975) Loof & Winiszewska-Slipińska, 1993
= *Iotonchus mucronatus* Eroshenko, 1975

P. paraamphigonicus (Eroshenko, 1975) **comb. n.**
= *Coomansus paraamphigonicus* (Eroshenko, 1975) Loof & Winiszewska-Slipińska, 1993
= *Iotonchus paraamphigonicus* Eroshenko, 1975
P. silvius (Eroshenko, 1975) **comb. n.**
= *Coomansus silvius* (Eroshenko, 1975) Jairajpuri & Khan, 1982
= *Clarkus silvius* Eroshenko, 1975
P. simmenensis (Kreis, 1924) **comb. n.**
= *Coomansus simmenensis* (Kreis, 1924) Loof & Winiszewska-Slipińska, 1993
= *Mononchus* (*Iotonchus*) *simmenensis* Kreis, 1924 (Schneider, 1939)
= *Iotonchus simmenensis* (Kreis, 1924) Schneider, 1939
P. zschokkei (Menzel, 1913) **comb. n.**
= *Coomansus zschokkei* (Menzel, 1913) Loof & Winiszewska-Slipińska, 1993
= *Mononchus zschokkei* Menzel, 1913
= *Mononchus* (*Iotonchus*) *zschokkei* Menzel, 1913 (Cobb, 1917)
= *Iotonchus zschokkei* (Menzel, 1913) Cobb, 1917
= *Clarkus scopulosus* Eroshenko, 1975
= *Coomansus scopulosus* (Eroshenko, 1975) Jairajpuri & Khan, 1982

Parkellus parkus Jairajpuri, Tahseen & Choi, 2001b
(Fig. 21E-H)

MEASUREMENTS

Female: L = 2.2-2.5 mm; a = 26-30; b = 3.5-3.8; c = 25-29; c′ = 1.5-1.8; V = 61-74.
Male: L = 2.07-2.49 mm; a = 27-31; b = 3.4-3.7; c = 23-29; c′ = 1.2-1.5.

DESCRIPTION

Lip region slightly offset, 43-48 × 14-18 μm; labial papillae slightly raised. Amphidial aperture oval. Buccal cavity strongly sclerotised, 55-62 × 28-32 μm. Dorsal tooth prominent, located in posterior third of buccal cavity, its apex at 35-39% of buccal cavity length from its base. Thin longitudinal ridge present on subventral walls. Female genital system amphidelphic. Vulva transverse slit, inconspicuous;

vulval papillae often present. Vagina with small *pars refringens*. Tail 85-90 μm long, conoid, ventrally curved, gradually tapering to a pointed terminus. Caudal glands rudimentary. Spinneret absent. Male with 82-92 μm long arcuate spicules. Gubernaculum 30-35 μm long with forked pincer-shaped proximal part with prominent sleeve. Lateral guiding pieces 22-25 μm long. Ventromedian supplements 21-25, regularly spaced.

DISTRIBUTION

Korea.

Key to species of *Parkellus* Jairajpuri, Tahseen & Choi, 2001

1 – Buccal cavity < 60 μm long 2
– Buccal cavity > 60 μm long 7

2 – Tail tip mucronate; gubernaculum thick, hooked proximally; ♀: L = 2.9 mm; a = 25; b = 4; c = 20; c′ = 2; V = 68; ♂: L = 1.85 mm; a = 24; b = 3.2; c = 25; supplements 20 (Russia-Far East) . . . . *mucronatus*
– Tail tip without mucro, gubernaculum not hooked proximally 3

3 – Tail terminus slender, acute or subacute 4
– Tail terminus conspicuously rounded 5

4 – Tail four anal body diam. long; dorsal tooth apex at 39-40% of buccal cavity length from its base; ♀: L = 2.8-3.4 mm; a = 24-27; b = 3.7-4.1; c = 16-19; c′ = 3.5-4; V = 63-72; ♂: L = 2.5 mm; a = 24; b = 3.6; c = 18; supplements 24 (Russia-Far East) . . . . . . *cobbi*
– Tail 2.5-3.4 anal body diam. long; dorsal tooth apex at 47-50% of buccal cavity length from its base; ♀: L = 2.5-3.2 mm; a = 23-27; b = 3.6-4.2; c = 14-18; c′ = 2.5-3.4; V = 62-71; ♂: L = 2.7-3.0 mm; a = 23-26; b = 3.6-3.8; c = 18-20; supplements 26-28 (Russia-Far East, Korea) . . . . *silvius*

5 – Tail cylindroid in posterior third with broadly rounded tip; body slender (a *ca* 40); ♀: L = 2.3-3.5 mm; a = 37-48; b = 3.8-4.8; c = 29-37; c′ = 2-2.5; V = 67; ♂: L = 2.9 mm; a = 43; b = 4.4; c = 26; supplements 23 (Switzerland, former Czechoslovakia, Romania, former Yugoslavia, Spain) . . . . *simmenensis*

– Tail uniformly narrowing in posterior-third with finely rounded tip; body not so slender (a *ca* 30) 6

6 – Buccal cavity 50-56 × 22-25 μm, tail 103-152 μm long in female; spicules 101-131 μm; ♀: L = 2.2-3.0 mm; a = 24-34; b = 3.7-4.3; c = 16-23; c′ = 2.1-3.2; V = 59-60; ♂: L = 2.0-3.0 mm; a = 25-35; b = 3.7-4.4; c = 17-25; supplements 19-24 (Switzerland, Austria, former Czechoslovakia, Hungary, Romania, Poland, Spain, Italy, Russia-Far East, Kazakhstan, Korea) . . . . *zschokkei*

– Buccal cavity 55-62 × 28-32 μm, tail 85-90 μm long in female; spicules 82-92 μm; ♀: L = 2.2-2.5 mm; a = 26-30; b = 3.5-3.8; c = 25-29; c′ = 1.5-1.8; V = 61-74; ♂: L = 2.07-2.49 mm; a = 27-31; b = 3.4-3.7; c = 23-29; c′ = 1.2-1.5; supplements 21-25 (Korea) . . . . *parkus*

7 – Body mostly well over 3 mm long; tail 150-190 μm long; ♀: L = 2.8-3.7 mm; a = 24-34; b = 4.2-4.8; c = 15-22; c′ = 2.2-3.2; V = 64-69; ♂: L = 2.5-3.4 mm; a = 23-33; b = 3.8-4.7; c = 17-33; supplements 19-26 (Switzerland, Poland) . . . . *menzeli*

– Body 2-2.7 mm long; tail 90-130 μm long 8

8 – Tail slightly curved; dorsal tooth rounded in contour; ♀: L = 2.0-2.4 mm; a = 17-22; b = 3.3-3.5; c = 18-20; c′ = 1.9; V = 70-72; ♂: unknown (Russia-Far East) . . . . *paramphigonicus*

– Tail strongly curved; dorsal tooth more or less pointed 9

9 – Buccal cavity shorter (66 μm); tail tip acute; ♀: L = 2.1 mm; a = 24; b = 3.2; c = 22; c′ = 1.7; V = 72; ♂: unknown (Russia-Far East) . . . . *acuticaudatus*

– Buccal cavity longer (70-75 μm); tail tip rounded; ♀: L = 2.2-2.7 mm; a = 20-24; b = 3.2-3.5; c = 17-20; c′ = 2.5; V = 65-75; ♂: L = 1.9-2.0 mm; a = 20-25; b = 3.2; c = 21; supplements 22 (Russia-Far East) . . . . *monticola*

Nematology Monographs & Perspectives, 2010, Vol. 7, 89-139

Chapter 5

Family Mylonchulidae Jairajpuri, 1969

Mylonchulus was initially proposed as a subgenus of the genus *Mononchus* Bastian, 1865 by Cobb (1916) and was subsequently raised to the status of an independent genus by Altherr (1953). It is characterised by the usually small body size, ventrally arcuate posture, several rows of transverse denticles on the subventral walls of the buccal cavity, non-tuberculate type of pharyngo-intestinal junction and the tail usually small and provided with well developed caudal glands and spinneret. This genus is widely distributed worldwide and is represented by over 60 species. Cobb (1916) had also described another genus, *Sporonchulus*, with denticles on the subventral walls. Andrássy (1958) added the genera *Granonchulus*, *Judonchulus* and *Brachonchulus* to this group, whereas Mulvey (1963b) added *Prionchuloides* and Mulvey and Jensen (1967) added the genus *Polyonchulus*. Jairajpuri (1969) considered *Mylonchulus* and related genera to form a relatively distinct group among Mononchida and accordingly proposed the family Mylonchulidae with two subfamilies: Mylonchulinae (for *Mylonchulus* and *Polyonchulus*) and Sporonchulinae (for *Sporonchulus*, *Granonchulus*, *Prionchuloides*, *Judonchulus* and *Brachonchulus*).

Andrássy (1972, 1976) added the genera *Margaronchulus* and *Oligonchulus*, respectively, to this group. Andrássy (1976) accepted Mylonchulidae but with only one subfamily, Mylonchulinae, containing five genera *viz.*, *Mylonchulus*, *Margaronchulus*, *Brachonchulus*, *Oligonchulus* and *Polyonchulus*. Jairajpuri and Khan (1982) accepted the two subfamilies proposed by Jairajpuri (1969) and grouped the genera *Mylonchulus*, *Paramylonchulus*, *Oligonchulus*, *Polyonchulus*, *Megaonchulus* and *Margaronchulus* under Mylonchulinae, and the genera *Sporonchulus*, *Granonchulus*, *Judonchulus*, *Prionchuloides* and *Brachonchulus* under Sporonchulinae. Siddiqi (1984a) considered *Sporonchulus* to belong to Mononchinae. Andrássy (1992) again accepted only a single subfamily, the Mylonchulinae, under Mylonchulidae with seven genera *viz.*, *Mylonchulus*, *Brachonchulus*, *Granonchulus*, *Mar-*

garonchulus, *Megaonchulus*, *Oligonchulus* and *Polyonchulus*. The genus *Sporonchulus* was placed under Mononchinae while *Prionchuloides* was transferred to Anatonchoidea.

The various genera belonging to this family are grouped herein into two subfamilies: the Mylonchulinae with the genera *Mylonchulus*, *Brachonchulus*, *Polyonchulus*, *Margaronchulus*, *Oligonchulus*, *Megaonchulus*, *Paramylonchulus*, *Crestonchulus* and *Margaronchuloides* **gen. n.**, and the Sporonchulinae with the genera *Sporonchulus*, *Granonchulus*, *Judonchulus*, *Prionchuloides* and *Actus*.

DIAGNOSIS

Medium-sized nematodes. Buccal cavity heavily sclerotised, goblet- or funnel-shaped, strongly tapering at base. Dorsal tooth large, claw-like, located in anterior part of buccal cavity and sharply pointed, directed anteriad, or moderately developed in anterior or posterior half of buccal cavity. Subventral walls with several rows of denticles arranged in transverse rows, or scattered, or both, denticles rarely absent. Tail generally short, conoid and arcuate, caudal glands and spinneret usually well developed, rarely absent.

TYPE SUBFAMILY

Mylonchulinae Jairajpuri, 1969

OTHER SUBFAMILY

Sporonchulinae Jairajpuri, 1969

Key to subfamilies of Mylonchulidae Jairajpuri, 1969

1 – Dorsal tooth large, claw-like, located in anterior part of buccal cavity; usually opposed by denticles arranged in transverse rows . . . . **Mylonchulinae**

– Dorsal tooth moderately developed, located in anterior or posterior half of buccal cavity, opposed by scattered denticles . . . . **Sporonchulinae**

Subfamily Mylonchulinae Jairajpuri, 1969

DIAGNOSIS

Dorsal tooth in anterior half of buccal cavity; denticles arranged in one or several transverse rows. Subventral teeth present or absent. Caudal glands and spinneret generally well developed.

TYPE GENUS

Mylonchulus Cobb, 1916

OTHER GENERA

Brachonchulus Andrássy, 1958
Polyonchulus Mulvey & Jensen, 1967
Margaronchulus Andrássy, 1972
Oligonchulus Andrássy, 1976
Megaonchulus Jairajpuri & Khan, 1982
Paramylonchulus Jairajpuri & Khan, 1982
Crestonchulus Siddiqi & Jairajpuri, 2002
Margaronchuloides **gen. n.**

Key to genera of Mylonchulinae Jairajpuri 1969

1 – Subventral walls with either two transverse, arcuate, smooth ribs or square-bracket-shaped crest 2
– Subventral walls with denticles in transverse rows with or without dentate longitudinal ridges or scattered denticles 3

2 – Subventral wall with square-bracket-shaped crest; no denticles or subventral teeth . . . . *Crestonchulus*
– Subventral walls with two transverse, arcuate ribs only, no denticles or teeth . . . . *Margaronchuloides* gen. n.

3 – Small denticles present only in transverse rows 4
– Dentate longitudinal ridge or scattered denticles also present in addition to transverse rows 7

4 – Transverse denticles numerous, forming rasp-like field 5
– Transverse denticles present only in one or two rows, not forming rasp-like field 6

5 – Buccal cavity comparatively broad; subventral walls with small teeth in addition to transverse rows of denticle; female genital system amphidelphic .... *Mylonchulus*
– Buccal cavity narrow; subventral walls without teeth; only transverse rows of denticles present; female genital system mono-prodelphic .... *Paramylonchulus*

6 – Female genital system mono-prodelphic; caudal glands reduced .. .. *Margaronchulus*
– Female genital system mono-opisthodelphic; caudal glands well developed.. *Oligonchulus*

7 – Subventral walls of buccal cavity with two serrate longitudinal ridges in addition to transverse rows of denticles 8
– Subventral walls of buccal cavity with scattered denticles and transverse rows of denticles .... *Brachonchulus*

8 – Tail in both sexes long, filiform, 8-20 anal body diam. long .... *Megaonchulus*
– Tail in both sexes short, 1-2 anal body diam. long ... *Polyonchulus*

Genus *Mylonchulus* Cobb, 1916

= *Mononchus* (*Mylonchulus* Cobb, 1916)
Pakmylonchulus Khan & Saeed, 1987
(Figs 26A-E; 27)

DIAGNOSIS

Body 0.5-2.9 mm long. Buccal cavity goblet- or funnel-shaped; dorsal wall thicker than ventral. Dorsal tooth very large, claw-like, obliquely directed anteriad with sharply pointed apex, located in anterior half of buccal cavity. Each subventral wall armed with several small rasp-like denticles more or less arranged in regular transverse rows. Anterior and/or posterior margin of rasp-field provided with sclerotised refractive rings. Two small subventral teeth present opposite base of dorsal tooth. Pharyngo-intestinal junction non-tuberculate. Female genital system amphidelphic. Spicules short, slender, arcuate; gubernaculum simple or bidentate with or without lateral accessory pieces. Tail of both sexes similar, short, usually *ca* 1-2 anal body diam., very rarely up to four times anal body diam., varying in form. Caudal glands usually well developed, grouped or in tandem. Spinneret terminal or subterminal.

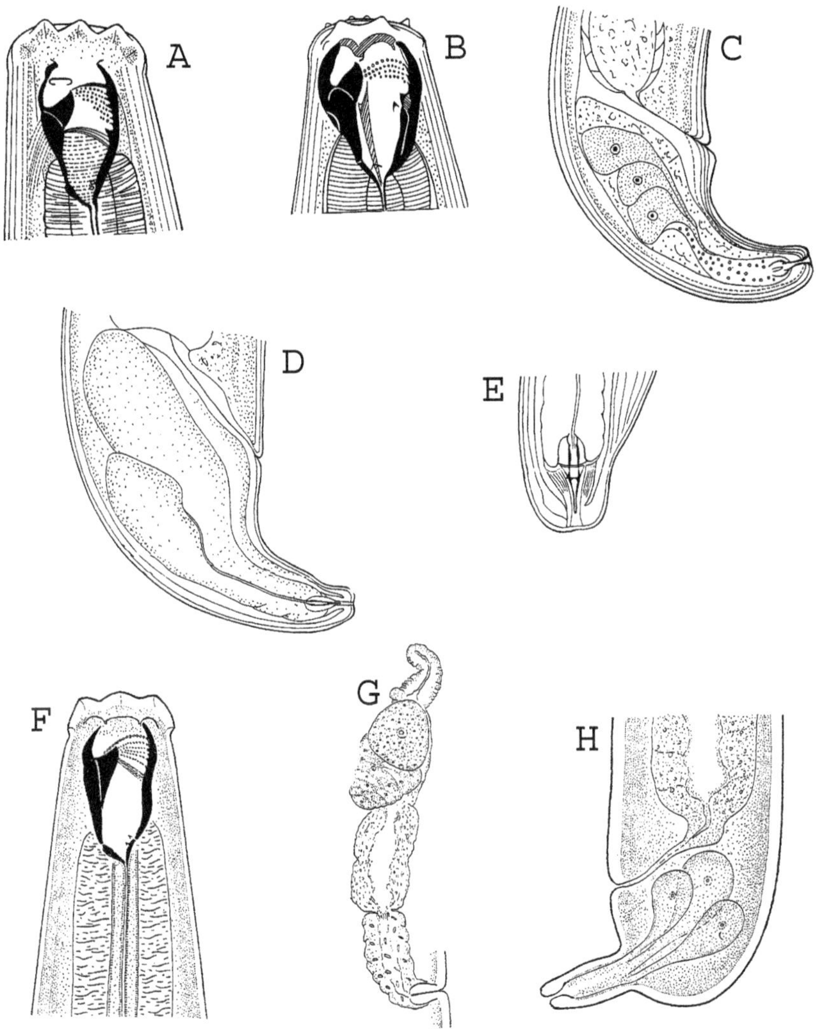

Fig. 26. *A-E:* Mylonchulus minor *(Cobb, 1893) Andrássy, 1958. A, B: Anterior region; C, D: Female posterior region; E: Female tail terminus. F-H:* Paramylonchulus index *(Cobb, 1906) Jairajpuri & Khan, 1982. F: Anterior region; G: Female genital system; H: Female posterior region. A-E after Mulvey and Jensen (1967); F-H after Jairajpuri (1970).*

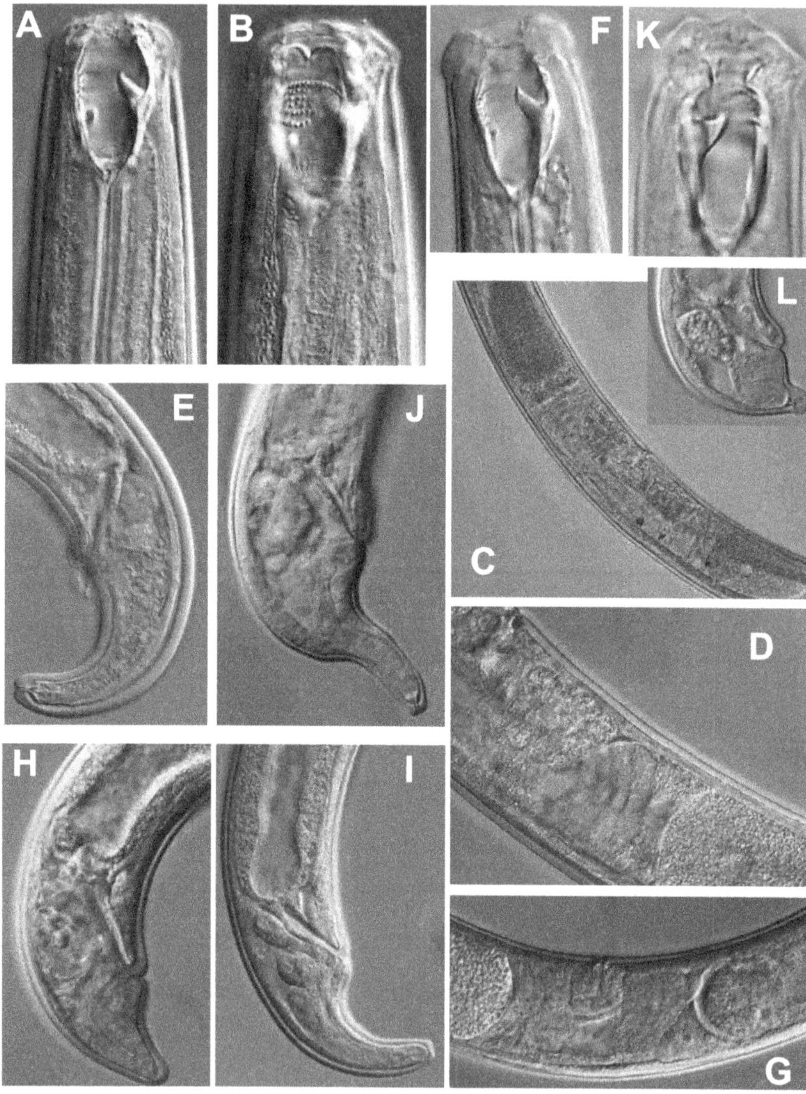

Fig. 27. *A-E:* Mylonchulus minor *Cobb, 1917. A, B: Anterior region; C, D: Female genital system; E: Female tail; F-H:* M. brachyuris *(Bütschli, 1873) Cobb, 1917. F: Anterior region; G: Female genital system; H: Female tail; I:* M. hawaiiensis *(Cassidy, 1931) Goodey, 1951, female tail region; J:* M. sigmaturus *Cobb, 1917, female tail region; K, L:* M. oceanicus *Andrássy, 1986. K: Anterior region; L: Female tail.*

TYPE SPECIES

Mylonchulus minor (Cobb, 1893) Cobb, 1916
= *Mononchus minor* Cobb, 1893
= *Mononchus* (*Mylonchulus*) *minor* Cobb, 1893 (Cobb, 1916)
= *Mononchus* (*Mylonchulus*) *brachyuris microdenticulatus minor* Cobb, 1893 (Micoletzky, 1922)
= *Mylonchulus sculentus* Jain, Saxena & Sharma, 1993 **syn. n.**

OTHER SPECIES

M. agilis Doucet, 1980
= *M. inflatus* Loof, 1993
M. aequatorialis Orcelli & Vinciguerra, 2007
M. amurus Khan & Jairajpuri, 1979
= *Pakmylonchulus amurus* (Khan & Jairajpuri, 1979) Khan & Saeed, 1987
M. ananasi Yeates, 1992
M. apapillatus Khan & Jairajpuri, 1979
M. arenicolus Clark, 1961c
M. brachyuris (Bütschli, 1873) Cobb, 1917
= *Mononchus brachyuris* Bütschli, 1873
= *Mononchus* (*Mylonchulus*) *brachyuris* Bütschli, 1873 (Cobb, 1917)
= *Mononchus* (*Mylonchulus*) *brachyuris microdenticulatus typicus* Micoletzky, 1922
= *Mylonchulus agriculturae* Coetzee, 1967a
= *Mylonchulus subterraneus* Schneider, 1940
= *Mylonchulus williamsi* Loof, 2006 **syn. n.**
M. brassicus Soni & Nama, 1980
M. brevicaudatus Cobb, 1917
= *Mononchus* (*Mylonchulus*) *brevicaudatus* Cobb, 1917
= *Mononchus* (*Mylonchulus*) *brachyuris macrodenticulatus brevicaudatus* Cobb, 1917 (Micoletzky, 1922)
M. bulbiferous Jensen & Mulvey, 1968
M. cavensis Schneider, 1940
= *Mononchus* (*Mylonchulus*) *cavensis* Schneider, 1940
M. cereris Coetzee, 1967a
M. contractus Jairajpuri, 1970b
M. curvicaudatus Mulvey & Jensen, 1967
M. dentatus Jairajpuri, 1970b

= *Mylonchulus wasimi* Tiasi, Chatterjee & Manna, 2008 **syn. n.**
M. doliolarius Andrássy, 1992
M. exacutus Jensen & Mulvey, 1968
M. gigas Gagarin, 1993
M. hawaiiensis (Cassidy, 1931) Goodey, 1951
= *Mononchus (Mylonchulus) hawaiiensis* Cassidy, 1931 (Goodey, 1951)
= *Mylonchulus bareilliensis* Sharma & Saxena, 1980
= *Mylonchulus indicus* Sharma & Saxena, 1980
M. incurvus Cobb, 1917
= *Mononchus (Mylonchulus) incurvus* Cobb, 1917
= *Mononchus (Mylonchulus) brachyuris macrodenticulatus incurvus* Cobb, 1917 (Micoletzky, 1922)
M. indicus Sharma & Saxena, 1980
M. insolitus Andrássy, 1968
M. jinjuensis Khan, Choi, Lee & Choi, 2000
M. kaszabi Andrássy, 1967
M. lacustris (Cobb *in* Cobb, 1915) Andrássy, 1958
= *Mononchus lacustris* Cobb *in* Cobb, 1915
= *Mononchus (Mylonchulus) lacustris* Cobb *in* Cobb, 1915 (Cobb, 1917)
= *Mononchus (Mylonchulus) brachyuris macrodenticulatus lacustris* Cobb *in* Cobb, 1915 (Micoletzky, 1922)
= *Mylonchulus sarmini* Azmi, 1991 **syn. n.**
= *Mylonchulus yassini* Elbadri, Khan, Moon, Lee & Choo, 2008 **syn. n.**
= *M. lacumurus* Singh, Lal & Saha, 2004 **syn. n.**
M. longus Altherr, 1972
= *Mylonchulus macrosoma* Altherr, 1976
M. maritimus Jiménez-Guirado & Murillo-Navarro, 2008
M. nainitalensis Jairajpuri, 1970a
M. neocontractus Patil & Khan, 1982
M. neoindicus Andrássy, 2009
= *M. indicus* Choudhury & Sivakumar, 1998 *nec M. indicus* Sharma & Saxena, 1980
M. obliquus Cobb, 1917
= *Mononchus (Mylonchulus) obliquus* Cobb, 1917
= *Mononchus (Mylonchulus) brachyuris macrodenticulatus obliquus* Cobb, 1917 (Micoletzky, 1922)
= *Mononchus (Mylonchulus) japonicus* Cobb, 1917

= *Mononchus* (*Mylonchulus*) *brachyuris microdenticulatus japonicus* Cobb, 1917 (Micoletzky, 1922)
= *Mylonchulus japonicus* Cobb, 1917
M. obtusicaudatus (von Daday, 1899) Cobb, 1916
= *Mononchus obtusicaudatus* von Daday, 1899 (Cobb, 1916)
M. oceanicus Andrássy, 1986
M. orbitus Jensen & Mulvey, 1968
= *Mylonchulus muradi* Jairajpuri, 1970b
M. oregonensis Andrássy, 1992
= *Mylonchulus striatus apud* Jensen & Mulvey, 1968
M. paitensis Yeates, 1992
M. parabrachyurus (Thorne, 1924) Schneider, 1939
= *Mononchus parabrachyurus* Thorne, 1924
= *Mononchus* (*Mylonchulus*) *parabrachyurus* (Schneider, 1939)
M. paracontractus Aboul-Eid & Ameen, 1982
M. polonicus (Stefanski, 1915) Cobb, 1917
= *Mononchus polonicus* Stefanski, 1915
= *Mononchus* (*Mylonchulus*) *polonicus* Stefanski, 1915 (Cobb, 1917)
= *Mononchus* (*Mylonchulus*) *brachyuris microdenticulatus polonicus* Stefanski, 1915 (Micoletzky, 1922)
= *Mononchus* (*Mylonchulus*) *montanus* Thorne, 1924 (Goodey, 1951)
= *Mylonchulus montanus* (Thorne, 1924) Goodey, 1951
= *Mylonchulus boveyi* Altherr *in* Altherr & Delamare Deboutteville, 1972
M. prodenticulatus Mulvey, 1961a
M. psammophilus Yeates, 1967
M. rosensis Khan, 1975
M. rotundicaudatus (Skwarra, 1921) Schneider, 1939
= *Mononchus rotundicaudatus* Skwarra, 1921
= *Mononchus* (*Mylonchulus*) *rotundicaudatus* Skwarra, 1921 (Schneider, 1939)
M. sagarensis Sinha, Baqri & Choudhury, 1989
M. sessus Jairajpuri, 1982
M. sexcristatus (Merzheevskaya, 1951) Mulvey, 1961a
M. sigmaturellus Mulvey, 1961a
= *M. abuelhasani* Elbadri, Khan, Moon, Lee & Choo, 2008 **syn. n.**
M. sigmaturus Cobb, 1917
= *Mononchus* (*Mylonchulus*) *sigmaturus* Cobb, 1917

= *Mononchus* (*Mylonchulus*) *brachyuris sigmaturus* Cobb, 1917 (Micoletzky, 1922)
= *Mononchus* (*Mylonchulus*) *brachyuris sigmaturoides* Schneider, 1939
M. solus Mulvey, 1961a
M. striatus (Thorne, 1924) Schneider, 1939
= *Mononchus striatus* Thorne, 1924
= *Mononchus* (*Mylonchulus*) *striatus* Thorne, 1924 (Schneider, 1939)
M. subsimilis Cobb, 1917
= *Mononchus* (*Mylonchulus*) *subsimilis* Cobb, 1917
= *Mononchus* (*Mylonchulus*) *cobbi* Micoletzky, 1922
M. subtenuis Cobb, 1917
= *Mononchus* (*Mylonchulus*) *subtenuis* Cobb, 1917
= *Mononchus* (*Mylonchulus*) *tenuis* Cobb, 1917, *nec* von Daday, 1908
= *Mononchus* (*Mylonchulus*) *brachyuris tenuis* Cobb, 1917 (Micoletzky, 1922)
M. taeguensis Khan, Choi, Lee & Choi, 2000
M. ubis Clark, 1961c
M. vasis Yeates, 1992
M. vulvalatus Ahmad, 2000
M. vulvapapillatus Altherr *in* Altherr & Delamare Deboutteville, 1972

Species inquirendae

M. denticulatus Cobb, 1917
M. madrasi Kannan, 1961c
M. micrurus Cobb, 1917
M. sparsus Cobb, 1916
M. subterraneus Schneider, 1940

Mylonchulus minor **(Cobb, 1893) Andrássy, 1958**
(Figs 26A-E; 27A-E)

Measurements

Female: L = 0.82-1.40 mm; a = 18-32; b = 3.0-3.8; c = 30-47; V = 54-67.

DESCRIPTION

Lip region 20-25 μm diam., 7-10 μm high. Amphidial aperture 3-4 μm wide, located at 9-10 μm from anterior end of body and 16-20 μm from base of buccal cavity. Buccal cavity 20-30 μm long, 12-18 μm broad. Dorsal tooth large, situated in anterior half of buccal cavity, its apex at 70-76% of length of buccal cavity length from its base. Subventral wall bearing five or six transverse rows of denticles. Subventral teeth present. Female genital system amphidelphic. Tail arcuate conoid with rounded terminus, 35-50 μm or 1.0-1.5 anal body diam. long. Caudal glands in tandem. Spinneret terminal. Male not known.

DISTRIBUTION

Egypt, Fiji, India, Iran, Java, Malaysia, Mauritius, Pakistan, Panama, South Africa, USA-Florida; Missouri, Venezuela.

REMARKS

The genus *Mylonchulus* is one of the most widely distributed genera among mononchs. Some of its species, *viz.*, *M. brachyuris*, *M. sigmaturus*, *M. minor* and, to some extent *M. lacustris* and *M. hawaiiensis*, have been recorded from many countries and probably represent more than one species.

Key to species of *Mylonchulus* Cobb, 1916

1 – Tail broadly rounded, hemispheroid 2
– Tail of different shape, not hemispheroid 5

2 – Spinneret absent; ♀: L = 1.4 mm; a = 27-34; b = 3.4; c = 41-62; c′ = 0.7; V = 61-69; ♂: L = 1.4 mm; a = 38; b = 3.3; c = 36; supplements 11 (USA-Oregon) . . . . *oregonensis*
– Spinneret present 3

3 – Spinneret terminal; ♀: L = 0.92-1.20 mm; a = 19-26; b = 3.4-3.9; c = 33.7-50.6; c′ = 0.7-0.9; V = 59-68; ♂: L = 0.96-1.27 mm; a = 22-28; b = 3.2-3.9; c = 28.4-39.9; supplements 10-12 . . . . *maritimus*
– Spinneret subterminal 4

4 – Spinneret subdorsal, tail bulbiferous, distinctly shorter than anal body diam.; ♀: L = 1.2-1.4 mm; a = 32-34; b = 3.0-3.4; c = 61-69; c′ = 0.6; V = 69-71; ♂: unknown (USA-Oregon) .... *bulbiferous*
– Spinneret subventral, tail not bulbiferous, as long as, or longer than, anal body diam.; ♀: L = 1.4-1.5 mm; a = 24-34; b = 3.4-4.2; c = 34-55; c′ = 1.1-1.5; V = 62-68; ♂: L = 1.4-1.9 mm; a = 30-34; b = 3.6-4.2; c = 45-50; supplements 14-16 (The Netherlands, Germany, Sweden, Italy, Estonia, Canada) ...... *rotundicaudatus*

5 – Caudal glands reduced, terminal opening absent 6
– Caudal glands and terminal opening present 10

6 – Tail arched, uniformly tapering 7
– Tail nearly straight, subdigitate or digitate 8

7 – Female tail short, as long as anal body diam.; spicules barely bent; ♀: L = 0.9-1.3 mm; a = 27-35; b = 3-4; c = 30-46; c′ = 0.8-1.0; V = 63-70; ♂: L = 0.9-1.3 mm; a = 36-40; b = 4.0-4.4; c = 37-42; supplements 10-11 (France, India, USA-Oregon, Brazil)..... .... *subsimilis*
– Female tail distinctly longer than one anal body diam.; spicules arched; ♀: L = 1.0-1.3 mm; a = 19-30; b = 3.3-4.6; c = 21-29; c′ = 1.3-1.7; V = 54-69; ♂: L = 1.0-1.3 mm; a = 22-30; b = 3.3-4.0; c = 20-27; supplements 10-11 (New Zealand).......... .... *psammophilus*

8 – Subventral teeth present; body length *ca* 1.5 mm; ♀: L = 1.4 mm; a = 25; b = 3.8; c = 35; c′ = 1.3; V = 68; ♂: unknown (Poland, Uzbekistan, Canada)..................................... *solus*
– Subventral teeth absent; body length *ca* 1 mm 9

9 – Transverse denticles in four rows; tail as long as 2-3 anal body diam., tip fine, almost pointed; supplements seven; ♀: L = 0.98 mm (juv.); a = 31; b = 4.0; c = 11; c′ = 3; ♂: L = 0.8 mm; a = 32; b = 3.9; c = 19: supplements seven (Congo Republic) .... *insolitus*
– Transverse denticles in six rows; tail *ca* as long as one anal body diam.; supplements ten; ♀: L = 1.1 mm; a = 25; b = 3.0; c = 37; c′ = 1-1.2; V = 74; ♂: L = 1.1 mm; a = 28; b = 3.2; c = 32; supplements ten (USA-Oregon) .... *exacutus*

10 – Spinneret subdorsal, *i.e.*, situated on dorsal side of tail tip..... 11
– Spinneret terminal, at centre of tail tip 24

11 – Vulva distinctly protruding like a blister on body surface; ♀: L = 1.24-1.48 mm; a = 30-33; b = 3.5-4.0; c = 25-28; c′ = 1.7-1.8; V = 59-60; ♂: unknown (Cameroon) .... *vulvalatus*
– Vulva not protruding like a blister 12

12 – Female tail as long as two anal body diam. 13
– Female tail 1.5 anal body diam. long or shorter 14

13 – Denticles in four transverse rows; body *ca* 1 mm long; ♀: L = 0.8-1.0 mm; a = 27-40; b = 3.0-4.4; c = 25-30; c′ = 2; V = 52-57; ♂: unknown (India) .... *neocontractus*
– Denticles in six transverse rows; body *ca* 1.5 mm long; ♀: L = 1.3-1.5 mm; a = 29-30; b = 3.7-3.8; c = 24-25; c′ = 2; V = 61-62; ♂: L = 1.6 mm; a = 33; b = 3.8; c = 25; supplements 10-14 (Poland, Uzbekistan, USA-Utah) .... *parabrachyurus*

14 – Advulval papillae (pre- and post-vulval) present; ♀: L = 1.4 mm; a = 23; b = 3.6; c = 30; c′ = 1.1; V = 62; ♂: unknown (USA-Massachusetts) .... *vulvapapillatus*
– Advulval papillae absent 15

15 – Tail tip dorsally bent; subventral teeth absent 16
– Tail tip not bent dorsally; subventral teeth mostly present 17

16 – Buccal cavity 26-29 × 16-17 μm; ♀: L = 1.1-1.2 mm; a = 24-26; b = 3.0-3.2; c = 34-35; c′ = 1-1.2; V = 63-64; ♂: unknown (USA-Hawaii) .... *oceanicus*
– Buccal cavity 20-21 × 8-10 μm; ♀: L = 0.89-1.07 mm; a = 24-28; b = 2.9-3.3; c = 24-32; c′ = 1.2-1.4; V = 61-64; ♂: unknown (Ecuador) .... *aequatorialis*

17 – Buccal cavity 30 μm or longer; body length *ca* 1.5 mm; ♀: L = 1.4-1.7 mm; a = 24-30; b = 2.9-3.2; c = 41-58; c′ = 0.8-1; V = 63-69; ♂: L = 1.4-1.7 mm; a = 26-32; b = 2.9-3.2; c = 40-42; supplements 14-15 (UK, Poland, Switzerland, Hungary, Italy, Russia, Georgia, India, Korea, Egypt, Nigeria, Canada, USA, St Lucia) .... *brevicaudatus*
– Buccal cavity 20 μm or shorter; body length < 1.5 mm 18

18 – Body very small, 0.6 mm long; buccal cavity unusually broad, barrel-shaped; ♀: L = 0.60-0.64 mm; a = 18-20; b = 3.0-3.1; c = 25-39; c′ = 1; V = 61-62; ♂: unknown (Vietnam) .... *doliolarius*
– Body longer, *ca* 1 mm; buccal cavity as usual; strongly tapering toward base 19

19 – Subventral teeth present, distinct 20
– Subventral teeth absent 23

20 – Tail 35-46 μm long; ♀: L = 0.9-1.4 mm; a = 20-35; b = 35; b = 3.1-4.0; c = 25-40; c′ = 1.2-1.5; V = 55-64; ♂: L = 1.0-1.4 mm; a = 27-32; b = 3.3-3.5; c = 27-36; supplements 8-12 (The Netherlands, Belgium, Germany, Denmark, UK, Sweden, Norway, Faeroe Islands, Greenland, Poland, Switzerland, Austria, former Czechoslovakia, Hungary, Spain, Romania, former Yugoslavia, Bulgaria, France, Italy, Moldavia, Estonia, Lithuania, Russia, Georgia, Iran, India, Korea, Krakatoa Island, Egypt, Nigeria, Uganda, South Africa, Mauritius, USA, Mexico, Costa Rica, Surinam, Venezuela, Argentina, Malaysia) . . . . *brachyuris*
– Tail 15-25 μm long 21

21 – Intestine characteristically narrowed at genital region; ♀: L = 0.5-1.1 mm; a = 19-35; b = 2.4-3.5; c = 30-45; c′ = 1-1.2; V = 52-65; ♂: L = 1.0 mm; a = 33; b = 3.3; c = 28; supplements nine (India, Dominica, Costa Rica) . . . . *contractus*
– Intestine not strikingly narrowed at genital region 22

22 – Tail tip oblique, body 1.2-1.4 mm long; ♀: L = 1.2-1.4 mm; a = 24-34; b = 3.2-3.5; c = 50; c′ = 0.8-1; V = 63-67; ♂: unknown (Germany, Japan) . . . . *obliquus*
– Tail tip rounded, not oblique; body 0.8-1.2 mm long; ♀: L = 0.8-1.2 mm; a = 24-30; b = 3.2-3.5; c = 40-46; c′ = 1; V = 62-64; ♂: L = 0.9 mm; a = 25; b = 3.4; c = 40; supplements 12 (India, Pakistan, Kenya) . . . . *nainitalensis*

23 – Intestine lying dorsal to female gonads; buccal cavity 15 μm long; ♀: L = 1.0-1.1 mm; a = 30-31; b = 3.3-3.4; c = 24-27; c′ = 1.4-1.7; V = 56-58; ♂: unknown (New Caledonia) . . . . *ananasi*
– Intestine overlapping female gonads; buccal cavity 20 μm long; ♀: L = 1.1-1.3 mm; a = 25-32; b = 3.3-3.8; c = 22-26; c′ = 1.5; V = 59-63; ♂: unknown (New Zealand) . . . . *ubis*

24 – Tail sigmoid, sharply bent ventrad with digitate posterior part inclining slightly dorsally and making dorsal contour somewhat concave 25
– Tail not sigmoid, either more or less arcuate or, if subdigitate, showing no concave dorsal contour or not sharply bent ventrad 30

25 – Tail 55-80 μm or 2-2.5 anal body diam. long; labial region distinctly broader than adjacent body; ♀: L = 1.6-2.6 mm; a = 28-50; b = 3.7-4.4; c = 22-38; c′ = 2-2.5; V = 62-66; ♂: L = 1.9-2.9 mm; a = 45-57; b = 3.9-4.6; c = 20-38; supplements 12-14 (Belgium, Romania, USA, Mexico)........... *sigmaturellus*
– Tail 25-50 μm or 0.8-1.5 anal body diam. long; labial region not or slightly broader than adjacent body 26

26 – Body > 1.5 mm long; spicules *ca* 60 μm long; ♀: L = 1.6-2.0 mm; a = 30-37; b = 3.2-3.6; c = 38-69; c′ = 1; V = 61-71; ♂: L = 1.6-2.0 mm; a = 31-45; b = 3.4-3.8; c = 30-42; supplements 10-12 (France) .... *agilis*
– Body shorter, 1-1.5 mm long; spicules *ca* 50 μm long........ 27

27 – Posterior finger-like portion of tail straight; ♀: L = 1.3-1.4 mm; a = 28-33; b = 3.3-3.6; c = 32-41; c′ = 1.2-1.4; V = 63-67; ♂: unknown (New Caledonia)......................... *paitensis*
– Posterior finger-like portion of tail slightly but perceptibly bent dorsad.. 28

28 – Rasp-like denticles numerous (*ca* 10-15 irregular rows); ♀: L = 1.0-1.5 mm; a = 30-37; b = 3.0-3.5; c = 30-35; c′ = 1.5; V = 61-65; ♂: unknown (India)......................... *dentatus*
– Rasp-like denticles less numerous (in 7-8 rows).............. 29

29 – Advulval papillae (pre- and post-vulval) present; ♀: L = 1.38-1.55 mm; a = 34-36; b = 3.5-3.6; c = 47-55; c′ = 0.9; V = 63-65; ♂: L = 1.25 mm; a = 40; b = 3.6; c = 37; c′ = 1.2; supplements 11 (Brunei)....................................... *sessus*
– Advulval papillae absent.; ♀: L = 1.0-1.6 mm; a = 23-35; b = 3.3-3.6; c = 35-50; c′ = 0.8-1; V = 62-68; ♂: L = 1.1-1.7 mm; a = 30-50; b = 3.4-3.9; c = 30-40; supplements 11 (The Netherlands, Germany, Poland, former Czechoslovakia, Hungary, Spain, France, Italy, Russia, Switzerland, Lithuania, Uzbekistan, Iran, India, Pakistan, Korea, Japan, Egypt, Nigeria, Mauritius, South Africa, Canada, USA, Mexico, Venezuela, Costa Rica, Argentina, Brazil, Colombia, Hawaii, Australia, New Zealand) .. .. *sigmaturus*

30 – Tail 2-4 anal body diam. long............................ 31
– Tail < 2 anal body diam. long............................. 35

31 – Small species, 0.8-1.3 mm long 32

– Larger species, to 2.5 mm long 33

32 – Tail three anal body diam. long, more or less right-angled; ♀: L = 1.3 mm; a = 30; b = 3.5; c = 26; c′ = 3; V = 66; ♂: L = 0.8 mm; a = 32; b = 3.4; c = 21; supplements ten (former Yugoslavia, El Salvador) . . . . *cavensis*

– Tail *ca* two anal body diam. long, arcuate; ♀: L = 1.0-1.1 mm; a = 35-38; b = 5.5; c = 23; c′ = 2.1-2.3; V = 53-55; ♂: unknown (Pakistan) . . . . *rosensis*

33 – Subventral denticles densely arranged, in *ca* ten irregular rows; buccal cavity 20-23 μm long; ♀: L = 1.3-1.6 mm; a = 34-33; b = 3.5-4.0; c = 25-34; c′ = 2.5; V = 59-65; ♂: unknown (India, Korea) . . . . *apapillatus*

– Subventral denticles not so dense, in 5-7 rows; buccal cavity 30-35 μm long 34

34 – Tail 60 μm long, anterior anal lip massive, overhanging; ♀: L = 1.9 mm; a = 43; b = 3.8; c = 33; c′ = 2.5; V = 72; ♂: L = 1.8 mm; a = 45; b = 4.0; c = 33; supplements 14 (The Netherlands, USA-Virginia) . . . . *subtenuis*

– Tail 80-120 μm long, anterior anal lip normal, not overhanging; ♀: L = 1.6-2.5 mm; a = 28-38; b = 3.2-3.7; c = 16-32; c′ = 2-4; V = 56-67; ♂: L = 1.5-1.8 mm; a = 27-32; b = 3.1-3.3; c = 1-21; supplements 13-14 (Poland, Hungary, Thailand, Egypt, South Korea, Canada, USA) . . . . *polonicus*

35 – Subventral teeth absent; ♀: L = 0.8-1.2 mm; a = 20-34; b = 2.8-3.4; c = 25-42; c′ = 1-1.5; V = 58-66; ♂: L = 0.8 mm; a = 25; b = 3.0; c = 42; supplements 13 (India) . . . . *amurus*

– Subventral teeth present 36

36 – Buccal cavity 35-55 μm long 37

– Buccal cavity 15-30 μm long 41

37 – Subventral teeth nearly basal, posterior refractive ring unusually strong; ♀ unknown; ♂: L = 2.2 mm; a = 4.4; b = 3.9; c = 42; supplements ten (Australia) . . . . *prodenticulatus*

– Subventral teeth more anterior, posterior refractive ring weak 38

38 – Tail slightly bent ventrally, conspicuously longer than anal body diam.; ♀: L = 1.8-1.9 mm; a = 21-30; b = 3.6; c = 21-34;

c′ = 1.3-1.5; V = 66-67; ♂: unknown (Paraguay, New Guinea, Costa Rica) . . . . *obtusicaudatus*

– Tail sharply bent ventrally, as long as, or shorter than, anal body diam. 39

39 – Posterior digitate part of tail conoid, distinctly narrowing to terminus; ♀: L = 1.4-2.3 mm; a = 25-31; b = 2.8-3.8; c = 37-47; c′ = 0.8-1.1; V = 63-68; ♂: L = 1.9-2.5 mm; a = 37-44; b = 3.1-7; c = 35-44; supplements 14-15 (Poland, France, Russia, India, Korea, Thailand, Mauritius, South Africa, Canada, USA, El Salvador, Jamaica, Brazil) . . . . *incurvus*

– Posterior digitate part of tail cylindroid, broadly rounded at terminus 40

40 – Advulval papillae present; body nearly 2 mm long; ♀: L = 1.8 mm; a = 29; b = 3.3; c = 50; c′ = 1; V = 63; ♂: L = 1.7-1.9 mm; a = 32-34; b = 3.4-3.7; c = 45-51; supplements 14-15 (Mongolia) . . . . *kaszabi*

– Advulval papillae absent; body 2.5-3 mm long; ♀: L = 2.4-2.8 mm; a = 22-35; b = 3.5-3.6; c = 30-36; c′ = 0.8-1.2; V = 64-69; ♂: unknown (Germany, Sweden) . . . . *longus*

41 – Tail 1.5-1.8 anal body diam. long 42

– Tail *ca* as long as anal body diam. 46

42 – Tail arcuate, cylindrical for almost entire length 43

– Tail sharply bent ventrad, consisting of an anterior wider and a posterior slender, digitate or subdigitate part 44

43 – Advulval papillae (pre- and post-vulval) present; ♀: L = 1.57-2.12 mm; a = 36-54; b = 3.8-4.5; c = 39-57; c′ = 1.3-1.9; V = 61-69; ♂: L = 1.61-1.65 mm; a = 41-47; b = 3.9-4.0; c = 35-38; c′ = 1.2-1.3; supplements 10-14 (Russia) . . . . *sexcristatus*

– Advulval papillae absent; ♀: L = 1.1-1.6 mm; a = 25-40; b = 3.3-4.3; c = 20-32; c′ = 1.5-2; V = 55-66; ♂: L = 1.1 mm; a = 34; b = 3.4; c = 27; supplements 9 + 2 (Georgia, India, Japan, Thailand, Java, Sumatra, Niger, Mauritius, Zaire, South Africa, Canada, USA, Mexico, Panama, Venezuela, Australia) . . *lacustris*

44 – Rasp-like denticles in four rows; ♀: L = 0.9-1.3 mm; a = 25-31; b = 3.2-3.6; c = 24-29; c′ = 1.3-1.8; V = 56-58; ♂: unknown (New Caledonia) . . . . *vasis*

– Rasp-like denticles in 6-7 rows 45

45 – Male supplements six; ♀: L = 0.7-1.0 mm; a = 21-26; b = 3.2-4.0; c = 22-26; c′ = 1.5; V = 54-57; ♂: L = 0.9 mm; a = 25; b = 3.4; c = 23; supplements six (India) .... *brassicus*

– Male supplements 10-12; ♀: L = 0.7-1.3 mm; a = 20-30; b = 2.9-4.0; c = 17-30; c′ = 1.5-1.8; V = 55-70; ♂: L = 0.9-1.0 mm; a = 25-39; b = 2.3-3.1; c = 35-44; supplements 10-12 (India, Egypt, Niger, Kenya, South Africa, El Salvador, Argentina, Hawaii, Costa Rica) .... *hawaiiensis*

46 – Tail obtuse with bluntly rounded tip 47

– Tail conoid with narrowly rounded tip 48

47 – Spinneret rather indistinct; supplements 12; ♀: L = 1.1-1.6 mm; a = 22-36; b = 3.4-4.0; c = 40-71; c′ = 0.8-1; V = 63-66; ♂: L = 1.1 mm; a = 34; b = 3.5; c = 34; supplements 12 (The Netherlands, Russia, Canada, USA, New Zealand) .... *striatus*

– Spinneret well developed; supplements 15; ♀: L = 1.0-1.5 mm; a = 24-30; b = 3.1-3.5; c = 39-46; c′ = 1; V = 60-65; ♂: L = 1.1-1.6 mm; a = 33-42; b = 3.5-3.9; c = 44; supplements 15 (Hungary, India, USA) .... *orbitus*

48 – Tail 30 μm long or shorter; ♀: L = 1.0-1.2 mm; a = 22-26; b = 3.1-3.2; c = 36-41; c′ = 0.8-1; V = 58-62; ♂: unknown (Niger, Singapore) .... *curvicaudatus*

– Tail 40-50 μm long 49

49 – Tail tip conoid-pointed; ♀: L = 1.2-1.4 mm; a = 30-34; b = 3.3-4.1; c = 39-44; c′ = 1; V = 62-67; ♂: L = 1.3-1.5 mm; a = 35-40; b = 3.4-4.0; c = 30-43; supplements 13 (South Africa) ... *cereris*

– Tail tip cylindroid.. 50

50 – Buccal cavity walls nearly parallel, barely tapering towards base; ♀: L = 1.4-1.5 mm; a = 29-31; b = 3.4-3.6; c = 30-32; c′ = 1; V = 63-64; ♂: L = 1.9 mm; a = 49; b = 4.0; c = 41; supplements 12 (Russia, Uzbekistan, New Zealand)............... *arenicola*

– Buccal cavity funnel-shaped, strongly tapering to base; ♀: L = 1.0-1.1 mm; a = 29-30; b = 3.5-3.8; c = 30-40; c′ = 1; V = 60; ♂: unknown (Fiji, Java, Panama, Malaysia, India, Iran) .... *minor*

Note

Not included in key because of unavailability of literature: *M. paracontractus*, *M. indicus*, *M. neoindicus* and *M. sagarensis*.

Genus *Brachonchulus* Andrássy, 1958
(Fig. 28B)

DIAGNOSIS

Body *ca* 1 mm long. Buccal cavity goblet-shaped. Dorsal tooth large, claw-like, in anterior half of buccal cavity and with anteriorly directed apex. Subventral teeth present. Denticles arranged in two groups: an anterior rasp-like field opposite to dorsal tooth composed of 5-6 rows of denticles, a posterior one consisting of irregularly scattered denticles. Refractive rings apparently absent. Pharyngo-intestinal junction simple. Female reproductive system amphidelphic. Tail short, conoid, as long as anal body diam. Caudal gland and orifice absent. Male not known.

TYPE AND ONLY SPECIES

Brachonchulus brachyuroides (Micoletzky, 1925) Andrássy, 1958
= *Mononchus brachyuroides* Micoletzky, 1925
= *Granonchulus brachyuroides* (Micoletzky, 1925) Mulvey, 1963b

Brachonchulus brachyuroides (Micoletzky, 1925) Andrássy, 1958
(Fig. 28B)

MEASUREMENTS

Female: L = 0.93-0.96 mm; a = 21.3-22.6; b = 3.0-3.13; c = 20.5-20.7; V = 58-59.

DESCRIPTION

Micoletzky (1925) described this species as *Mononchus brachyuroides* based on two female specimens collected from Surinam. He differentiated this species from *M. brachyuris* (now *Mylonchulus brachyuris*) by the structure of the buccal cavity. This has, besides the dorsal tooth opposite to the rasp-like area, a field of coarse basal denticles. These are present only on the left subventral wall close to the beginning of the respective subdorsal pharyngeal sector.

Micoletzky considered this species to represent a transition between two subgenera erected by Cobb, *Mylonchulus* (with subventral rasp) and *Sporonchulus* (with irregular basal denticles). Andrássy (1958) proposed

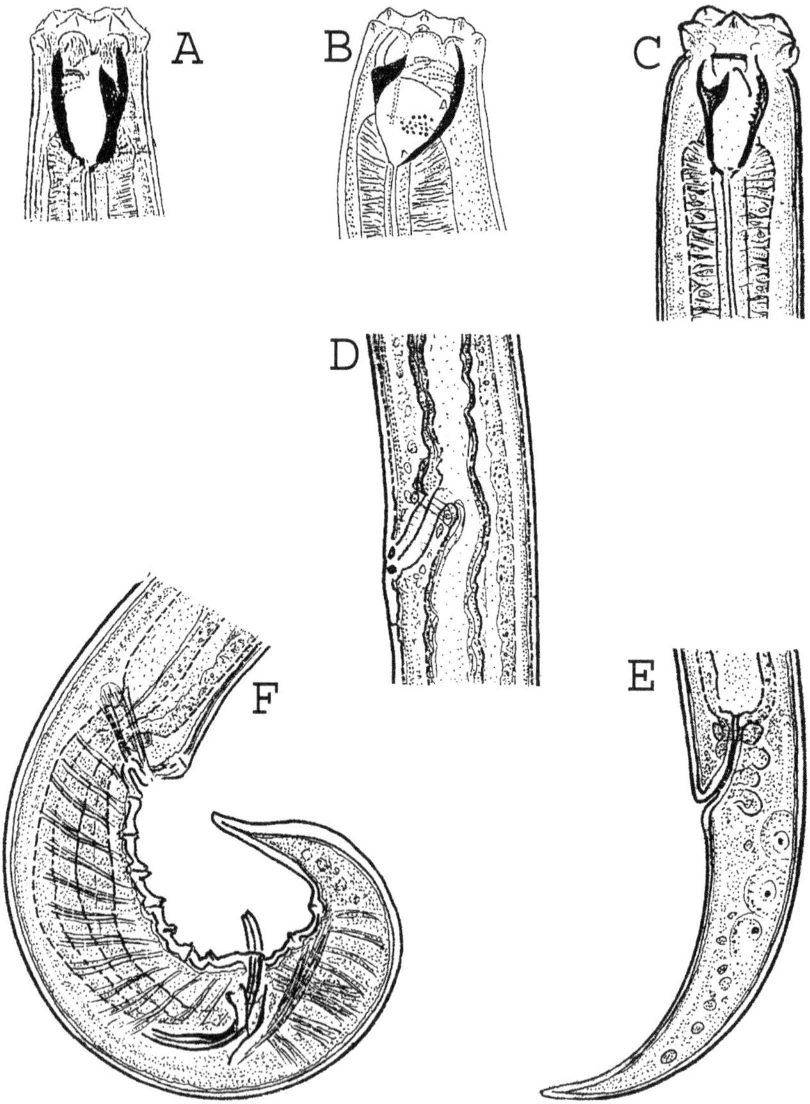

Fig. 28. *A:* Oligonchulus reversus *(Cobb, 1917) Andrássy, 1976, anterior region; B:* Brachonchulus brachyuroides *(Micoletzky, 1925) Andrássy, 1958, anterior region; C-F:* Crestonchulus crestus *Siddiqi & Jairajpuri, 2002. C: Anterior region; D: Vaginal region; E: Female tail; F: Male tail. A after Cobb (1917); B after Micoletzky (1925); C-F after Siddiqi and Jairajpuri (2002).*

a new genus *Brachonchulus* for this species. This species has never been described since its original brief description and its taxonomic status is doubtful.

Genus *Margaronchulus* Andrássy, 1972
(Fig. 29A-F)

DIAGNOSIS

Body small, 0.9-1.0 mm long. Buccal cavity comparatively small, broader proximally than usual in family. Dorsal tooth not too large, in anterior half of buccal cavity with apex directed anteriad. Buccal armature very poor, subventral teeth absent, rasp-like denticles present in a single row. Each subventral wall provided with a short transverse ridge. Pharyngo-intestinal junction non-tuberculate. Female genital system mono-prodelphic, without post-uterine sac. Tail long, filiform. Caudal glands reduced; spinneret lacking. Male not known.

TYPE AND ONLY SPECIES

Margaronchulus mulveyi Andrássy, 1972

Margaronchulus mulveyi Andrássy, 1972
(Fig. 29A-F)

MEASUREMENTS

Female: L = 0.9-1.0 mm; a = 31-43; b = 3.5-4.2; c = 3.8-4.6; c′ = 11.5-13.9; V = 57-60.

DESCRIPTION

Lip region distinctly offset, wider than adjoining body, 15-19 μm diam. and 7-8 μm high. Amphidial aperture slit-like, at 9-10 μm from anterior end. Buccal cavity 15-18 × 9-11 μm. Apex of dorsal tooth at 13-14 μm from base of buccal cavity, opposed by a single row of denticles. Female genital system monoprodelphic. Vulva transverse, vagina 7-8 μm deep; *pars refringens vaginae* present. *Pars dilatata* part of oviduct consisting of four rows of four cells each (quadricollumela). Tail long filiform, whip-like, 200-235 μm long. Caudal glands in tandem. Spinneret terminal. Male not known.

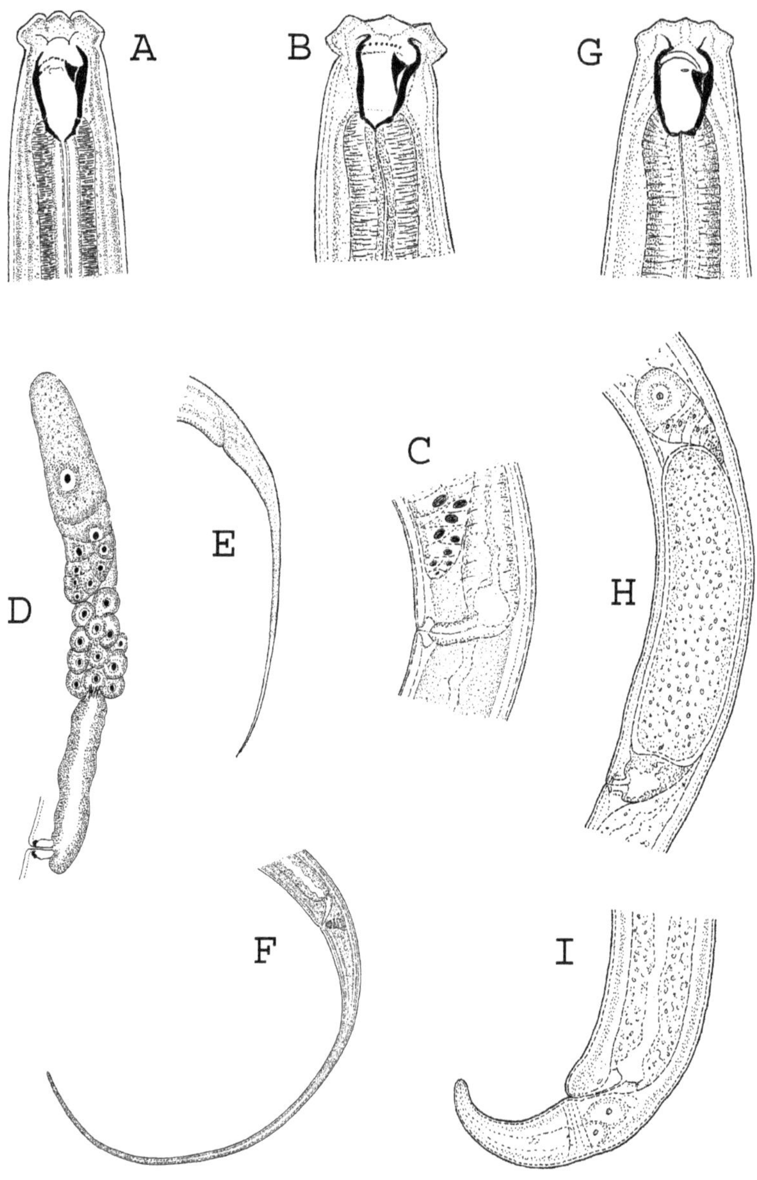

Fig. 29. *A-F:* Margaronchulus mulveyi *Andrássy, 1972. A, B: Anterior region; C: Vulval region; D: Female genital system; E, F: Female posterior region; G-I:* Margaronchuloides adenticulatus *(Andrássy, 1985) comb. n. G: Anterior region; H: Female genital system; I: Female posterior region. A, D, F after Ahmad (2000); B, C, E after Andrássy (1972); G-I after Andrássy (1985).*

DISTRIBUTION

Cameroon, Congo Republic.

Genus *Oligonchulus* Andrássy, 1976
(Fig. 28A)

DIAGNOSIS

Body *ca* 1 mm long. Cuticle finely striated, striae resolving into small dots. Buccal cavity goblet-shaped. Dorsal tooth large and oblique with pointed apex located quite close to beginning of buccal capsule. Subventral teeth absent. Transverse denticles only in two rows and very few (*ca* six in number on each side). Refractive rings absent. Female genital system mono-opisthodelphic, without anterior uterine sac; vulva located anterior to mid-body. Eggs usually longer than broad. Tail arcuate, cylindroid in its posterior half, twice as long as anal body diam. Caudal glands and spinneret present. Male not known.

TYPE AND ONLY SPECIES

Oligonchulus reversus (Cobb, 1917) Andrássy, 1976
= *Mononchus* (*Mylonchulus*) *reversus* Cobb, 1917
= *Mylonchulus reversus* Cobb, 1917

Oligonchulus reversus (Cobb, 1917) Andrássy, 1976
(Fig. 28A)

MEASUREMENTS

Female: L = 1.1 mm; a = 29; b = 3.8; c = 25; c′ = 2; V = 40.

REMARKS

Cobb (1917) recorded this species from Rio de Janeiro, Brazil, and it has not been recorded since. As per Cobb's description, "This very interesting abnormal form possesses striae that appear resolvable into rows of dots. This species is readily distinguishable by the sparseness of pharyngeal denticles, only a few of which can be seen and the form of sexual organ which is mono-opisthodelphic without anterior uterine sac. Eggs relatively large size, about 5-6 times as long as body width. Tail more or less cylindroid in posterior half, its terminus one-fourth to one-fifth as wide as its base. Spinneret terminal."

DISTRIBUTION

Brazil.

Genus *Polyonchulus* Mulvey & Jensen, 1967
(Fig. 30)

DIAGNOSIS

Body length *ca* 1 mm. Buccal cavity *ca* twice as long as broad. Dorsal tooth large, claw-like, situated in anterior half of buccal cavity with anteriorly directed apex. Subventral teeth absent but two longitudinal serrate ridges present instead. Rasp-like subventral denticles in 2-7 rows, refractive rings lacking. Pharyngo-intestinal junction non-tuberculate. Female genital system amphidelphic or mono-prodelphic without post-uterine sac. Vulva at 66-70% of body length. Male known in one species; spicules arcuate, without lateral accessory pieces. Supplements small, six in number. Tail of both sexes similar, conoid arcuate or subdigitate, 1-2.5 anal body diam. long. Caudal glands and spinneret either present or reduced.

TYPE SPECIES

Polyonchulus cobbi Mulvey & Jensen, 1967

OTHER SPECIES

P. clavicaudatus (Schuurmans Stekhoven & Teunissen, 1938) Jairajpuri, 1970b
= *Mononchus clavicaudatus* Schuurmans Stekhoven & Teunissen, 1938
= *Mononchus* (*Mylonchulus*) *clavicaudatus* Schuurmans Stekhoven & Teunissen, 1938 (Goodey, 1951)
= *Mylonchulus clavicaudatus* Schuurmans Stekhoven & Teunissen, 1938 (Goodey, 1951)

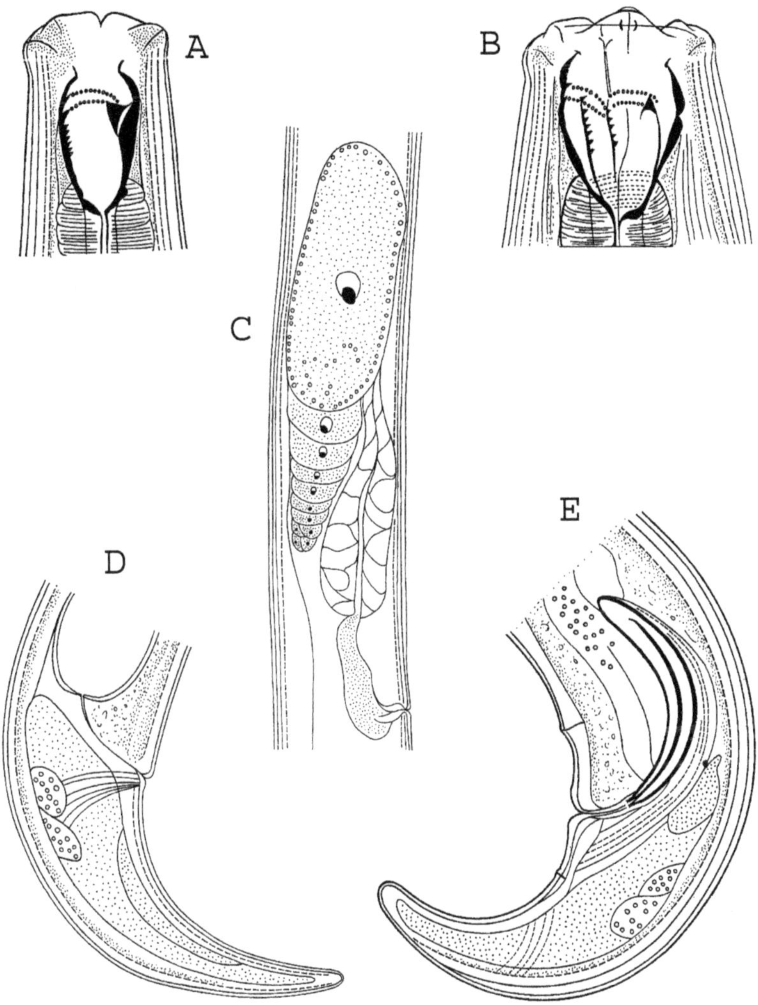

Fig. 30. Polyonchulus cobbi *Mulvey & Jensen, 1967. A, B: Anterior region; C: Female genital system; D: Female posterior region; E: Male posterior region. After Mulvey and Jensen (1967).*

Polyonchulus cobbi Mulvey & Jensen, 1967
(Fig. 30)

MEASUREMENTS

Female: L = 0.9-1.1 mm; a = 26-34; b = 3.0-3.7; c = 13.8-18.8; V = 67-70.

Male: L = 1.2 mm; a = 33; b = 3.5; c = 21.

DESCRIPTION

Lip region slightly offset. Amphids obscure. Buccal cavity 22-25 μm long and 13-15 μm diam. Dorsal tooth massive, opposed by two longitudinal ribs, each bearing five or six small tooth-like structures with two rows of large transverse denticles present anterior to these. Female reproductive system mono-prodelphic. Post-uterine sac absent. *Pars refringens vaginae* absent. Tail conoid arcuate. Caudal glands and spinneret absent. Male spicule 45 μm long, gubernaculum weakly developed, lateral accessory pieces absent. Ventromedian supplements six. Tail conoid arcuate, without caudal gland and spinneret.

DISTRIBUTION

Nigeria.

Key to species of *Polyonchulus* Mulvey & Jensen, 1967

1 – Female genital system mono-prodelphic, transverse denticles in two rows; tail arcuate conoid with reduced gland; ♀: L = 0.9-1.1 mm; a = 25-34; b = 3.0-3.7; c = 14-19; c′ = 2.5; V = 67-70; ♂: L = 1.2 mm; a = 33; b = 3.5; c = 21; supplements six (Nigeria) . . . . *cobbi*

– Female genital system amphidelphic; transverse rows of denticles in seven rows; tail subdigitate, dorsally bent with well developed glands; ♀: L = 0.95 mm; a = 27; b = 3; c = 36; c′ = 1; V = 66; ♂: unknown (Zaire) . . . . *clavicaudatus*

Genus *Megaonchulus* Jairajpuri & Khan, 1982
(Fig. 31)

DIAGNOSIS

Body 1.4-1.7 mm long, very slender. Buccal cavity goblet-shaped, strongly narrowing towards base. Dorsal tooth large, triangular with apex directed anteriad. Subventral teeth absent but two fine, serrate, longitudinal ribs present instead. Transverse denticles rasp-like, arranged

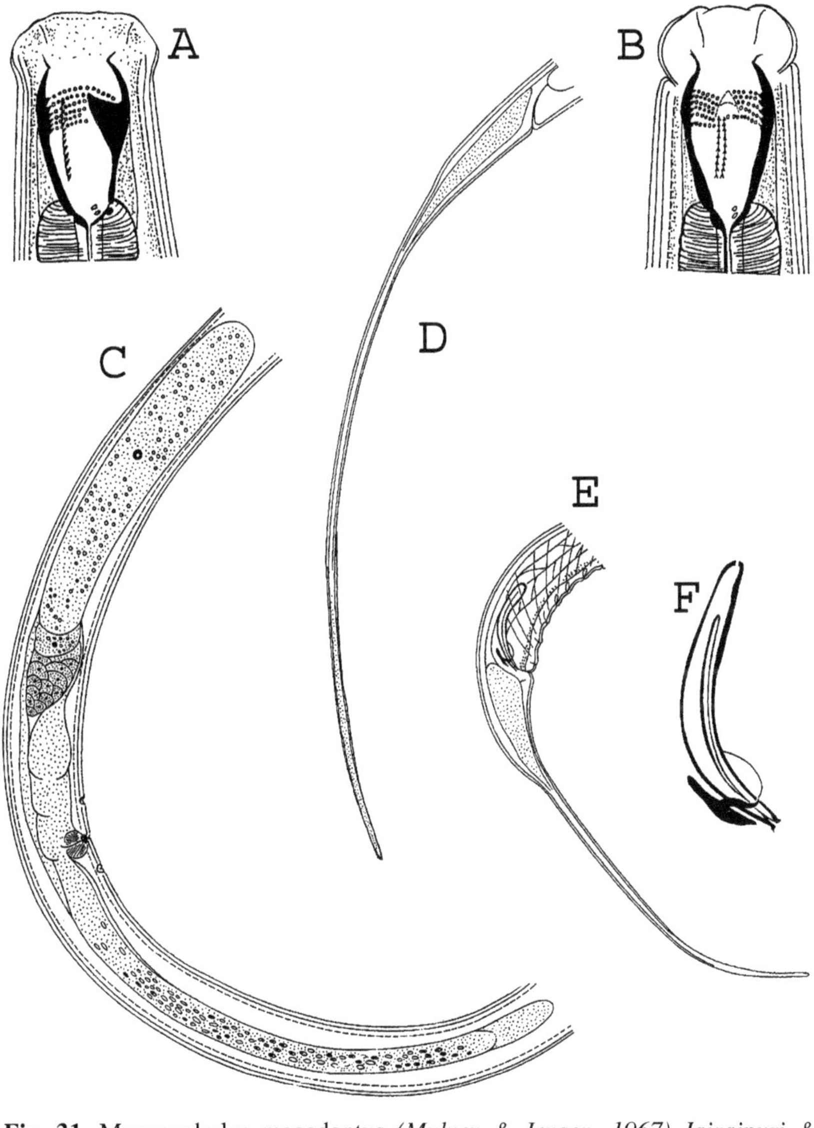

Fig. 31. Megaonchulus megadontus *(Mulvey & Jensen, 1967) Jairajpuri & Khan, 1982. A, B: Anterior region; C: Female genital system; D: Female posterior region; E: Male posterior region; F: Spicule, gubernaculum and lateral accessory piece. After Mulvey and Jensen (1967).*

in *ca* six rows. No refractive rings. Pharyngo-intestinal junction non-tuberculate. Female genital system mono-prodelphic, post-uterine sac present, unusually long. Vulva at mid-body. Spicules comparatively short, without lateral accessory pieces. Supplements eight. Tail of both sexes similar, filiform, 8-20 anal body diam. long. Caudal glands and spinneret absent.

TYPE AND ONLY SPECIES

Megaonchulus megadontus (Mulvey & Jensen, 1967) Jairajpuri & Khan, 1982
= *Polyonchulus megadontus* Mulvey & Jensen, 1967

Megaonchulus megadontus (Mulvey & Jensen, 1967) Jairajpuri & Khan, 1982
(Fig. 31)

MEASUREMENTS

Female: L = 1.7 mm; a = 55; b = 4.7; c = 4.6; V = 50.
Male: L = 1.4-1.5 mm; a = 45-52; b = 4.2-5.0; c = 6.3-6.5.

DESCRIPTION

Lip region distinctly offset. Buccal cavity thick-walled, 25 μm long and 14 μm diam. Dorsal tooth massive, opposed by two longitudinal ribs, each bearing 9-10 small denticles. In addition, two groups of small transverse denticles, each with six rows regularly arranged. Amphidial apertures obscure. Female genital system mono-prodelphic with reflexed ovary and long post-uterine sac. Vulval papillae present. Tail 360 μm long, filiform, without caudal glands or spinneret. Male with 22-25 × 12-13 μm buccal cavity. Spicules 45 μm long. Gubernaculum *ca* one-fourth of spicule length. Lateral accessory pieces absent. Ventromedian supplements eight. Tail 210-230 μm long, filiform, without caudal glands or spinneret.

DISTRIBUTION

Nigeria.

Genus *Paramylonchulus* Jairajpuri & Khan, 1982
(Figs 26F-H; 32)

DIAGNOSIS

Body 0.6-1.2 mm long. Buccal cavity comparatively narrow, tapering at base. Dorsal tooth large to massive, situated in anterior half of buccal cavity. Subventral walls bearing from three to many rows of denticles. Subventral teeth absent. Pharyngo-intestinal junction non-tuberculate. Female genital system mono-prodelphic. Spicules short; gubernaculum simple. Tail variable in shape. Caudal glands and spinneret well developed.

TYPE SPECIES

Paramylonchulus index (Cobb, 1906) Jairajpuri & Khan, 1982
= *Mylonchulus index* (Cobb, 1906) Cobb, 1917
= *Mononchus index* Cobb, 1906
= *Mononchus* (*Mylonchulus*) *index* Cobb, 1906 (Cobb, 1917)

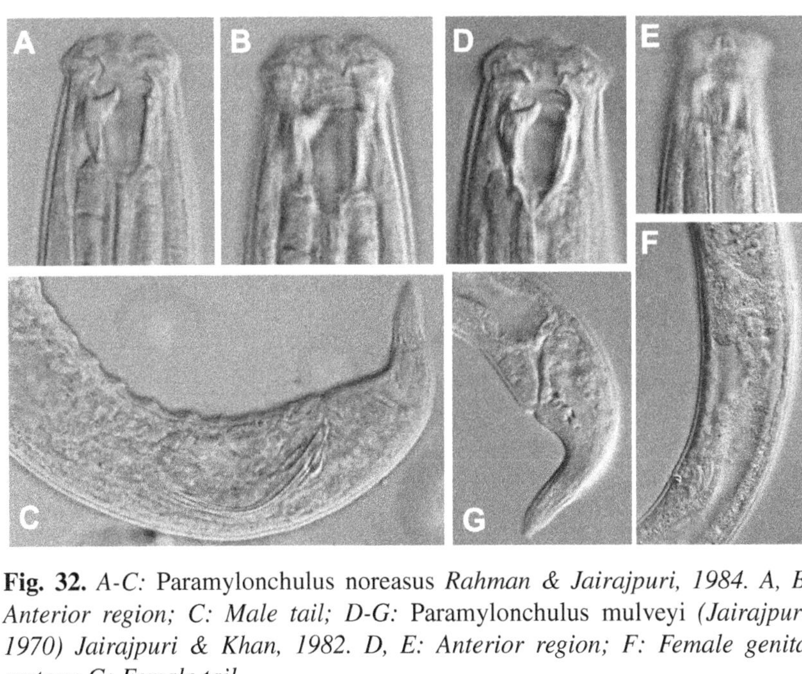

Fig. 32. *A-C:* Paramylonchulus noreasus *Rahman & Jairajpuri, 1984. A, B: Anterior region; C: Male tail; D-G:* Paramylonchulus mulveyi *(Jairajpuri, 1970) Jairajpuri & Khan, 1982. D, E: Anterior region; F: Female genital system; G: Female tail.*

OTHER SPECIES

P. andrassyi (Loof, 1993) **comb. n.**
= *Mylonchulus andrassyi* Loof, 1993
P. californicus (Jairajpuri, 1970b) Jairajpuri & Khan, 1982
= *Mylonchulus californicus* Jairajpuri, 1970b
P. caespitosus (Razzhivin, 1971) Jairajpuri & Khan, 1982
= *Mylonchulus caespitosus* Razzhivin, 1971
P. cassicus Dhanachand, Rombati & Anandi, 1995
= *Mylonchulus cassicus* (Dhanachand, Rombati & Anandi, 1995) Mohilal, Pramodini & Romoni, 2006
P. indicus Choudhury & Sivakumar, 1998
= *Mylonchulus indiensis* Andrássy, 2009
P. japonicus Dhanachand, Rombati & Anandi, 1995
= *Mylonchulus japonicus* (Dhananchand, Rombati & Anandi, 1995) Mohilal, Pramodini & Romoni, 2006
= *Mylonchulus politus* Susulovsky, 2000 **syn. n.**
= *Paramylonchulus politus* (Susulovsky, 2000) **comb. n., syn. n.**
P. longisacculus Popovici, 1990
= *Mylonchulus longisacculus* (Popovici, 1990) Loof, 1992
P. mashhoodi (Khan & Jairajpuri, 1979) Jairajpuri & Khan, 1982
= *Mylonchulus mashhoodi* Khan & Jairajpuri, 1979
= *Mylonchulus hortulanus* Khan, 1992 **syn. n.**
= *Paramylonchulus hortulanus* (Khan, 1992) **comb. n., syn. n.**
P. mulveyi (Jairajpuri, 1970b) Jairajpuri & Khan, 1982
= *Mylonchulus mulveyi* Jairajpuri, 1970b
= *Paramylonchulus lepidus* Lal & Khan, 1988
= *Paramylonchulus indistinctus* Anandi, Mohilal & Dhanachand, 1997
P. noreasus Rahman & Jairajpuri, 1984
= *Mylonchulus noreasus* (Rahman & Jairajpuri, 1984) Loof, 1992
P. orientalis (Andrássy, 1992) **comb. n.**
= *Mylonchulus orientalis* Andrássy, 1992
P. paraindex (Ahmad, Baniyamuddin & Jairajpuri, 2005) **comb. n.**
= *Mylonchulus paraindex* Ahmad, Baniyamuddin & Jairajpuri, 2005
P. silvaticus (Razzhivin, 1971) Jairajpuri & Khan, 1982
= *Mylonchulus silvaticus* Razzhivin, 1971
P. subterraneus (Schneider, 1940) Jairajpuri & Khan, 1982
= *Mylonchulus subterraneus* Schneider, 1940

Paramylonchulus index (Cobb, 1906) Jairajpuri & Khan, 1982
(Fig. 26F-H)

MEASUREMENTS

Female: L = 0.6-0.9 mm; a = 23-28; b = 2.7-3.4; c = 21-26; c′ = 1.0-1.2; V = 70-77.

DESCRIPTION

Lip region 15-20 μm diam., 6-8 μm high. Amphidial aperture 3-5 μm wide, located at 9-12 μm from anterior end of body and 12-15 μm from base of buccal cavity. Buccal cavity 16-21 μm long, 8-13 μm diam. Dorsal tooth large, its apex at 12-17 μm or 75-80% of buccal cavity length from its base. Subventral walls bearing five transverse rows of denticles. Female genital system mono-prodelphic; post-uterine sac absent. Sphincter present at oviduct-uterus junction but poorly developed. Vulva transverse. *Pars refringens vaginae* absent. Tail hemispheroid but with a ventral finger-like projection, 25-40 μm or *ca* one anal body diam. long. Caudal glands grouped. Spinneret terminal.

DISTRIBUTION

Austria, former Czechoslovakia, El Salvador, Georgia, USA-Hawaii, India, Mauritius, Poland, Singapore, Thailand.

Key to species of *Paramylonchulus* Jairajpuri & Khan, 1982

1 – Post-uterine sac present, 1-8 vulval body diam. long 2
 – Post-uterine sac < vulval body diam. or completely absent 7

2 – Caudal glands reduced, spinneret absent; tail conoid, ventrally bent; ♂: L = 1.5-1.6 mm; a = 28-30; b = 3.6-3.9; c = 41-47; c′ = 1; V = 67-68; ♂: L = 1.3-1.4 mm; a = 28-30; b = 3.4-3.8; c = 45-49; supplements 12 (Kazakhstan) . . . . *silvaticus*
 – Caudal glands well developed, spinneret present; tail digitate, somewhat dorsally bent 3

3 – Body 2 mm long; post-uterine sac very long, *ca* eight body diam.; ♀: L = 2.0 mm; a = 40-45; b = 3.5-3.7; c = 25; c′ = 2.1-2.4; V =

72-74; ♂: L = 1.9-2.2 mm; a = 35-36; b = 3.4-3.8; c = 22-27; supplements 13 (Romania) .... *longisaculus*
– Body 1.5 mm or shorter; post-uterine sac shorter, 1-4 body diam. 4

4 – Tail S-shaped, dorsally concave with long posterior cylindrical part (this latter *ca* one anal body diam. long); ♀: L = 1.0-1.3 mm; a = 26-39; b = 3.1-4.0; c = 19-29; c′ = 1.8-2.2; V = 72-78; ♂: L = 1.0-1.3 mm; a = 27-36; b = 3.2-5.0; c = 20-27; supplements 7-10 (Austria, Hungary, Italy) .... *andrassyi*
– Tail not S-shaped, not concave dorsally, with shorter posterior cylindrical part 5

5 – Vulval lips sclerotised, post-uterine branch *ca* four body diam. long 6
– Vulval lips not sclerotised, post-uterine branch *ca* two body diam. long; ♀: L = 1.1-1.5 mm; a = 30-47; b = 3.1-3.8; c = 32-47; c′ = 1.4; V = 75-80; ♂: L = 1.1 mm; a = 35; b = 3.4; c = 26; supplements nine (USA-California) .... *californicus*

6 – Buccal cavity 23 × 12 μm; post-uterine sac 167 μm; spicules 56 μm; ♀: L = 1.4 mm; a = 37; b = 3.4; c = 30; c′ = 1.6; V = 79; ♂: L = 1.5 mm; a = 3.5; c = 31; supplements eight (India).. .... *noreasus*
– Buccal cavity 16-19 × 7-9 μm; post-uterine sac 110 μm; spicules 33 μm; ♀: L = 0.99-1.24 mm; a = 37-41; b = 3.1-3.5; c = 24-27; c′ = 1.6-1.7; V = 76-78; ♂: L = 1.23 mm; a = 57; b = 3.5; c = 28; c′ = 1.7; spicules 33 μm; supplements ten (India). .*japonicus*

7 – Caudal glands inconspicuous, spinneret absent; body > 1 mm; ♀: L = 1.2-1.4 mm; a = 27-36; b = 3.7-5.0; c = 31-33; c′ = 2; V = 66-70; ♂: unknown (Kazakhstan) .... *caespitosus*
– Caudal glands and spinneret conspicuous; smaller species, 0.6-1 mm 8

8 – Tail short, one anal body diam. or so; rasp-like denticles in 5-6 transverse rows 9
– Tail longer, 1.5-2 anal body diam.; rasp-like denticles in 2-4 transverse rows 11

9 – Caudal spinneret subdorsal; ♀: L = 0.98 mm; a = 32; b = 3.1; c = 49; c′ = 0.9; V = 73; ♂: unknown (Vietnam) .... *orientalis*
– Caudal spinneret terminal 10

10 – Short post-uterine sac present; vulval lips sclerotised; ♀: L = 0.71-1.01 mm; a = 29-31; b = 2.9-3.4; c = 20-28; c′ = 1.3-1.6; V = 70-71; ♂: unknown (Singapore) .... *paraindex*

– Post-uterine sac completely absent; vulval lips not sclerotised; ♀: L = 0.6-0.9 mm; a = 23-38; b = 2.9-3.4; c = 21-26; c′ = 1-1.2; V = 70-77; ♂: unknown (Austria, former Czechoslovakia, Poland, Georgia, India, Thailand, Mauritius, El Salvador, USA-Hawaii) .... *index*

11 – Short post-uterine sac present; tail > 2 anal body diam.; ♀: L = 0.92-1.03 mm; a = 30-36; b = 3.1-3.4; c = 15-16; c′ = 2.4-2.9; V = 69-85; ♂: L = 1.21-1.29 mm; a = 33-39; b = 3.2-3.6; c = 15.2-17.5; c′ = 2.6-2.8; supplements six (India) .... *cassicus*

– Post-uterine sac completely absent; tail usually < 2 anal body diam. 12

12 – Sclerotised pieces in vulva elongate; tail tip strongly narrowing, almost pointed; ♀: L = 0.7-1.0 mm; a = 21-37; b = 3.0-3.4; c = 17-24; c′ = 2-2.2; V = 74-78; ♂: L = 0.9 mm; a = 26; b = 2.9; c = 22; supplements eight (India, El Salvador, St Lucia) .... *mulveyi*

– Sclerotised pieces in vulva dot-like; tail tip rounded; ♀: L = 0.8-1.1 mm; a = 24-39; b = 2.8-4.0; c = 20-33; c′ = 1.5-2.1; V = 68-76; ♂: unknown (India, New Caledonia) .... *mashhoodi*

Note

Not included in key because of unavailability of literature: *P. indicus*.

Genus *Crestonchulus* Siddiqi & Jairajpuri, 2002
(Fig. 28C-F)

DIAGNOSIS

Body comparatively long and slender. Cuticle thin and appearing smooth. Cephalic region sharply offset from body by a constriction. Lips large, partially separated, very prominent. Buccal cavity prominent, narrowing towards base. Dorsal tooth massive, inserted on anterior half of thick-walled dorsal metarhabdion, with apex directed anteriad. Mid-ventral junction of two ventral metarhabdial plates minutely denticulate (serrated). Each subventral wall provided with a sclerotised and characteristic, square-bracket-shaped, crest situated latero-ventrally at *ca* level of dorsal tooth apex. Subventral teeth and transverse rows of denticles absent. Pharyngo-intestinal junction non-tuberculate. Female genital system pseudo-monoprodelphic with long post-uterine sac. Spicules slender, arcuate with gubernaculum and lateral accessory pieces. Supplements 6-9 in number. Tail conoid in both sexes, with narrow terminus. Caudal glands and spinneret absent.

TYPE AND ONLY SPECIES

Crestonchulus crestus Siddiqi & Jairajpuri, 2002

Crestonchulus crestus Siddiqi & Jairajpuri, 2002

(Fig. 28C-F)

MEASUREMENTS

Female: L = 1.14-1.40 mm; a = 40-48; b = 3.4-3.9; c = 16-20; c′ = 3.3-4.3; V = 64-71.

Male: L = 1.13-1.32 mm; a = 42-48; b = 3.4-3.9; c = 16.5-19.4; c′ = 3.0-4.2.

DESCRIPTION

Lip region knob-like, sharply offset from body by a constriction, 20-22 μm diam. with partially separated lips. Buccal cavity prominent, narrowing towards its base, 20-22 μm long and 11-12 μm diam. Dorsal tooth massive, in anterior half of buccal cavity, its apex anteriorly directed. Subventral walls minutely denticulate (serrated). In addition, each subventral wall provided with a sclerotised, square-bracket-shaped, crest situated latero-ventrally and obliquely to body axis at *ca* level of apex of dorsal tooth. Female genital system pseudo-monoprodelphic; posterior branch represented by an elongate sac extending to more than half distance from vulva to anus, filled with sperm. Vulva transverse oval, body surface posterior to it often wrinkled and with a ventral body pore. Vagina elongate, leading inward and anteriad for *ca* three-quarters of body diam.; *pars refringens vaginae* appearing as two irregular dots in lateral view. Tail conoid, ventrally arcuate with minutely rounded tip, 64-79 μm long. Caudal glands absent, very rarely three weakly developed glands present. Spinneret absent. Male with elongate-slender and rather strongly arcuate, spicules 40-46 μm long, gubernaculum 17-21 μm long and lateral accessory pieces conspicuous, distally globular in lateral view. Ventromedian supplements 6-9, mammiform. Tail similar to female but more strongly ventrally curved, 64-71 μm long. Caudal glands and spinneret absent.

DISTRIBUTION

Cameroon.

Genus *Margaronchuloides* gen. n.
(Fig. 29G-I)

DIAGNOSIS

Body small, 0.7 mm. Buccal cavity small, comparatively broad proximally. Dorsal tooth large, in anterior half of buccal cavity, apex directed anteriad. Buccal armature poor, subventral teeth absent, each subventral wall with two transverse, arcuate smooth ribs, denticles absent. Pharyngo-intestinal junction non-tuberculate. Female genital system mono-prodelphic, without post-uterine sac. Tail short, conoid, ventrally arcuate with finely rounded tip. Caudal glands rudimentary, spinneret absent. Male not known.

TYPE AND ONLY SPECIES

M. adenticulatus (Andrássy, 1985) **comb. n.**
= *Margaronchulus adenticulatus* Andrássy, 1985

DIFFERENTIAL DIAGNOSIS

The new genus is close to *Margaronchulus* but differs in having two transverse, arcuate, non-denticulate ribs at level of dorsal tooth apex instead of a single, transverse row of denticles.

Margaronchuloides adenticulatus (Andrássy, 1985) comb. n.
(Fig. 29G-I)

MEASUREMENTS

Female: L = 0.73-0.75 mm; a = 26-28; b = 3.6-3.7; c = 22-24; c′ = 2.3-2.4; V = 68-69.

DESCRIPTION

Lip region well offset, 17-18 μm diam.; lips prominent with rounded papillae. Amphidial apertures at level of dorsal tooth apex. Buccal cavity 15-16 × 11 μm, *Mylonchulus*-like in general appearance but comparatively broader at proximal end. Dorsal tooth large, with apex directed anteriad and situated at 78-79% of buccal cavity length from its base, opposed by two transverse, arcuate ribs on each subventral wall running parallel for most part but convergent dorsally. These ribs are smooth and continuous, never split into denticles. Vulval lips flat. Female genital system monoprodelphic. Uterus with very small posterior sac. Vagina sclerotised. Tail short, conoid, ventrally curved with finely rounded tip. Caudal glands rudimentary, spinneret mostly absent, rarely present, but if so, very weak.

DISTRIBUTION

Congo Republic.

Subfamily Sporonchulinae Jairajpuri, 1969

DIAGNOSIS

Dorsal tooth moderately developed, midway or in anterior half of buccal cavity. Denticles scattered, not arranged in transverse rows.

TYPE GENUS

Sporonchulus Cobb, 1917
= *Mononchus* (*Sporonchulus* Cobb, 1917)

OTHER GENERA

Granonchulus Andrássy, 1958
Judonchulus Andrássy, 1958
Prionchuloides Mulvey, 1963b
Actus Baqri & Jairajpuri, 1974

Key to genera of Sporonchulinae Jairajpuri, 1969

1 – Dorsal tooth in middle region of buccal cavity 2
– Dorsal tooth in anterior region of buccal cavity 3
2 – Each subventral wall bearing finely denticulate longitudinal ridge and a number of minute, irregularly dispersed denticles opposed to dorsal tooth .... *Prionchuloides*
– Each subventral wall only bearing large number of scattered denticles opposed to dorsal tooth .... *Judonchulus*
3 – Single transverse row of denticles present plus scattered denticles in longitudinal rows .... *Granonchulus*
– No transverse row of denticles; only irregular denticles in longitudinal rows 4
4 – Subventral denticles in two irregular rows, 6-12 in number .. *Actus*
– Subventral denticles in four irregular rows, 20-25 in number .... *Sporonchulus*

Genus *Sporonchulus* Cobb, 1917

= *Sporonchuloides* Mohandas & Prabhoo, 1982
(Figs 33; 34A-D)

DIAGNOSIS

Body 1.1-1.8 mm long. Buccal cavity barrel-shaped, dorsal tooth moderately developed, in anterior part of buccal cavity, directed anteriad and opposed by numerous, *ca* 25, small denticles more or less arranged in four irregular rows. Pharyngo-intestinal junction non-tuberculate. Female genital system amphidelphic or mono-prodelphic; vulva situated at 55-65% of body length. Male known in only one species, possessing 12 supplements. Tail ventrally curved, conoid, short, 1.5-2.5 anal body diam. long. Caudal glands and spinneret well developed or reduced.

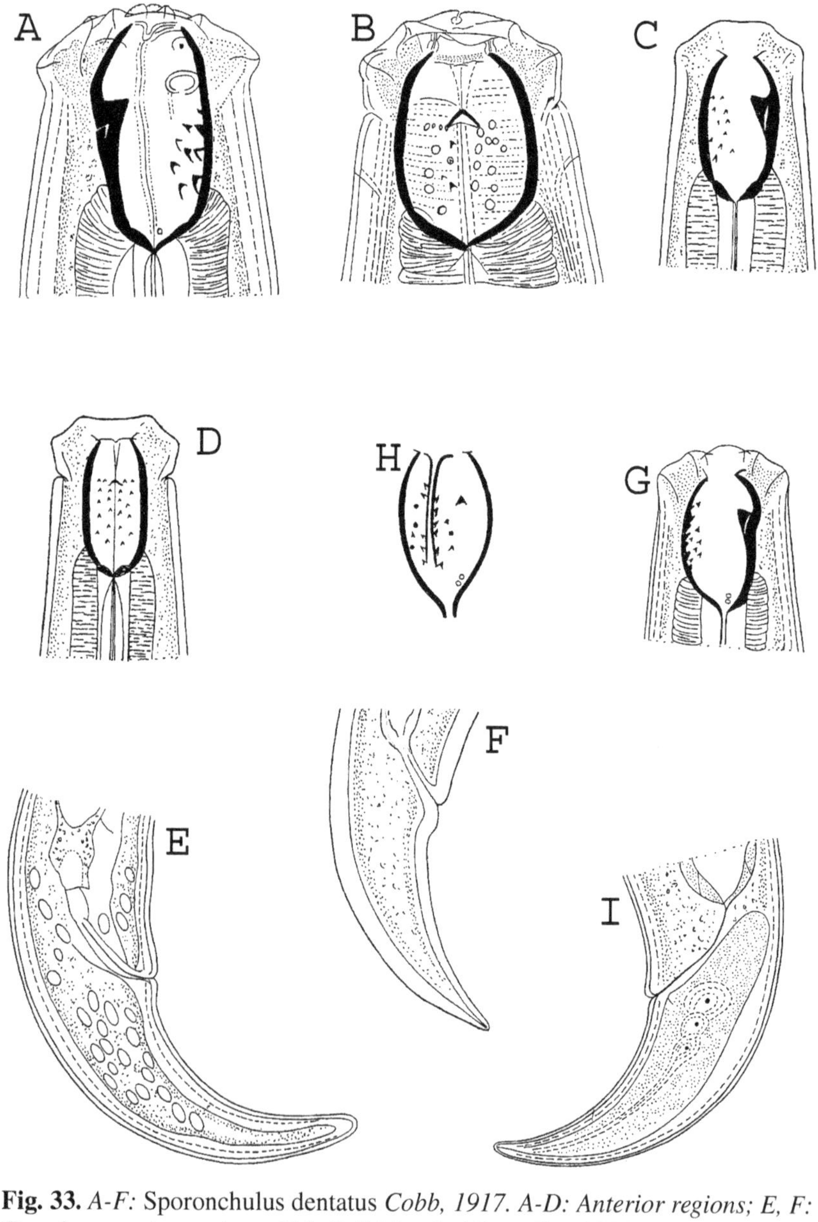

Fig. 33. *A-F:* Sporonchulus dentatus *Cobb, 1917. A-D: Anterior regions; E, F: Female posterior regions; G-I:* S. ibitiensis *(Carvalho, 1951) Andrássy, 1958. A, B, E, G-I after Mulvey (1963); C, D, F after Williams (1958).*

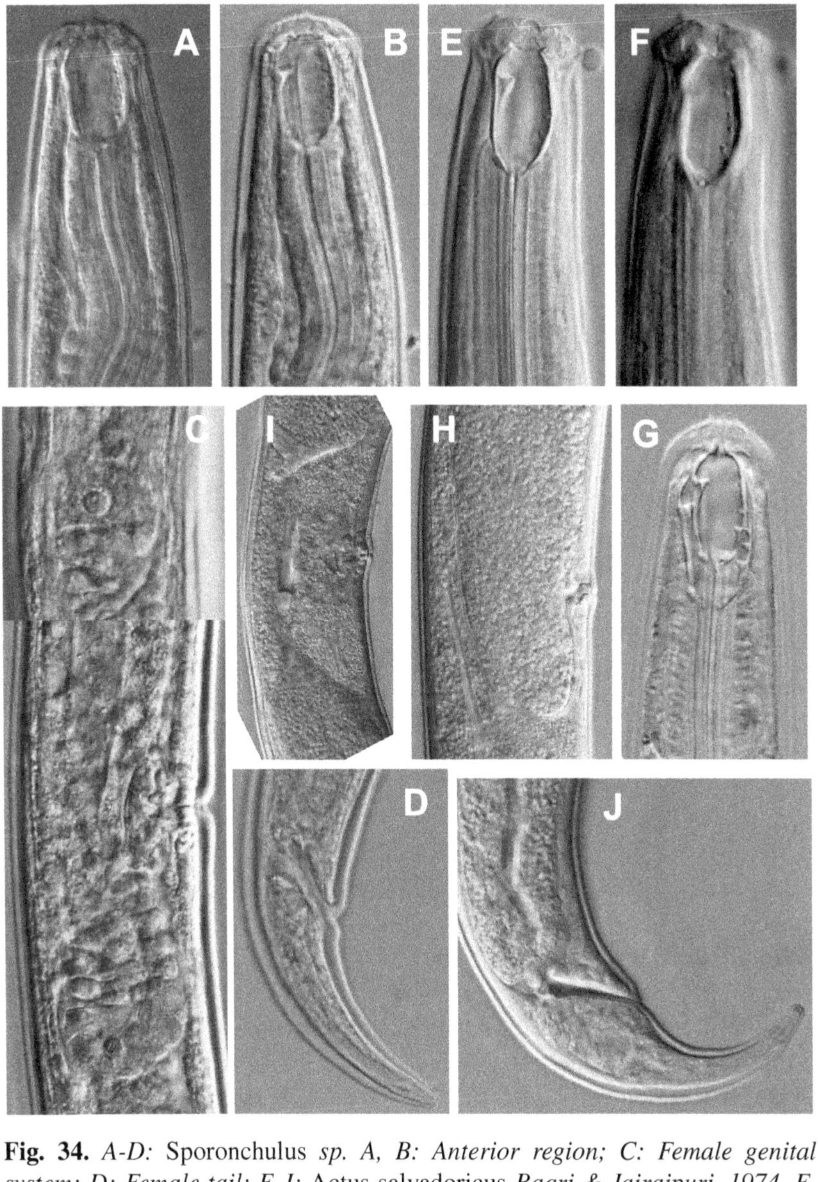

Fig. 34. *A-D:* Sporonchulus *sp. A, B: Anterior region; C: Female genital system; D: Female tail; E-J:* Actus salvadoricus *Baqri & Jairajpuri, 1974. E, F: Anterior region; G: Moulting specimen; H, I: Vulval region; J: Female tail region.*

TYPE SPECIES

Sporonchulus dentatus Cobb, 1917
= *Mononchus* (*Sporonchulus*) *dentatus* Cobb, 1917

OTHER SPECIES

S. coronatus (Carvalho, 1956b) Andrássy, 1958
= *Mononchus coronatus* Carvalho, 1956b
= *Sporonchuloides coronatus* (Carvalho, 1956b) Mohandas & Prabhoo, 1982
S. ibitiensis (Carvalho, 1951) Andrássy, 1958
= *Mononchus ibitiensis* Carvalho, 1951
= *Sporonchuloides ibitiensis* (Carvalho, 1951) Mohandas & Prabhoo, 1982
S. vagabundus Jairajpuri, 1971

Sporonchulus dentatus Cobb, 1917
(Fig. 33A-F)

MEASUREMENTS

Female: L = 1.1-1.5 mm; a = 26-30; b = 3.4-3.9; c = 10-23; c′ = 2.0-2.5; V = 62-64.

DESCRIPTION

Lip region offset, wider than adjoining body. Buccal cavity barrel-shaped, 32-35 × 18 μm. Dorsal tooth moderately developed, anterior to middle of buccal cavity, its apex at 67-73% of buccal cavity length from its base, opposed by numerous irregularly arranged denticles on subventral walls. Pharyngo-intestinal junction non-tuberculate. Female genital system amphidelphic. Tail conoid, ventrally arcuate with rounded terminus. Caudal glands and spinneret absent. Male not known.

REMARKS

Cobb (1917) based his description of *S. dentatus* on a young female from Brazil. Later, he found specimens of this species from Angola. Williams (1958) collected it from a sugarcane field in Mauritius and provided a brief description and figures of the female anterior and posterior regions. Mulvey (1963b) provided measurements of a single specimen from Venezuela. Since then this species has not been redescribed. There is a need to restudy all nominal species of this genus.

DISTRIBUTION

Angola, Brazil-Bahia, Mauritius, Venezuela, West Africa.

Key to species of *Sporonchulus* Cobb, 1917

1 – Female prodelphic; ♀: L = 1.1-1.4 mm; a = 17-22; b = 3.3-4.0; c = 20-26; c′ = 1.5; V = 60-65; ♂: unknown (Brazil) . . *coronatus*
– Female amphidelphic 2
2 – Caudal glands and terminal opening conspicuous; ♀: L = 1.2-1.8 mm; a = 22-36; b = 3.7-4.8; c = 21-27; c′ = 2-2.5; V = 55-63; ♂: unknown (India, St Lucia) . . . . *vagabundus*
– Caudal glands and terminal opening inconspicuous 3
3 – Buccal cavity 20-28 μm long; tail tip sharply pointed; ♀: L = 1.2-1.8 mm; a = 22-36; b = 3.7-4.8; c = 21-27; c′ = 2.5; V = 55-63; ♂: L = 1.2-1.3 mm; a = 27-30; b = 3.8-4.0; c = 23-24; supplements 11-12 (India, Sri Lanka, Nigeria, Zaire, Brazil, Argentina, Malaysia, Cameroon) . . . . *ibitiensis*
– Buccal cavity 32-35 μm long; tail tip finely rounded; ♀: L = 1.1-1.5 mm; a = 26-30; b = 3.4-3.9; c = 10-23; c′ = 2-2.5; V = 62-64; ♂: unknown (Angola, Mauritius, Venezuela, Brazil) . . . . . *dentatus*

Genus *Granonchulus* Andrássy, 1958
(Fig. 35)

DIAGNOSIS

Body 1.0-2.2 mm long. Buccal cavity somewhat different from *Mylonchulus* type, *i.e.*, more ovoid, not tapering so strongly towards base. Dorsal tooth of moderate size situated in anterior half of buccal cavity and directed anteriad. Subventral teeth and refractive rings absent. Subventral denticles present in two arrangements: rasp-like in single row and restricted in number (5-10 on each side); scattered in

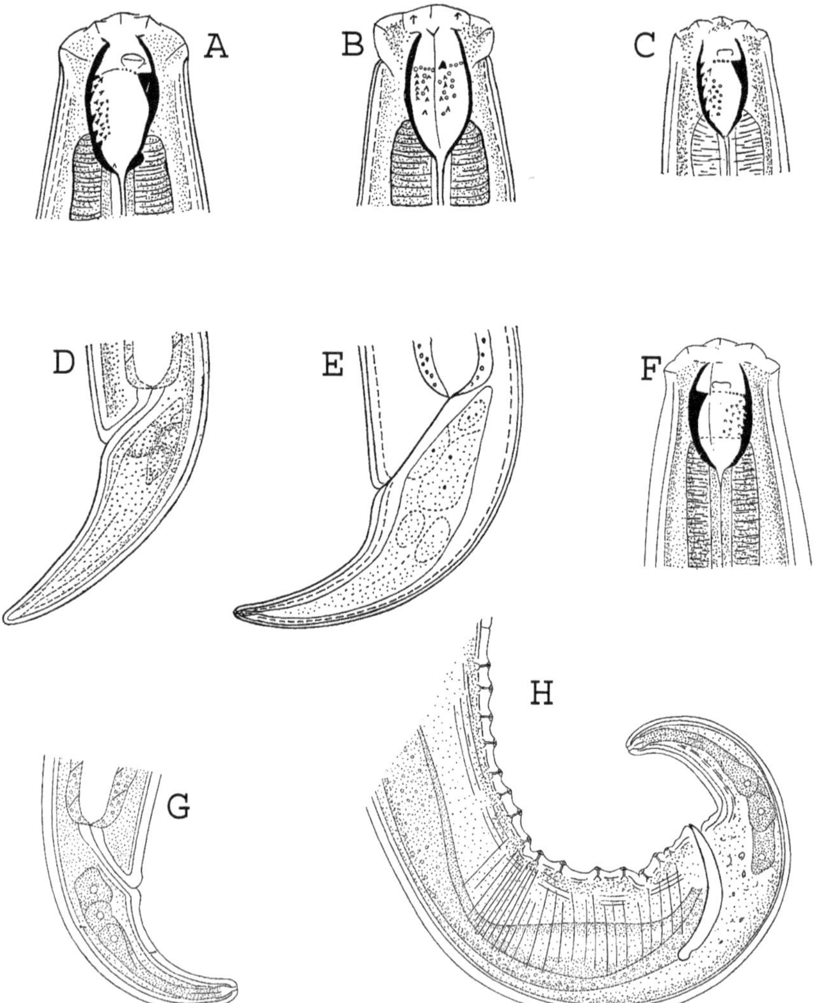

Fig. 35. *A-E:* Granonchulus decurrens *(Cobb, 1917) Andrássy, 1958. A-C: Anterior regions; D, E: Female posterior region; F-H:* Granonchulus subdecurrens *Coetzee, 1966. A, B, E after Mulvey (1963); C, D, F-H after Coetzee (1963).*

irregular longitudinal rows, located just posterior to former. Pharyngo-intestinal junction non-tuberculate. Female genital system amphidelphic. Supplements numerous, 15-21. Tail in both sexes similar, hemispheroid or conoid, 1-2.5 anal body diam. long. Caudal glands mostly well developed.

TYPE SPECIES

Granonchulus decurrens (Cobb, 1917) Andrássy, 1958
= *Mononchus* (*Sporonchulus*) *decurrens* Cobb, 1917

OTHER SPECIES

G. helicus Sharma & Saxena, 1981
G. schulzi (Meyl, 1955a) Andrássy, 1958
= *Mononchus* (*Sporonchulus*) *schulzi* Meyl, 1955a
= *Sporonchulus schulzi* Meyl, 1955a
= *Granonchulus auberti* Altherr *in* Altherr & Delamare Deboutteville, 1972
G. subdecurrens Coetzee, 1966

Granonchulus decurrens (Cobb, 1917) Andrássy, 1958

(Fig. 35A-E)

MEASUREMENTS

Female: L = 1.0-1.3 mm; a = 25-27; b = 3.6-4.0; c = 20-26; c′ = 2.0-2.5; V = 57-61.

DESCRIPTION

Lip region offset. Buccal cavity 23 × 13 μm, *ca* twice as long as broad. Dorsal tooth situated in anterior third of buccal cavity, opposed by denticles arranged in two groups. Anterior transverse row consisting of 5-6 very small denticles, scattered denticles in irregular longitudinal rows just posterior to transverse row. Amphidial apertures situated just anterior to dorsal tooth apex. Pharyngo-intestinal junction non-tuberculate. Female genital system amphidelphic. Tail conoid, ventrally arcuate. Caudal glands and spinneret absent.

REMARKS

Cobb (1917) first described and illustrated this species on the basis of two juvenile specimens collected from Florida, USA. Mulvey (1963b) provided a description based on three female specimens collected, again from Florida, by Dr A.C. Tarjan.

DISTRIBUTION

South Africa, USA-Florida.

Key to species of *Granonchulus* Andrássy, 1958

1 – Tail hemispheroid, *ca* as long as anal body diam.; body length nearly 2 mm; ♀: L = 1.7-2.0 mm; a = 23-30; b = 3.9-5.5; c = 36-61; c′ = 0.8-1; V = 54-65; ♂: L = 1.7-2.2 mm; a = 23-41; b = 4.3-5.2; c = 39-60; supplements 17-21 (Germany, The Netherlands, Poland, USA) . . . . *schulzi*
– Tail conoid, *ca* two anal body diam. long; body length nearly 1 mm 2

2 – Caudal glands reduced, spinneret absent; tail tip somewhat pointed; ♀: L = 1.0-1.3 mm; a = 25-27; b = 3.6-4.0; c = 20-26; c′ = 2-2.5; V = 57-61; ♂: unknown (South Africa, USA) . . . . *decurrens*
– Caudal glands normally developed; spinneret present; tail tip more rounded 3

3 – Tail arcuate; ♀: L = 1.0-1.4 mm; a = 20-34; b = 3.8-5.0; c = 16-24; c′ = 2; V = 53-62; ♂: L = 1.1 mm; a = 23; b = 4; c = 17; supplements 15 (Nigeria, South Africa) . . . . *subdecurrens*
– Tail almost straight; ♀: L = 1.1-1.2 mm; a = 19-22; b = 4.0-4.5; c = 19; c′ = 2; V = 57-61; ♂: unknown (India) . . . . *helicus*

Genus *Judonchulus* Andrássy, 1958
(Fig. 36)

DIAGNOSIS

Body quite slender, 1.4-5.1 mm long. Buccal cavity elongate, large, armed with a large dorsal tooth situated midway or more posterior in buccal cavity, large number of scattered denticles present opposite dorsal tooth. Pharyngo-intestinal junction non-tuberculate. Female genital system monoprodelphic or amphidelphic. Male known for only one species. Supplements 9-10 in number. Tail elongate, 4-7 anal body diam. long, almost cylindrical. Caudal glands and spinneret present.

TYPE SPECIES

Judonchulus recessus (Cobb, 1917) Andrássy, 1958

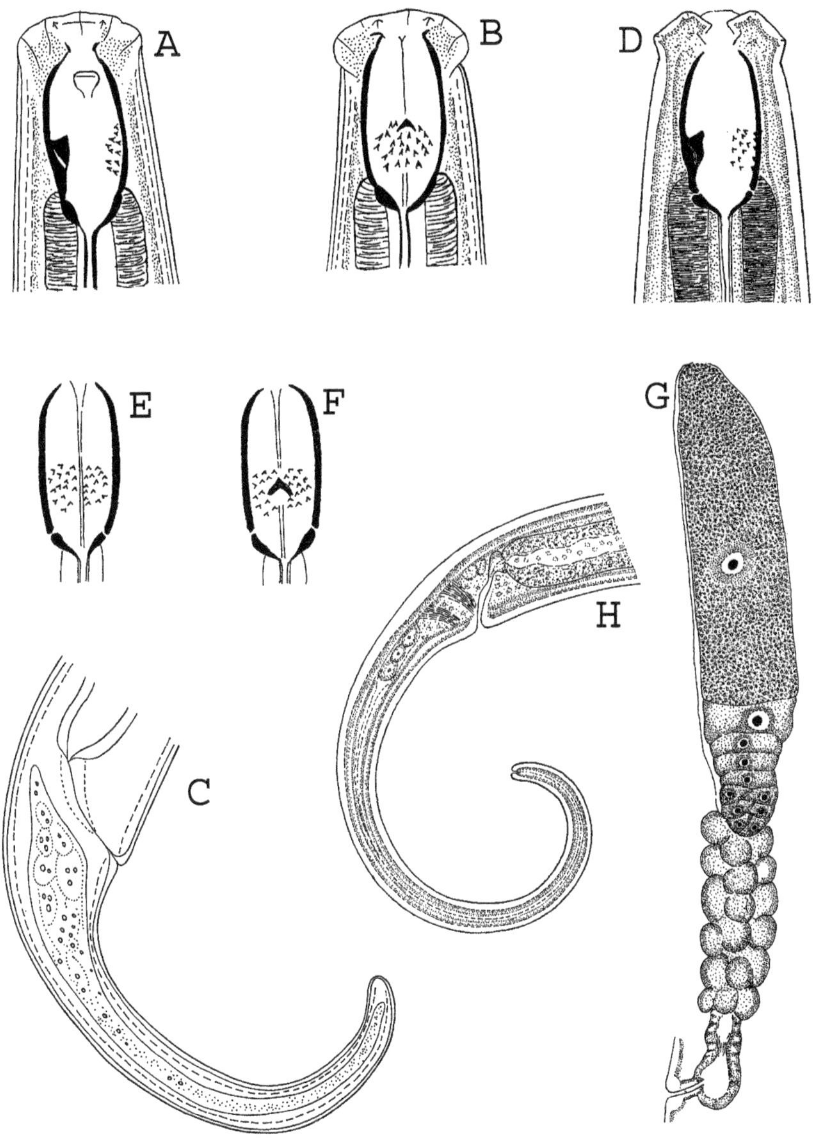

Fig. 36. *A-C:* Judonchulus recessus *(Cobb, 1917) Andrássy, 1958. A, B: Anterior region; C: Female posterior region; D-H:* Judonchulus monhystera *Jairajpuri, Ahmad & Sturhan, 1998. D: Anterior region; E, F: Buccal cavities; G: Female genital system; H: Female posterior region. A-C after Mulvey (1963); D-H after Jairajpuri* et al. *(1998).*

= *Mononchus* (*Sporonchulus*) *recessus* Cobb, 1917
= *Sporonchulus recessus* Cobb, 1917

OTHER SPECIES

J. magnus (Altherr, 1977) Andrássy, 1993a
= *Sporonchulus magnus* Altherr, 1977
J. monhystera Jairajpuri, Ahmad & Sturhan, 1998

Judonchulus recessus (Cobb, 1917) Andrássy, 1958
(Fig. 36A-C)

MEASUREMENTS

Female: L = 1.4-1.6 mm; a = 32-40; b = 3.3-3.7; c = 11-14; c′ = 4-5; V = 63-65.

DESCRIPTION

Body slender. Lip region slightly offset. Buccal cavity *ca* twice as long as broad, 33-35 × 18-19 μm. Dorsal tooth large, situated midway or further posterior in buccal cavity, small subventral denticles irregularly arranged opposite dorsal tooth. Amphidial apertures 5-6 μm wide, located *ca* 85% from base of buccal cavity. Pharyngo-intestinal junction non-tuberculate. Female genital system amphidelphic. Tail tapering suddenly immediately posterior to anus then continuing conoid, ventrally arcuate. Caudal glands small, inconspicuous, spinneret present or absent. Male not known.

REMARKS

Cobb (1917) described this species as *Mononchus* (*Sporonchulus*) *recessus* from Florida, USA. Andrássy (1958) proposed a new genus, *Judonchulus*, for this species and *M.* (*S.*) *micoletzkyi* (Meyl, 1954). Mulvey (1963b) did not accept this genus and returned the type species, *J. recessus*, to *Sporonchulus* and proposed a new genus, *Prionchuloides*, with *M.* (*S.*) *micoletzkyi* as its type. Andrássy (1993a) did not accept Mulvey's synonymy of *Judonchulus* and also transferred *Sporonchulus magnus* (Altherr, 1977) to *Judonchulus*. Jairajpuri *et al.* (1998) accepted Andrássy's views and also added a new species, *J. monhystera*, from New Zealand.

Key to species of *Judonchulus* Andrássy, 1958

1 – Female genital system mono-prodelphic; ♀: L = 1.25-1.43 mm; a = 39-46; b = 3.4-3.8; c = 6-7; c′ = 9-11; V = 60-63; ♂: unknown (New Zealand) . . . . *monhystera*
– Female genital system amphidelphic 2

2 – Body very long, 4-5 mm in length, extremely slender; ♀: L = 4.6-5.1 mm; a = 75-95; b = 3.9-4.0; c = 20-37; c′ = 7; V = 64; ♂: L = 4.4 mm; a = 70; b = 3.7; c = 51; supplements 9-10 (Brazil) . . . . *magnus*
– Body much smaller, <2 mm long, not so slender; ♀: L = 1.4-1.6 mm; a = 32-40; b = 3.3-3.7; c = 11-14; c′ = 4-5; V = 63-65; ♂: unknown (USA-Florida) . . . . *recessus*

Genus *Prionchuloides* Mulvey, 1963b
(Fig. 37A)

DIAGNOSIS

Buccal cavity barrel-shaped, tapering at base. Dorsal tooth located midway in buccal cavity, opposed by two groups of denticles on subventral walls, one group arranged in two longitudinal rows and other irregularly placed. Pharyngo-intestinal junction non-tuberculate. Female genital system amphidelphic. Tail conoid, caudal glands and spinneret obscure.

TYPE AND ONLY SPECIES

Prionchuloides micoletzkyi (Meyl, 1954) Mulvey, 1963b
= *Mononchus* (*Sporonchulus*) *micoletzkyi* Meyl, 1954
= *Judonchulus micoletzkyi* (Meyl, 1954) Andrássy, 1958

Prionchuloides micoletzkyi (Meyl, 1954) Mulvey, 1963b
(Fig. 37A)

MEASUREMENTS

Female: L = 1.7 mm; a = 22.1; b = 4.0; c = 14; V = 62.

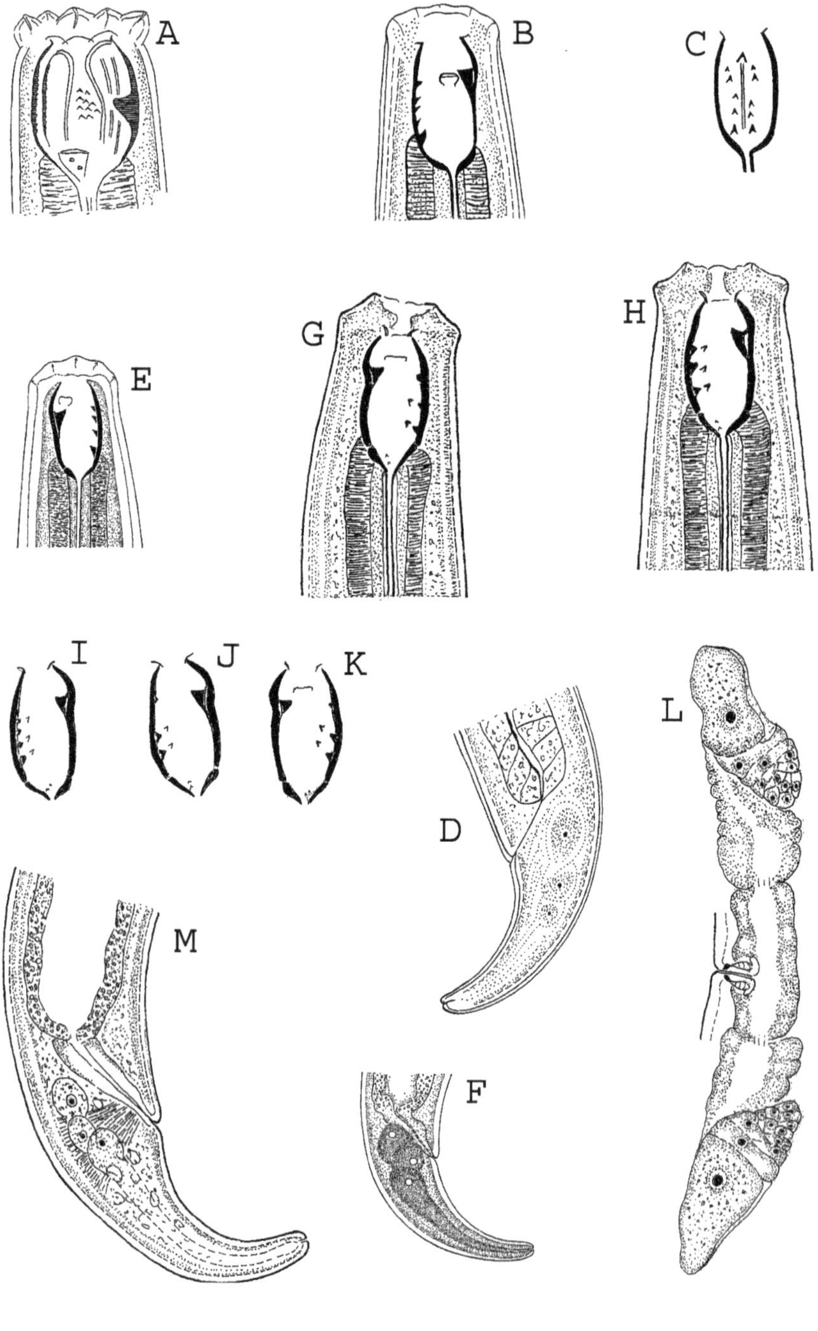
A
B
C
E
G
H
I
J
K
D
L
M
F

DESCRIPTION

Buccal cavity a little longer than broad, 40 × 34 μm. Dorsal tooth strong, located in middle of buccal cavity, opposed by two longitudinal rows of 12-13 very small denticles arranged along a rib. At level of dorsal tooth, *ca* ten scattered denticles also present. Female genital system amphidelphic. Tail ventrally arcuate, short.

REMARKS

Meyl (1954) described this species as *Mononchus* (*Sporonchulus*) *micoletzkyi* based on a single female. Andrássy (1958) placed it in the genus *Judonchulus* along with *M.* (*S.*) *recessus*. Mulvey (1963b) synonymised *Judonchulus*, transferred *M.* (*S.*) *recessus* to *Sporonchulus* and proposed a new genus *Prionchuloides* for *M.* (*S.*) *micoletzkyi*. This species has never been recorded since its original description.

DISTRIBUTION

Italy-Ischia Island.

Genus *Actus* Baqri & Jairajpuri, 1974

(Figs 34E-J; 37B-M)

DIAGNOSIS

Body small, 0.8-1.5 mm long. Buccal cavity ovoid, *ca* twice as long as broad, tapering at base. Dorsal tooth located in anterior part of buccal cavity, tooth apex directed anteriad, opposed by two rows of denticles which are not arranged on longitudinal ridges. Pharyngo-intestinal junction non-tuberculate. Female reproductive system amphidelphic. Tail short, 0.8-2.6 anal body diam. long, ventrally curved. Caudal glands and spinneret present. Male not known.

Fig. 37. *A:* Prionchuloides micoletzkyi *(Meyl, 1954) Mulvey, 1963, anterior region; B-D:* Actus minutus *(Mulvey, 1963) Baqri & Jairajpuri, 1974. B: Anterior region; C: Buccal cavity; D: Female tail; E-M:* Actus salvadoricus *Baqri & Jairajpuri, 1974. E, G, H: Anterior region; F, M: Female tail region; I-K: Buccal cavity; L: Female genital system. A after Meyl (1954); B-D after Mulvey (1963); E-M after Baqri and Jairajpuri (1974).*

TYPE SPECIES

Actus minutus (Mulvey, 1963b) Baqri & Jairajpuri, 1974
= *Sporonchulus minutus* Mulvey, 1963b

OTHER SPECIES

A. baqrus Mohilal & Dhanachand, 1997
A. conoidus Dhanachand, Romoni & Pramodini, 2006
A. nanurus Siddiqi, 1984a
A. neocaledonensis Yeates, 1992
A. salvadoricus Baqri & Jairajpuri, 1974

Actus minutus (Mulvey, 1963b) Baqri & Jairajpuri, 1974

(Fig. 37B-D)

MEASUREMENTS

Female: L = 0.83-0.85 mm; a = 21-22; b = 3.1-3.4; c = 17.6-19.1; V = 62-65.

DESCRIPTION

Lip region offset by depression. Lips partially separated, papillae slightly raised. Amphidial aperture slit-like, at level of dorsal tooth apex. Buccal cavity 28 × 14-17 μm or *ca* twice as long as broad, with moderately thickened walls. Dorsal tooth medium-sized, opposed by two longitudinal rows of relatively large denticles (not arranged along longitudinal rib). Posteriormost two denticles nearly basal and larger than other eight. Female reproductive system amphidelphic. Vulva a transverse slit. Vagina with small *pars refringens*. Tail conoid, ventrally arcuate, 44-47 μm long. Caudal glands three. Terminal opening obscure. Male not known.

DISTRIBUTION

Canada, Mariana and Carolina Islands.

Key to species of *Actus* Baqri & Jairajpuri, 1974

1 – Tail 18-33 μm, *ca* as long as anal body diam.; ♀: L = 0.91-1.1 mm; a = 22-28; b = 3.2-4.3; c = 33-59; c′ = 0.8-1.3; V = 60-60; ♂: unknown (Fiji) . . . . *nanurus*
– Tail 45-72 μm, 2-2.5 anal body diam. long 2

2 – Buccal cavity > 30 μm (31-36 μm); ♀: L = 1.1-1.5 mm; a = 24-31; b = 3.4-3.9; c = 19-25; c′ = 1.8-2.6; V = 58-63; ♂: unknown (New Caledonia) . . . . *neocaledonensis*
– Buccal cavity < 30 μm 3

3 – Vulval papillae present; ♀: L = 1.10-1.14 mm; a = 28-30; b = 3-3.8; c = 20-22; c′ = 2.0-2.4; V = 61-62; ♂: unknown . . . . *baqrus*
– Vulval papillae absent 4

4 – Buccal cavity short, 21-22 μm long; ♀: L = 1.21-1.39 mm; a = 31-36; b = 4.1-4.5; c = 20-24; c′ = 2.4-2.6; V = 59-61; ♂: unknown. . . . . *conoidus*
– Buccal usually > 25 μm long 5

5 – Buccal cavity 28 μm long; anterior edge of dorsal tooth straight, perpendicular to body axis; caudal spinneret obscure; ♀: L = 0.83-0.85 mm; a = 21-22; b = 3.1-3.4; c = 17.6-19; V = 62-65; ♂: unknown (Canada, Marianna and Carolina Islands) . . . . *minutus*
– Buccal cavity 24-27 μm long; anterior edge of dorsal tooth concave, tooth anteriorly directed; caudal spinneret distinct; ♀: L = 1.0-1.2 mm; a = 20-24; b = 3.5-4.3; c = 17-20; c′ = 1.8-2.4; V = 61-63; ♂: unknown (El Salvador, Japan) . . . . *salvadoricus*

Note

Andrássy (1993a) synonymised *A. salvadoricus* with *A. minutus* but Olia *et al.* (2008) recently revalidated this species.

Nematology Monographs & Perspectives, 2010, Vol. 7, 141-155

Chapter 6

Family Cobbonchidae Jairajpuri, 1969

Cobb (1917) described under the genus *Mononchus*, subgenus *Mononchus*, two species of mononchs, *M. radiatus* and *M. palustris*, having a dorsal tooth in the anterior half of the buccal cavity opposed by almost equally developed teeth, one on each subventral wall and situated in the posterior half of the buccal cavity. Andrássy (1958) erected a new genus, *Cobbonchus*, for these two species and designated *Cobbonchus palustris* (= *Mononchus palustris*) as type species of the genus. More species were added to this genus by Clark (1961b), Coetzee (1965, 1968), Baqri *et al.* (1978), Rahman and Jairajpuri (1984), Tsalolikhin (1988), Yeates (1992), Andrássy (1993a), Dhanam and Jairajpuri (1998), Orselli and Vinciguerra (2007), *etc.*

Jairajpuri (1969) proposed a new family, Cobbonchidae, for this genus whilst Jairajpuri and Khan (1982) proposed a new genus, *Comiconchus* Jairajpuri & Khan, 1982, similar to *Cobbonchus*, and also considered *Tigronchus* Kirjanova & Krall, 1969 to belong to this family. Andrássy (1993a) accepted Cobbonchinae only as a subfamily under Mononchidae with two genera, *Cobbonchus* and *Comiconchus*. In 1996, he added another new genus, *Tricaenonchus*, to this group. Recently, Andrássy (2009) proposed a new genus, *Cobbonchulus*, for an unusual *Cobbonchus* species, *C. longicaudatus* Jairajpuri, Ahmad & Sturhan, 1998, described from New Zealand. This species is characteristic in having subventral teeth much smaller than its dorsal tooth, a buccal cavity shape more like *Mylonchulus* species in form, and a long filiform tail. This proposal makes a total of four genera that are grouped together in the family Cobbonchidae.

DIAGNOSIS

Dorsal and subventral walls of buccal cavity armed either with a tooth of nearly equal size or dorsal tooth massive and the subventral teeth much smaller; all three teeth with anteriorly directed apex. Dorsal tooth generally located in anterior part, subventrals in posterior part, of buccal cavity; rarely all three teeth lying at same level in mid-buccal cavity or subventral teeth lying in anterior half of buccal cavity and dorsal tooth in posterior half. Tail short, conoid, rarely elongate-conoid or long filiform; caudal glands and spinneret usually well developed.

TYPE AND ONLY SUBFAMILY

Cobbonchinae Jairajpuri, 1969

TYPE GENUS

Cobbonchus Andrássy, 1958

OTHER GENERA

Comiconchus Jairajpuri & Khan, 1982
Tricaenonchus Andrássy, 1996
Cobbonchulus Andrássy, 2009

Key to genera of Cobbonchidae Jairajpuri, 1969

1 – Dorsal tooth in anterior half of buccal cavity; subventral teeth in posterior half or rarely at level with dorsal tooth 2
– Dorsal tooth in posterior half of buccal cavity; subventral teeth in anterior half . . . . *Tricaenonchus*

2 – Dorsal tooth massive, *Mylonchulus* type; subventral teeth much smaller than dorsal; tail long filiform . . . . *Cobbonchulus*
– Dorsal moderately developed, subventral teeth slightly smaller than dorsal; tail short conoid to elongate conoid 3

3 – Subventral teeth more or less posterior to dorsal tooth; tail short-conoid, with conspicuous glands . . . . *Cobbonchus*
– Subventral teeth at level of dorsal tooth; tail elongate-conoid, with reduced glands . . . . *Comiconchus*

Genus *Cobbonchus* Andrássy, 1958
(Fig. 38A-E)

DIAGNOSIS

Body 0.8-3.8 mm long. Buccal cavity oblong, armed with a large dorsal and two similar, but generally somewhat smaller, subventral teeth; dorsal tooth located in anterior part, subventral teeth in posterior part of buccal cavity, or all closer to each other in mid-buccal cavity region (subventral teeth in this case also a little more posterior than dorsal). Pharyngo-intestinal junction non-tuberculate. Female genital system amphidelphic or mono-prodelphic, vulva at 58-80% of body length. Male known in 50% of species; supplements 6-18 in number. Tail in both sexes similar, short, 0.5-2.3 anal body diam. long, conoid to digitate, or blunt, ventrally arcuate. Caudal glands and terminal orifice conspicuous.

TYPE SPECIES

Cobbonchus palustris (Cobb, 1917) Andrássy, 1958
= *Mononchus palustris* Cobb, 1917

OTHER SPECIES

C. abrupticaudatus (Altherr, 1960) Goodey, 1963
= *Mononchus abrupticaudatus* Altherr, 1960
C. aequatorialis Andrássy, 2008
C. altitudinarum Yeates, 1992
C. artemisiae Coetzee, 1968
C. australis Clark, 1963
C. charlesi Coetzee, 1966
C. chauliodus Clark, 1961b
C. citri Dhanam & Jairajpuri, 1999
C. coetzeeae Andrássy, 1970
C. collaris Andrássy, 1985
C. couleensis Yeates, 1992
C. curvicaudatus Tsalolikhin, 1988
C. dianae Coetzee, 1965
C. eurystoma Coetzee, 1965
C. heynsi Coetzee, 1965

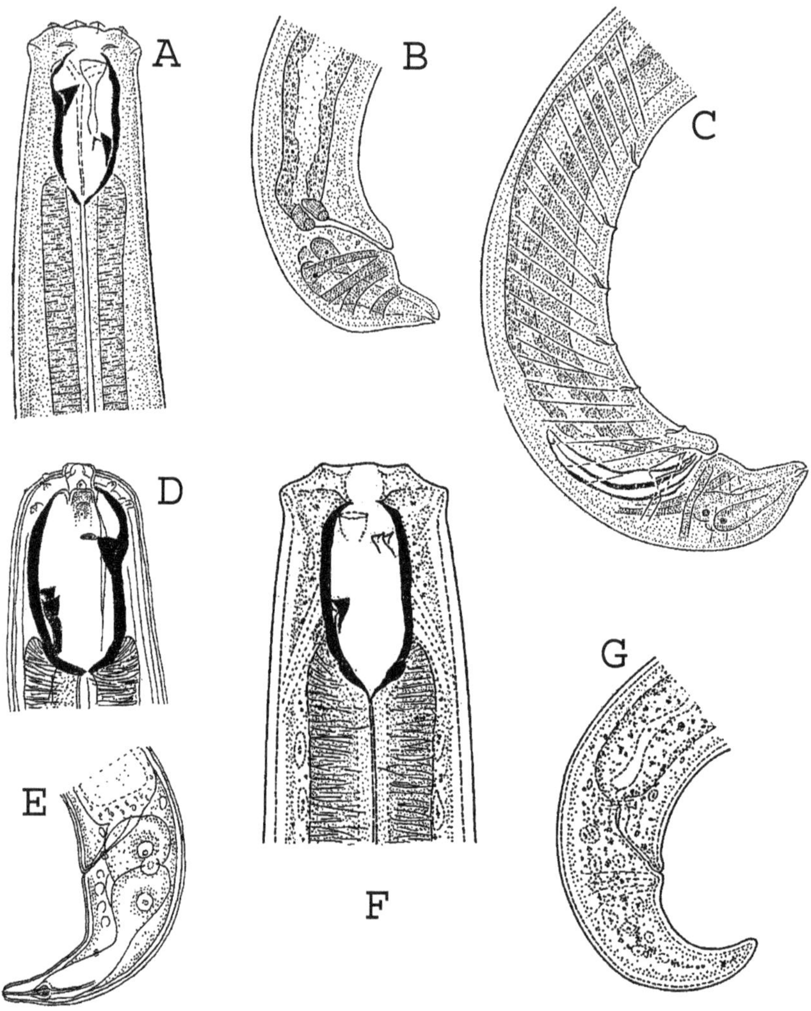

Fig. 38. *A-C:* Cobbonchus indicus *Baqri* et al., *1978. A: Anterior region; B: Female tail; C: Male posterior region; D, E:* Cobbonchus palustris *(Cobb, 1917) Andrássy, 1958. D: Anterior region; E: Female tail; F, G:* Tricaenonchus caucasicus *Andrássy, 1996. F: Anterior region; G: Female tail region. A-C after Baqri* et al. *(1978); D, E after Clark (1960); F, G after Andrássy (1996).*

C. impositus Mohilal & Dhanachand, 1997
C. inclinatus Dhanam & Jairajpuri, 1998
C. indicus Baqri, Baqri & Jairajpuri, 1978
= *C. subcaudatus* Dhanachand, Romoni & Pramodini, 2006 **syn. n.**

= *C. neoindicus* Choudhury & Sivakumar, 1998 **syn. n.**
C. macrampula Orselli & Vinciguerra, 2007
C. mauritianus (Williams, 1958) Clark, 1960a
= *Mononchus mauritianus* Williams, 1958
= *Cobbonchus dzumaci* Yeates, 1992
C. megalus Coetzee, 1966
C. ockerti Coetzee, 1965
C. orstomi Yeates, 1992
C. papillatus Dhanam & Jairajpuri, 1998
C. paraindicus Rahman & Jairajpuri, 1984
C. pleiades Andrássy, 1993a
C. pounamua Clark, 1961b
= *C. regulus* Altherr, 1963
C. radiatus (Cobb, 1917) Andrássy, 1958
= *Mononchus radiatus* Cobb, 1917
C. rotundicaudatus Coetzee, 1968
= *Cobbonchus ockerti apud* Coetzee, 1966 *nec* Coetzee, 1965
C. soutoi (Carvalho, 1956a) Andrássy, 1993a
= *Mononchus soutoi* Carvalho, 1956a
= *Miconchus soutoi* (Carvalho, 1956a) Andrássy, 1958
C. thesigei Coetzee, 1968

Cobbonchus palustris (Cobb, 1917) Andrássy, 1958

(Fig. 38D, E)

MEASUREMENTS

Female: L = 1.4 mm; a = 37; b = 3.33; c = 33.3; V = 62.

DESCRIPTION

Lip region not offset, rounded. Amphidial apertures small, at level of dorsal tooth apex. Buccal cavity large, almost twice as long as broad, well indented in region of teeth, base broad. Dorsal tooth apex at 68% of buccal cavity length from its base; subventral tooth apices at 43%. Female genital system amphidelphic; both branches almost equally developed. Tail short, ventrally arcuate, rather conoid. Caudal glands broad, saccate; ampulla large, duct terminal. Male not known.

DISTRIBUTION

Former Czechoslovakia, USA-Wisconsin.

Key to species of *Cobbonchus* Andrássy, 1958

1 – Female genital system prodelphic, posterior ovary absent or rudimentary 2
– Female genital system amphidelphic, both ovaries normal 11

2 – Large species, >3 mm long 3
– Smaller species, 1-2.4 mm long 4

3 – Dorsal tooth in anterior fifth of buccal cavity; large, reaching centre of buccal cavity; ♀: L = 3.3 mm; a = 51; b = 3.1; c = 41; c′ = 1.5; V = 70; ♂: unknown (New Zealand). . . . *chauliodus*
– Dorsal tooth in anterior third of buccal cavity, not as large; ♀: L = 3.4-3.9 mm; a = 36-50; b = 3-4; c = 33-40; c′ = 2; V = 72-76; ♂: L = 3.8 mm; a = 49; b = 4; c = 49; supplements 16 (South Africa) . . . . *megalus*

4 – Dorsal tooth and subventral teeth at similar level, former in anterior third of buccal cavity or more posterior, latter close to mid-buccal cavity 5
– Dorsal tooth and subventral teeth more separated, former in anterior fifth/fourth, latter in posterior third of buccal cavity 8

5 – Buccal cavity broad, barrel-shaped, dorsal tooth comparatively small; tail short, as long as one anal body diam.; ♀: L = 1.0 mm; a = 24; b = 3.5; c = 25; c′ = 1; V = 71; ♂: L = 1.1 mm; a = 27; b = 3.0; c = 21; supplements 14 (South Africa) . . . . . . *eurystoma*
– Buccal cavity oblong, not so broad, dorsal tooth large; tail distinctly longer than one anal body diam. 6

6 – Subventral teeth nearly as big as dorsal tooth; tail two anal body diam. long; ♀: L = 1.5-1.8 mm; a = 37-43; b = 3.2-3.8; c = 23-24; c′ = 2; V = 70-74; ♂: L = 1.6-1.7 mm; a = 41-44; b = 3.7-4.0; c = 21-24; supplements 15 (South Africa) . . . . *dianae*
– Subventral teeth conspicuously smaller than dorsal tooth; tail 1.5 anal body diam. long 7

7 – Body 2 mm long; posterior ovary present but very small; ♀: L = 2.1 mm a = 46; b = 3.0; c = 38; c′ = 1.5; V = 74; ♂: L =

1.9 mm; a = 42; b = 3.0; c = 31; supplements 18 (India, South Africa) . . . . *artemisiae*
– Body < 1.5 mm long; posterior ovary completely absent; ♀: L = 1.1-1.4 mm; a = 37-44; b = 3-4; c = 33-39; c′ = 1.3-1.5; V = 66-75; ♂: L = 1.3-1.4 mm; a = 40-46; b = 4; c = 32-38; supplements 16-18 (South Africa) . . . . *charlesi*

8 – Tail 80-90 μm long, posterior narrowed part as long as, or longer than, anal body diam.; ♀: L = 1.9-2.4 mm; a = 40-42; b = 3.5-3.6; c = 23-25; c′ = 1.8-2.1; V = 72-75; ♂: L = 1.9-2.3 mm; a = 40-51; b = 3.4-3.8; c = 22-29; supplements 6-8 (New Caledonia) . . . . *couleensis*
– Tail 30-45 μm long, posterior narrowed part conspicuously shorter than anal body diam. 9

9 – Buccal cavity unusually narrow with parallel walls; tail < anal body diam.; ♀: L = 0.9-1.1 mm; a = 26-36; b = 3-4; c = 55-67; c′ = 0.7-0.8; V = 77-80; ♂: L = 0.9-1.3 mm; a = 40-41; b = 4.2-4.3; c = 40; supplements 12-14 (South Africa) . . . . *ockerti*
– Buccal cavity barrel-shaped with convex walls; tail somewhat longer than anal body diam. 10

10 – Dorsal tooth apex at 80% of buccal cavity length from its base; ♀: L = 1.0 mm; a = 23; b = 3.6; c = 24; c′ = 1.2; V = 71; ♂: unknown (South Africa) . . . . *heynsi*
– Dorsal tooth apex at 65-70% of buccal cavity length from its base; ♀: L = 1.3-1.5 mm; a = 35-42; b = 2.9-3.3; c = 31-36; c′ = 1.4-1.6; V = 73-75; ♂: L = 1.3-1.4 mm; a = 39-43; b = 3.1-3.4; c = 29-32; supplements 12-13 (New Caledonia) . . . . *orstomi*

11 – Apex of dorsal tooth at middle of buccal cavity, all three teeth nearly at same level 12
– Apex of dorsal tooth in anterior third/fourth of buccal cavity, subventral teeth conspicuously distanced from former 17

12 – Vulva unusually anterior, at 35-36% of body length, each gonad *ca* two body diam. long; ♀: L = 2.3-2.8 mm; a = 28-33; b = 3.1-3.4; c = 30; c′ = 1.8-2; V = 35-36; ♂: unknown (Brazil) . . . . . . *soutoi*
– Vulva in usual position, well posterior to mid-body; each gonad > three body diam. long 13

13 – Tail characteristic in shape, broadly conoid in anterior half, then rapidly narrowing, digitiform, ending in a mammiform tip; ♀: L =

1.67-1.87 mm; a = 28-32; b = 2.8-3.1; c = 31-37; c′ = 1.4-1.5; V = 65-69; ♂: unknown (Ecuador) . . . . *aequatorialis*

– Tail blunt with broadly rounded tip, or tapering with a narrow rounded tip 14

14 – Tail blunt with broadly rounded tip; ♀: L = 1.9 mm; a = 33; b = 3.4; c = 38; c′ = 1.2; V = 65; ♂: unknown (Brazil) . . . . *curvicaudatus*

– Tail tapering with a narrow rounded tip 15

15 – Tail finger-like in posterior two-thirds, strongly bent ventrad, anterior anal lip overhanging; each gonad 7-8 body diam. long; ♀: L = 2.2 mm; a = 4.7; b = 4; c = 42; c′ = 2.3; V = 65; ♂: unknown (South Africa) . . . . *coetzeeae*

– Tail not finger-like in posterior part, straight or dorsally bent; each gonad 3-4 body diam. long 16

16 – Tail < anal body diam. long; body length *ca* 1 mm; ♀: L = 1.0-1.2 mm; a = 29-31; b = 3.2-3.5; c = 48-64; c′ = 0.5-0.7; V = 65-70; ♂: unknown (New Caledonia) . . . . *altitudinarum*

– Tail > anal body diam. long; body length *ca* 1.5 mm; ♀: L = 1.4-1.6 mm; a = 30-33; b = 4; c = 33-42; c′ = 1.2; V = 70; ♂: unknown (South Africa) . . . . *thesigeri*

17 – Body *ca* 3 mm long; ♀: L = 2.9 mm; a = 44; b = 3.7; c = 34; c′ = 1.5-1.7; V = 59; ♂: L = 2.2-2.9 mm; a = 46-53; b = 3.4-4.0; c = 27-34; supplements 16-18 (New Zealand). . . . *pounamua*

– Body 2 mm or shorter 18

18 – Tail up to *ca* two anal body diam. long, uniformly narrowing . . 19

– Tail < 2 anal body diam. long, conoid with a ventral flexure then cylindroid or uniformly cylindroid 22

19 – Lip region rounded, as broad as adjacent body; ♀: L = 1.4 mm; a = 37; b = 3.3; c = 33; c′ = 2.2; V = 62; ♂: unknown (former Czechoslovakia, USA) . . . . *palustris*

– Lip region angular, broader than adjacent body 19

20 – Vulva at 74% of body length; buccal cavity barely 1.5 labial diam. long; ♀: L = 0.8 mm; a = 35; b = 3.3; c = 20; c′ = 2; V = 74; ♂: unknown (Uzbekistan, USA) . . . . *radiatus*

– Vulva at 60% of body length; buccal cavity nearly twice labial diam. long 21

21 – Tail conspicuously pressed on dorsal side, anterior anal lip large, overhanging; ♀: L = 1.38 mm; a = 34; b = 3.6; c = 26; c′ = 2.2; V = 58; ♂: unknown (Tanzania) . . . . *pleiades*

– Tail not pressed on dorsal side (uniformly contoured), anterior anal lip small, not overhanging; ♀: L = 1.60-1.62 mm; a = 42-44; b = 3.3-3.4; c = 24-25; c′ = 2.2-2.3; V = 60; ♂: unknown (Brazil) . . . . *collaris*

22 – Advulval papillae present 23

– Advulval papillae absent 24

23 – Buccal cavity 19-28 × 14 μm; dorsal tooth apex at 74-85% of buccal cavity length from its base; ♀: L = 1.0-1.1 mm; a = 24-26; b = 3.0-3.2; c = 48-53; c′ = 0.7; V = 62-67; ♂: L = 1.2 mm; a = 32; b = 3.3; c = 45; spicules 58 μm; supplements 16 (India) . . . . *paraindicus*

– Buccal cavity 34-47 × 17-20 μm; dorsal tooth apex at 58-65% of buccal cavity length from its base; ♀: L = 1.6-2.3 mm; a = 39-58; b = 2.7-3.5; c = 44-69; c′ = 1.1-1.9; V = 62-73; ♂: L = 1.4-1.5 mm; a = 43-50; b = 3.3-3.5; c = 41-63; c′ = 1.0-1.3; spicules 41-52 μm; supplements seven (India) . . . . *papillatus*

24 – Female tail only half as long as anal body diam., tip of male tail broadly rounded; ♀: L = 1.0 mm; a = 29; b = 4; c = 51; c′ = 0.5; V = 72; ♂: L = 0.9-1.2 mm; a = 30-37; b = 4; c = 43-52; supplements 16-17 (South Africa) . . . . *rotundicaudatus*

– Female tail 1-2 anal body diam. long, tip of male tail narrowly rounded 25

25 – Tail uniformly narrowing, ventrally arcuate 26

– Tail convex conoid with a ventral flexure, then rather digitate . . 28

26 – Dorsal tooth apex in anterior third of buccal cavity; subventral teeth apex near middle of buccal cavity 27

– Dorsal tooth apex in anterior fourth of buccal cavity; subventral teeth apex in posterior third; ♀: L = 1.1-1.3 mm; a = 32; b = 3.2; c = 24; c′ = 1.8; V = 63; ♂: unknown (Mauritius) . . *mauritianus*

27 – Caudal glands ending with well developed ampulla; vulva at 64-65%; ♀: L = 1.17-1.44 mm; a = 25-29; b = 2.6-3.2; c = 22-23; c′ = 1.5-1.8; V = 64-65; ♂: unknown (Italy) . . . . . . *macrampulla*

– Ampulla of caudal glands not so well developed; vulva at 53-59%; ♀: L = 1.36-1.50 mm; a = 30-50; b = 3.1-3.3; c = 26.6-35.7; c′ = 1.6-2.5; V = 53-59 (India) . . . . *impositus*

28 – Dorsal tooth apex in anterior fourth of buccal cavity; subventral teeth in posterior third; ♀: L = 1.1 mm; a = 37; b = 3.4; c = 49; c′ = 1-1.2; V = 66; ♂: L = 1.1 mm; a = 45; b = 3.4; c = 43; supplements six (India) . . . . *indicus*

– Dorsal tooth apex in anterior third of buccal cavity; subventral teeth near middle of buccal cavity 29

29 – Lip region rounded, continuous; ♀: L = 1.1-1.3 mm; a = 28-38; b = 3.1-3.6; c = 29-35; c′ = 1.7; V = 63-64; ♂: unknown (Cameroon) . . . . *abrupticaudatus*

– Lip region angular, wider than adjoining body 30

30 – Vagina inclined anteriad with angular bend; ♀: L = 1.3-1.6 mm; a = 35-48; b = 2.9-3.7; c = 35-54; c′ = 1.2-1.6; V = 63-68; ♂: unknown (India) . . . . *inclinatus*

– Vagina not inclined anteriad, straight 31

31 – Buccal cavity 34-35 × 18-20 μm; supplements 14-18; ♀: L = 1.9 mm; a = 34-46; b = 3.3-3.9; c = 31-34; c′ = 1; V = 63-66; ♂: L = 1.4-1.9 mm; a = 36-39; b = 3.4-3.6; c = 35-30 (New Zealand, Campbell Islands) . . . . *australis*

– Buccal cavity 25 × 15 μm; supplements six; ♀: L = 1.3 mm; a = 48; b = 3.4; c = 49; c′ = 1.2; V = 66; ♂: L = 1.1-1.3 mm; a = 37-44; b = 3.3-3.4; c = 42-45; c′ = 0.9-1.2 (India) . . . . *citri*

Genus *Comiconchus* Jairajpuri & Khan, 1982
(Fig. 39E-H)

DIAGNOSIS

Body 3-4 mm long. Buccal cavity barrel-shaped, *ca* twice as long as broad. Dorsal and both subventral teeth of approximately equal size, located near middle of buccal cavity, their apices directed anteriorly. Pharyngo-intestinal junction non-tuberculate. Female genital system amphidelphic. Tail elongate conoid, 3-4 anal body diam. long. Caudal glands reduced, spinneret absent. Males with slender spicules, gubernaculum and 22-24 ventromedian supplements.

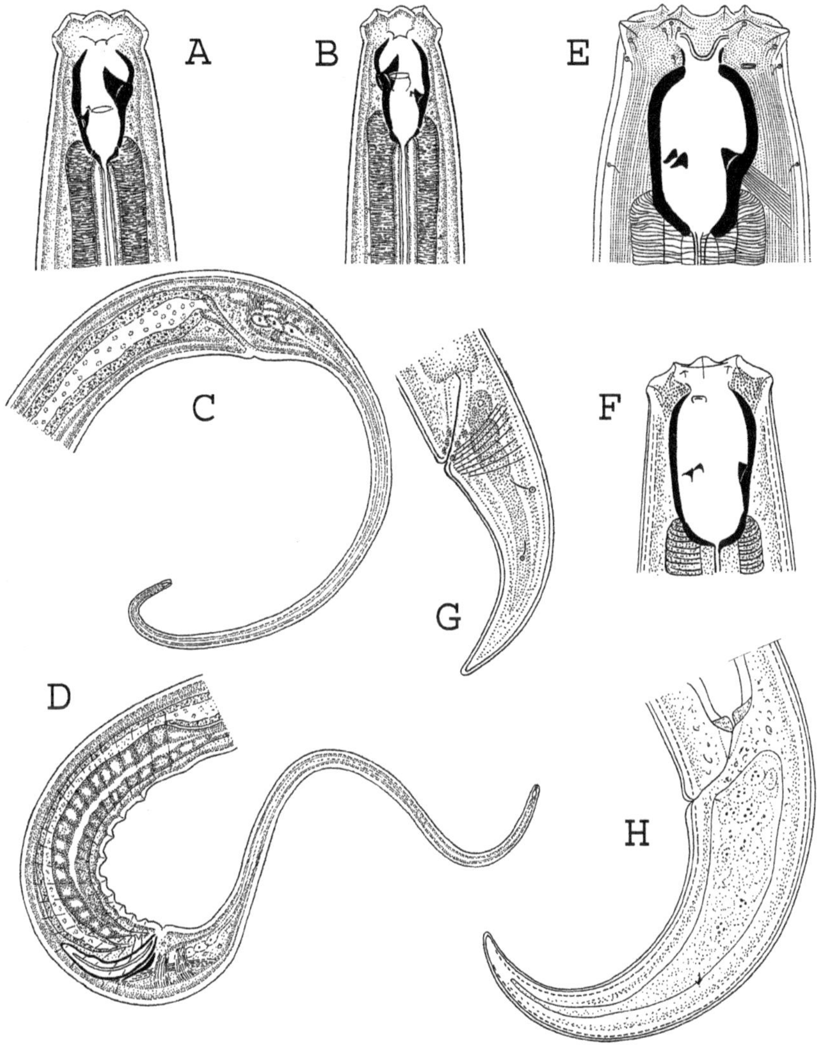

Fig. 39. *A-D:* Cobbonchulus longicaudatus *(Jairajpuri, Ahmad & Sturhan, 1998) Andrássy, 2009. A, B: Anterior region; C: Female posterior region; D: Male posterior region; E-H:* Comiconchus trionchus *(Thorne, 1924) Jairajpuri & Khan, 1982. E, F: Anterior region; G, H: Female tail. A-D after Jairajpuri* et al. *(1998); E, G after Thorne (1924); F, H after Mulvey (1962).*

TYPE SPECIES

Comiconchus trionchus (Thorne, 1924) Jairajpuri & Khan, 1982
= *Mononchus trionchus* Thorne, 1924
= *Mononchus* (*Iotonchus*) *trionchus* Thorne, 1924 (Micoletzky, 1927)
= *Miconchus trionchus* (Thorne, 1924) Andrássy, 1958

OTHER SPECIES

C. zduni Susulovsky, 2001

Comiconchus trionchus **(Thorne, 1924) Jairajpuri & Khan, 1982**
(Fig. 39E-H)

MEASUREMENTS

Female: L = 2.6-4.1 mm; a = 31-36; b = 3.9-5.1; c = 15-24.5; c′ = 3.5; V = 60-64.5.
Male: L = 2.4 mm; a = 30; b = 3.9; c = 26.

DESCRIPTION

Lip region offset by broad shallow depression with prominent labial and cephalic papillae. Amphidial apertures near base of lips. Buccal cavity 58-70 × 32-40 μm. Dorsal tooth large, located just posterior to middle of buccal cavity, apex at *ca* 45-50% of buccal cavity length from its base. Subventral teeth nearly of same size as dorsal, located at level of dorsal tooth. Female genital system amphidelphic. Male with long slender spicules 106 μm, gubernaculum 40 μm long and 24 ventromedian supplements. Tail elongate conoid, 3-4 anal body diam. long. Caudal glands reduced. Spinneret absent.

DISTRIBUTION

Canada, Russia, Ukraine, USA-Utah.

Key to species of *Comiconchus* Jairajpuri & Khan, 1982

1 – Buccal cavity with comparatively thin walls and small teeth; spicules relatively stout; gubernaculum with strong swelling; ♂:

L = 2.41-2.84 mm; a = 30-31; b = 4.1-4.3; c = 25-34; spicules 91-95 μm; ventromedian supplements 22; ♀: unknown (Ukraine) . . . . *zduni*

– Buccal cavity with thick walls and large teeth; spicules rather slender; gubernaculum without strong swelling; ♀: L = 2.6-4.1 mm; a = 29-36; b = 3.9-5.1; c = 15-27; c′ = 1.7-3.5; V = 60-72; ♂: L = 2.4 mm; a = 30; b = 3.9; c = 26; spicules 106 μm; ventromedian supplements 24 (USA-Utah, Russia, Canada, Ukraine) . . . *trionchus*

Genus *Tricaenonchus* Andrássy, 1996
(Fig. 38F, G)

DIAGNOSIS

Lip region offset. Buccal cavity oblong, armed with three, nearly equal, teeth of which dorsal tooth lying in posterior half of buccal cavity and subventral teeth in anterior half, far from former. Pharyngo-intestinal junction non-tuberculate. Female genital system amphidelphic; vulva post-equatorial. Tail conoid, ventrally arcuate, shorter than two anal body diam. Caudal glands and terminal spinneret absent. Male not known.

TYPE AND ONLY SPECIES

Tricaenonchus caucasicus Andrássy, 1996

Tricaenonchus caucasicus Andrássy, 1996
(Fig. 38F, G)

MEASUREMENTS

Female: L = 2.70-2.83 mm; a = 40-42; b = 4.3-4.4; c = 33-36; c′ = 1.6-1.7; V = 71-73.

DESCRIPTION

Lip region 42 μm diam., somewhat wider than adjoining body. Lips barely separate; labial papillae conoid. Amphidial apertures close to beginning of buccal cavity. Buccal cavity 46 × 22 μm, somewhat longer than labial diam., oval, moderately roomy with fairly thick and parallel walls. Dorsal and subventral teeth in unusual arrangement with dorsal tooth lying in posterior half of buccal cavity, both subventral teeth in anterior half and fairly far anteriad from former. Apex of dorsal tooth located at 45-49% and apices of subventral teeth at 80-82% of buccal cavity length from its base. Female genital system amphidelphic. Vulva a transverse slit; vagina well sclerotised. Advulval papillae absent. Tail 78-81 μm long, conoid, ventrally arcuate with rounded tip. Caudal glands and spinneret absent.

DISTRIBUTION

Russia-Caucasus.

Genus *Cobbonchulus* Andrássy, 2009
(Fig. 39A-D)

DIAGNOSIS

Body 1.5-1.6 mm long. Lip region distinctly offset. Buccal cavity elongate, distinctly narrow at base. Dorsal tooth large, massive, anteriorly directed, somewhat like *Mylonchulus*, its apex at 76-84% of buccal cavity length from its base. Subventral teeth distinctly smaller than dorsal with apex near middle of buccal cavity. Pharyngo-intestinal junction non-tuberculate. Female genital system amphidelphic. Tail long filiform. Caudal glands inconspicuous; terminal opening weak, non-cuticularised.

TYPE AND ONLY SPECIES

Cobbonchulus longicaudatus (Jairajpuri, Ahmad & Sturhan, 1998) Andrássy, 2009
= *Cobbonchus longicaudatus* Jairajpuri, Ahmad & Sturhan, 1998

Cobbonchulus longicaudatus (Jairajpuri, Ahmad & Sturhan, 1998) Andrássy, 2009
(Fig. 39A-D)

MEASUREMENTS

Female: L = 1.47-1.59 mm; a = 62-64; b = 4.2-4.4; c = 8.8-10; c′ = 8-8.2; V = 61-64.
Male: L = 1.52-1.65 mm; a = 63-69; b = 4.4-4.5; c = 8-10; c′ = 7-8.

DESCRIPTION

Body ventrally arcuate, tapering markedly in posterior region and with long filiform tail. Lip region clearly demarcated, 14-15 μm diam., 5 μm high. Position of amphidial aperture variable in two sexes, near middle of dorsal tooth in male and at level with tip of subventral teeth in female. Buccal cavity 17-19 $\times$ 9-10 μm, dorsal tooth massive, 5-6 μm long, anteriorly directed, its apex at 76-84% of buccal cavity length from its base. Subventral teeth distinctly smaller than dorsal, 2.0-2.5 μm long, their apices at 48-59% of buccal cavity length from its base. Pharyngo-intestinal junction non-tuberculate. Female genital system amphidelphic, both sexual branches almost equally developed. Vulval papillae present but weakly developed. Sphincter at oviduct-uterus junction well developed. Tail 7-8 anal body diam. long, gradually tapering, almost filiform, similar in both sexes. Caudal glands inconspicuous, arranged in tandem with terminal non-cuticularised opening. Spicules 29-32 μm long, gubernaculum 11-14 μm and lateral guiding pieces 6-8 μm. Supplements 9-11.

DISTRIBUTION

New Zealand.

Nematology Monographs & Perspectives, 2010, Vol. 7, 157-191

Chapter 7

Family Anatonchidae Jairajpuri, 1969

Jairajpuri (1969) proposed the family Anatonchidae for the genera *Anatonchus*, *Miconchus* and *Crassibucca*. Later, *Tigronchus* Kirjanova & Krall, 1969; *Tigronchoides* Ivanova & Dzhuraeva, 1971; *Paracrassibucca* Baqri & Jairajpuri, 1974 and *Tectonchus* Tsalolikhin, 1974 were added to this group. Andrássy (1976) proposed the new subfamily Miconchinae under the family Mononchidae for *Miconchus*, *Crassibucca* and *Paracrassibucca*, as well as for the genus *Cobbonchus*. Under Anatonchidae, Andrássy (1976) left only the genera *Anatonchus*, *Tigronchus* and *Tigronchoides*, and further suggested the possible congeneric status of the latter two with *Anatonchus* – the so-called 'tiger-teeth' in *Tigronchus* were nothing but simple splits between the mural plates (interparietalia) of the buccal capsule at their point of junction. Jairajpuri and Khan (1982) accepted two families, Anatonchidae Jairajpuri, 1969 and Iotonchidae Jairajpuri, 1969, with the subfamilies Anatonchinae Jairajpuri, 1969 and Miconchinae Andrássy, 1976 under Anatonchidae, and the subfamilies Iotonchinae Jairajpuri, 1969 and Hadronchinae Khan & Jairajpuri, 1980 under Iotonchidae.

Siddiqi (1984a) accepted Andrássy's view, synonymised both *Tigronchus* and *Tigronchoides* with *Anatonchus* and proposed a new genus, *Truxonchus* Siddiqi, 1984a. Siddiqi (1984b) also proposed *Nullonchus* Siddiqi, 1984b, an interesting new genus of toothless mononchs, and also made comments on the phylogenetic relationships among mononchs. He recognised Anatonchidae as a family under Mononchoidea with three subfamilies, Anatonchinae, Iotonchinae and Nullonchinae. Andrássy (1993b) did not accept Nullonchinae, and recognised the three subfamilies, Anatonchinae, Iotonchinae and Miconchinae under Anatonchidae. He also revalidated the genus *Tigronchoides* and accepted three genera, *viz.*, *Anatonchus*, *Truxonchus* and *Tigronchoides* under Anatonchinae. The genus *Nullonchus* was placed in the subfamily Iotonchinae. We are also of the opinion that *Nullonchus* fits better with the genus *Iotonchus* and should be placed with this group. In the present account the classification of Jairajpuri and Khan (1982) is largely followed.

DIAGNOSIS

Buccal cavity broad and flattened at base. Dorsal and subventral teeth almost equal in size and similar in position, their apices pointing anteriad or posteriad. Pharyngo-intestinal junction tuberculate. Tail conoid or long cylindroid, may be filiform. Caudal glands poorly developed, may be absent.

TYPE SUBFAMILY

Anatonchinae Jairajpuri, 1969

OTHER SUBFAMILY

Miconchinae Andrássy, 1976

Key to subfamilies of Anatonchidae Jairajpuri, 1969

1 – Subventral teeth retrorse; dorsal tooth also usually retrorse . . . . **Anatonchinae**
– Subventral teeth and dorsal tooth anteriorly directed . . . . **Miconchinae**

Subfamily Anatonchinae Jairajpuri, 1969

DIAGNOSIS

Buccal cavity roomy. Teeth of equal size and shape, located in anterior, middle or posterior half of buccal cavity. Dorsal tooth retrorse, rarely anteriorly directed (*Micatonchus*), subventral teeth always retrorse.

TYPE GENUS

Anatonchus Cobb, 1916
= *Tigronchus* Kirjanova & Krall, 1969

OTHER GENERA

Tigronchoides Ivanova & Dzhuraeva, 1971
Truxonchus Siddiqi, 1984a
Micatonchus Jairajpuri, Tahseen & Choi, 2001

Key to genera of Anatonchinae Jairajpuri, 1969

1 – Dorsal tooth retrorse 2
– Dorsal tooth anteriorly directed . . . . *Micatonchus*

2 – Teeth located posterior to middle of buccal cavity . . . . . *Truxonchus*
– Teeth lying midway or farther anterior in buccal cavity 3

3 – Teeth, especially in female, located in anterior end of buccal cavity; dorsal wall of buccal cavity becoming thinner and somewhat angular in mid-region . . . . *Tigronchoides*
– Teeth located midway in buccal cavity; dorsal wall of buccal cavity not becoming conspicuously thinner and angular in mid-region . . . . *Anatonchus*

Genus *Anatonchus* Cobb, 1916

= *Tigronchus* Kirjanova & Krall, 1969
(Figs 40; 41A-E)

DIAGNOSIS

Body 1.5-4.1 mm long. Cuticle smooth or occasionally finely annulated. Buccal cavity either oblong or spheroid, in latter case very roomy. Teeth retrorse, of equal shape and level, located at mid-region of buccal cavity, hinging by haft on buccal walls and ending in arrow-like tips. Pharyngo-intestinal junction tuberculate, tubercles especially well developed. Female genital system amphidelphic. Advulval papillae generally present. Spicules long and slender, gubernaculum simple, lateral accessory pieces present or absent. Tail elongate to filiform, 2-14 anal body diam. long, similar in both sexes. Caudal glands and spinneret usually present.

TYPE SPECIES

Anatonchus tridentatus (de Man, 1876) De Coninck, 1939
= *Mononchus tridentatus* de Man, 1876

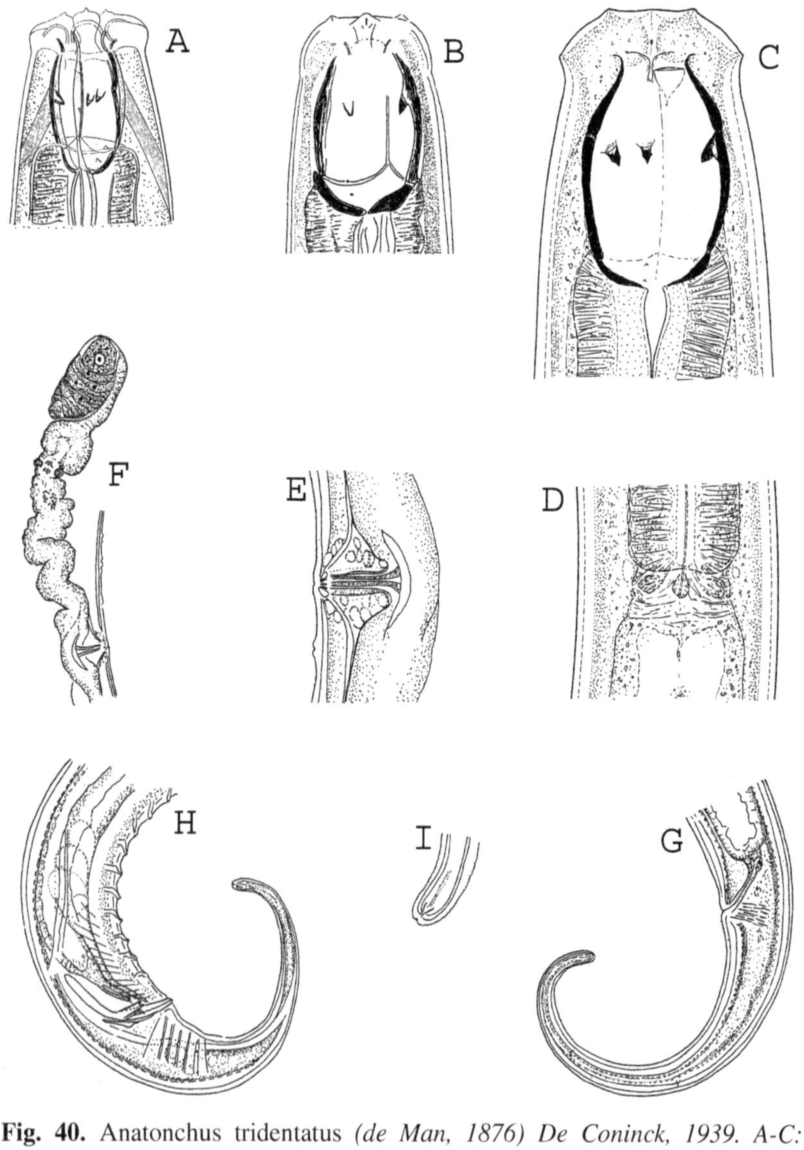

Fig. 40. Anatonchus tridentatus *(de Man, 1876) De Coninck, 1939. A-C: Anterior region; D: Pharyngo-intestinal junction; E: Vulval region; F: Anterior female genital system; G: Female tail; H: Male tail; I: Tail tip. A after Cobb (1917); B, E-I after Peneva* et al. *(1999); C, D after Andrássy (1993).*

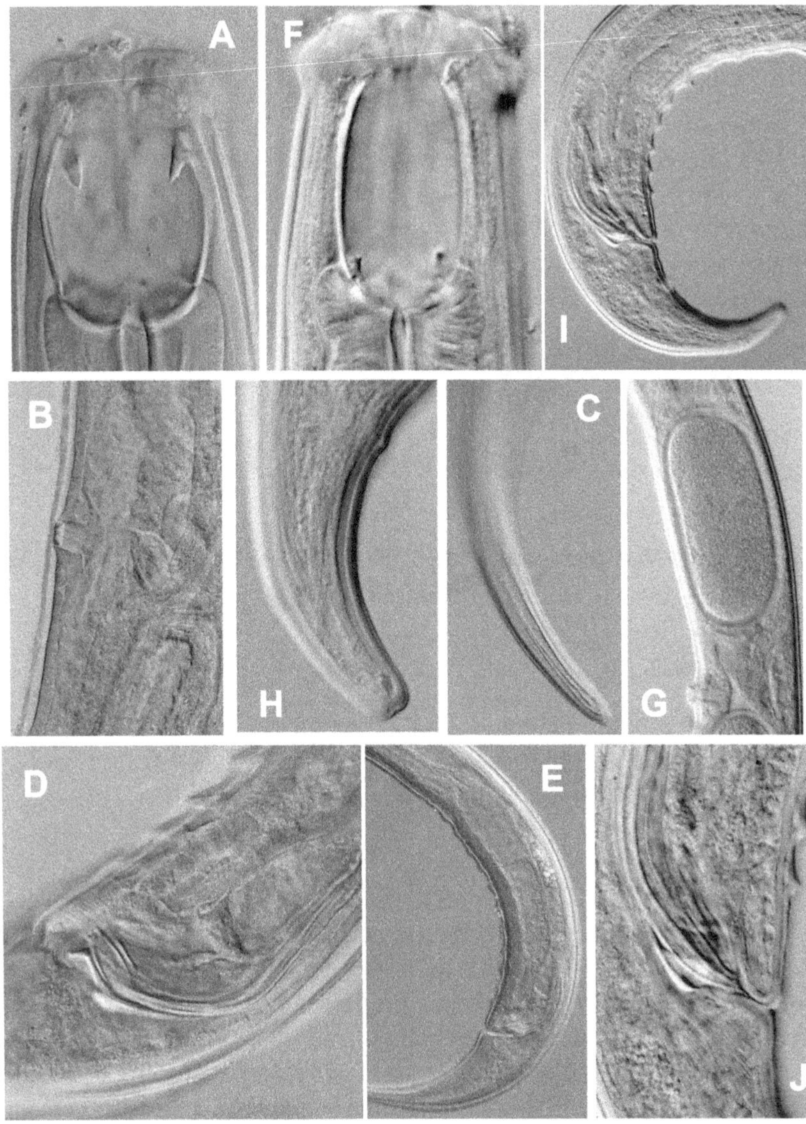

Fig. 41. *A-E:* Anatonchus *sp. A: Anterior region; B: Vulval region; C: Female tail terminus; D: Male cloacal region; E: Male posterior region; F-J:* Miconchus studeri *(Steiner, 1914) Andrássy, 1958. F: Anterior region; G: Female genital system; H: Female tail; I: Male tail region; J: Male cloacal region.*

OTHER SPECIES

A. acutus Altherr, 1974
A. ferox Siddiqi, 1984c
A. hortensis Andrássy, 1973
A. kafii Olia, Choudhury, Ahmad & Jairajpuri, 2004
A. kashmirensis Jairajpuri & Khan, 1982
A. mamillatus Altherr, 1968
A. narsikulovi Dzhuraeva & Kankina, 1997
A. sympathicus Andrássy, 1993b

SPECIES INQUIRENDAE

A. kreisi Meyl, 1961
Tigronchus taurinus Kirjanova & Krall, 1969

Anatonchus tridentatus **(de Man, 1876) De Coninck, 1939**
(Fig. 40)

MEASUREMENTS

Female: L = 2.8-3.6 mm; a = 27-40; b = 4.5-5.2; c = 8-12; c′ = 5-8; V = 60-65.
Male: L = 2.5-2.7 mm; a = 30-40; b = 4-5; c = 8-12.

DESCRIPTION

Lip region distinctly offset. Buccal cavity comparatively thin-walled, barrel-shaped, 42-63 × 31-45 μm or *ca* 1.5 times as long as wide with three medially placed retrorse teeth. Proparietale long, almost two-thirds length of interparietale. Teeth rooted on anterior ends of interparietalia, *i.e.*, located in first half of buccal cavity, their apices at 56-58% of buccal cavity from its base. Pharyngo-intestinal junction tuberculate. Female genital system amphidelphic. Sphincter between uterus and oviduct strongly developed. Cuticle in region of vulva with few irregularly spaced fine granular papillae. Male spicules 83-110 μm long, stout with bifid tip, 15-17 midventral supplements. Tail elongate, ventrally bent, 180-430 μm long in female and 130-320 μm in male, with small caudal glands and terminal spinneret.

REMARKS

In *A. tridentatus*, the teeth of the first-stage juveniles are anteriorly directed, becoming retrorse during the first moult but do not assume their normal position until the last moult. In *Truxonchus dolichurus* (=*A. dolichurus*) the three teeth of the juvenile stages point anteriad until the final moult.

DISTRIBUTION

Austria, Belgium, Bohemia, Estonia, France, Georgia, Germany, Ghana, Hungary, Iran, Italy, Mexico, The Netherlands, Poland, Romania, Russia, Slovakia, Spain, Switzerland, Turkey, USA, UK, Uzbekistan, former Yugoslavia.

Key to species of *Anatonchus* Cobb, 1916

1 – Tail 10-14 anal body diam. long; teeth with unusually long hafts; ♀: L = 3.6-4.1 mm; a = 45-53; b = 4.6-5.1; c = 4.9-7.6; c′ = 10-14; V = 58-65; ♂: L = 3.2 mm; a = 50; b = 4.7; c = 5-8; supplements 14 (Fiji Islands) . . . . *ferox*

– Tail 2-8 anal body diam. long; teeth with short hafts 2

2 – Bigger animals, 2.5-3.6 mm; eggs spherical, generally 6-8 in uterus; ♀: L = 2.5-3.6 mm; a = 23-40; b = 4.1-5.2; c = 8-12; c′ = 5-8; V = 60-70; ♂: L = 2.4-2.8 mm; a = 27-40; b = 4-5; c = 8-14; supplements 15-17 (The Netherlands, Belgium, Germany, UK, Poland, Switzerland, Austria, Bohemia, Slovakia, Hungary, Romania, former Yugoslavia, Spain, France, Italy, Estonia, Russia, Turkey, Georgia, Uzbekistan, Iran, Ghana, USA, Mexico, Bulgaria) . . . . *tridentatus*

– Smaller animals, 1.5-2.3 mm; eggs oblong, 1-2 in uterus 3

3 – Tail very sharp at tip, without spinneret; ♀ unknown; ♂: L = 1.6 mm; a = 34; b = 4.2; c = 9.6; c′ = 4; supplements 11 (Germany) . . . . *acutus*

– Tail finely rounded at tip, with spinneret 4

4 – Buccal cavity oblong, much longer than wide 5

– Buccal cavity spherical, *ca* as wide as long 7

5 – Female tail 2.5-4 anal body diam. long, male tail conoid, uniformly tapering; vulva transverse; ♀: L = 1.5-2.0 mm; a = 25-37; b = 3.8-

4.8; c = 11-20; c′ = 2.5-4.0; V = 63-71; ♂: L = 1.5-1.99 mm; a = 28-36; b = 3.7-4.6; c = 12-20; supplements 10-14 (Hungary, Romania, former Yugoslavia, France, Bulgaria) . . . . *hortensis*

– Female tail 5-6 anal body diam. long 6

6 – Vulva longitudinal; vaginal sclerotisation in two pieces; tail strongly ventrally arcuate; ♀: L = 2.0-2.3 mm; a = 28-33; b = 4.2-4.5; c = 8.7-10; c′ = 5.2-5.6; V = 62-63; ♂: L = 1.9-2.4 mm; a = 30-37; b = 4.2-4.5; c = 10-12; supplements 11-13 (Germany, Hungary) . . . . *sympathicus*

– Vulva transverse; vaginal sclerotisation in three pieces; tail slightly ventrally curved; ♀: L = 2.3-2.7 mm; a = 31-34; b = 4.2-4.4; c = 10.2-10.5; c′ = 5.0-5.5; V = 65; ♂: unknown . . . . *kafii*

7 – Tail 300 μm or 6-7 anal body diam. long; ♀: L = 2.1 mm; a = 27; b = 4.5; c = 6.7; c′ = 6-7; V = 62; ♂: L = 2.1 mm; a = 35; b = 4.6; c = 11.5; supplements 14 (Germany) . . . . *mamillatus*

– Tail 180 μm or 4-5 anal body diam. long; ♀: L = 1.8 mm; a = 30; b = 3.6; c = 11; c′ = 4.5; V = 67; ♂: unknown (India) . . . . *kashmirensis*

Note

Not included in key because of unavailability of literature: *A. narsikulovi*.

Genus *Tigronchoides* Ivanova & Dzhuraeva, 1971
(Fig. 42)

DIAGNOSIS

Medium-sized to large body, 1.7-5 mm long. Cuticle smooth, especially finely annulated on tail. Buccal cavity roomy to very roomy, often nearly globular, dorsal wall (interparietale) transversely striated, becoming thinner in mid-region and bulging. Teeth retrorse, equal in shape and location, in female lying at anterior end of buccal cavity and hinged by conspicuous hafts on walls, in male located somewhat more posteriorly and having shorter hafts. Apices of teeth arrow-like. Pharyngo-intestinal junction tuberculate. Female genital system amphidelphic or mono-prodelphic, vulva at 59-77% of body length, often provided with advulval papillae. Spicules with forked accessory pieces. Supplements

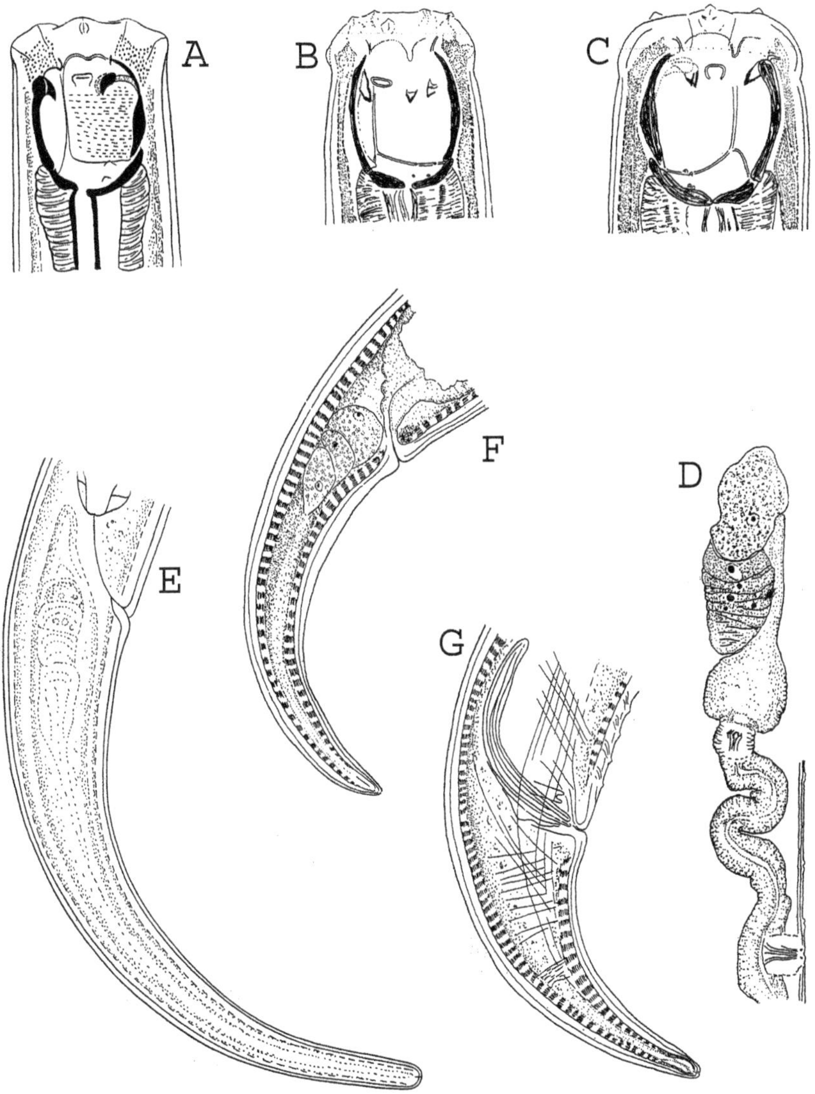

Fig. 42. Tigronchoides ginglymodontus *(Mulvey, 1961) Andrássy, 1993. A-C: Anterior region; D: Anterior female genital system; E, F: Female tail; G: Male tail. A, E after Mulvey (1961); B-D, F, G after Peneva* et al. *(1999).*

9-18. Tail of both sexes similar in shape, conoid to filiform, 3-20 anal body diam. long. Caudal glands and spinneret mostly present.

TYPE SPECIES

Tigronchoides ginglymodontus (Mulvey, 1961b) Andrássy, 1993b
=*Anatonchus ginglymodontus* Mulvey, 1961b
=*Anatonchus killickae* Clark, 1963
=*Tigronchoides varidentus* Ivanova & Dzhuraeva, 1971
=*Anatonchus varidentus* (Ivanova & Dzhuraeva, 1971) Siddiqi, 1984a

OTHER SPECIES

T. amiciae (Coomans & Lima, 1965) Andrássy, 1993b
=*Anatonchus amiciae* Coomans & Lima, 1965
T. australicus (Winiszewska-Slipińska, 1989) Andrássy, 1993b
=*Anatonchus australicus* Winiszewska-Slipińska, 1989
T. filicaudatus (Altherr, 1971) Andrássy, 1993b
=*Anatonchus filicaudatus* Altherr, 1971
T. istvani (Winiszewska-Slipińska, 1989) Andrássy, 1993b
=*Anatonchus istvani* Winiszewska-Slipińska, 1989
T. monohystera (Altherr, 1977) Andrássy, 1993b
=*Anatonchus monohystera* Altherr, 1977
T. quercus Peneva, Neilson & Nedelchev, 1999
T. sukuli (Baqri, Das & Ahmad, 1981) Andrássy, 1993b
=*Anatonchus sukuli* Baqri, Das & Ahmad, 1981

Tigronchoides ginglymodontus (Mulvey, 1961b) Andrássy, 1993b
(Fig. 42)

MEASUREMENTS

Female: L = 1.7-3.2 mm; a = 20-42; b = 3.7-5.2; c = 8-17; c′ = 3.5-6; V = 59-72.
Male: L = 2.5-3.4 mm; a = 32-44; b = 4.0-5.1; c = 11-19.

DESCRIPTION

Lip region slightly offset and slightly wider than adjoining body, 45-58 μm broad and 14-22 μm high. Amphids funnel-shaped, apertures 7-8 μm wide at level of anterior end of buccal cavity. Buccal cavity barrel-shaped, nearly as broad as long, 42-62 × 40-53 μm; walls of buccal cavity anterior to base of teeth (praeparietale) weakly developed. Three medium-sized teeth hinged to anterior wall of buccal cavity, 67-79% from base of buccal cavity. Pharyngo-intestinal junction tuberculate. Vulva a transverse slit. Pre-vulval papillae 0-4, 0-7 post-vulval papillae. Vagina not sclerotised distally. Female genital system amphidelphic. Male buccal cavity 50-54 × 40-42 μm. Spicules 100-110 μm long, arcuate; gubernaculum 27.5-35 μm long, lateral accessory pieces 17 μm. Supplements 12-17, regularly spaced. Tail elongate conoid, ventrally arcuate, 160-307 μm long in female and 190 μm long in male. Caudal glands three, leading through a terminal duct.

DISTRIBUTION

Bulgaria, New Zealand, Poland, Romania, Spain, Tadzhikistan, USA, former Yugoslavia.

Key to species of *Tigronchoides* Ivanova & Dzhuraeva, 1971

1 – Female monodelphic with post-uterine sac 2
– Female amphidelphic 3

2 – Buccal cavity wider than long; tail straight; body *ca* 3 mm long; ♀: L = 3.1 mm; a = 40; b = 4; c = 22; c′ = 3; V = 75; ♂: unknown (Brazil) . . . . *monohystera*
– Buccal cavity longer than wide; tail arcuate; body *ca* 2 mm long; ♀: L = 1.7-2.0 mm; a = 30-35; b = 3.9-4.3; c = 12-16; c′ = 3-4; V = 73-77; ♂: L = 1.7-2.0 mm; a = 32-38; b = 4.1-4.7; c = 14-16; supplements 13-17 (former Yugoslavia, Italy) . . . . *amiciae*

3 – Large species, 4-5 mm long 4
– Smaller species, 2-3 mm long 5

4 – Tail filiform, 12-20 anal body diam. long; spicules 80-90 μm long; ♀: L = 3.1-4.8 mm; a = 40-50; b = 5.0-5.7; c = 4-5; c′ = 12-20; V = 60-64; ♂: L = 3.3-4.6 mm; a = 42-56; b = 4.4-5.6; c = 5.0-7.6; supplements 13-14 (Romania) . . . . *filicaudatus*
– Tail shorter, 7-8 anal body diam. long; spicules 140-150 μm long; ♀: L = 4.9 mm; a = 47; b = 5; c = 8; c′ = 7-8; V = 64; ♂: L = 4.4-4.7 mm; a = 43-46; b = 5.1-5.2; c = 12; supplements 14-15 (India) . . . . *sukuli*

5 – Advulval papillae and caudal spinneret lacking; tail strongly curved, hook-like; ♀: L = 2.0 mm; a = 35; b = 5.7; c = 7.4; c′ = 7; V = 61; ♂: L = 1.7-1.9 mm; a = 36-40; b = 4.5-5.4; c = 12-13; supplements 9-12 (Australia) .... *australicus*
– Advulval papillae and caudal spinneret present; tail not hook-like 6

6 – Teeth hanging from anterior end of buccal capsule; ♀: L = 1.7-3.2 mm; a = 20-42; b = 3.7-5.2; c = 8-17; c′ = 3.5-6.0; V = 59-72; ♂: L = 2.5-3.2 mm; a = 32-44; b = 4.0-5.1; c = 11-19; supplements 12-17 (Poland, Romania, former Yugoslavia, Spain, Tadzhikistan, USA, New Zealand, Bulgaria) ...... *ginglymodontus*
– Teeth located in anterior fourth of buccal capsule 7

7 – Buccal cavity 67 μm long in female, 51-61 μm in male; tail 184 μm long in female, 120-148 μm in male; ♀: L = 2.8 mm; a = 33; b = 4.7; c = 15; c′ = 3.4; V = 72; ♂: L = 2.1-2.6 mm; a = 27-33; b = 4.5-4.9; c = 17-19; supplements 16-18 (Poland)....... *istvani*
– Buccal cavity 51-58 μm long in female, 42-48 μm in male; tail 198-267 μm long in female, 161-204 μm in male; ♀: L = 2.4-2.9 mm; a = 25-30; b = 4.3-4.5; c = 10.4-12.4; c′ = 3.37-4.9; V = 61-66; ♂: L = 2.1-2.6 mm; a = 22-30; b = 3.9-4.6; c = 12.7-14.2; c′ = 2.9-3.3; supplements 12-15 (Bulgaria) .... *quercus*

Genus *Truxonchus* Siddiqi, 1984a
(Fig. 43A-D)

DIAGNOSIS

Body 2.0-6.5 mm long. Cuticle smooth. Buccal cavity either oblong, *ca* half as wide as long, or very broad and nearly as wide as long. One dorsal and two subventral retrorse teeth, equal in shape and position, located posterior to middle of buccal cavity (originating on or near posterior ends of interparietalia). Pharyngo-intestinal junction tuberculate. Female genital system amphidelphic. Vulva at 49-68% of body length. Spicules fairly massive, arcuate, with forked lateral pieces. Supplements 12-20 in number. Tail in both sexes similar, arcuate-conoid, exceptionally straight, 2-14 anal body diam. long. Caudal glands and spinneret present or absent.

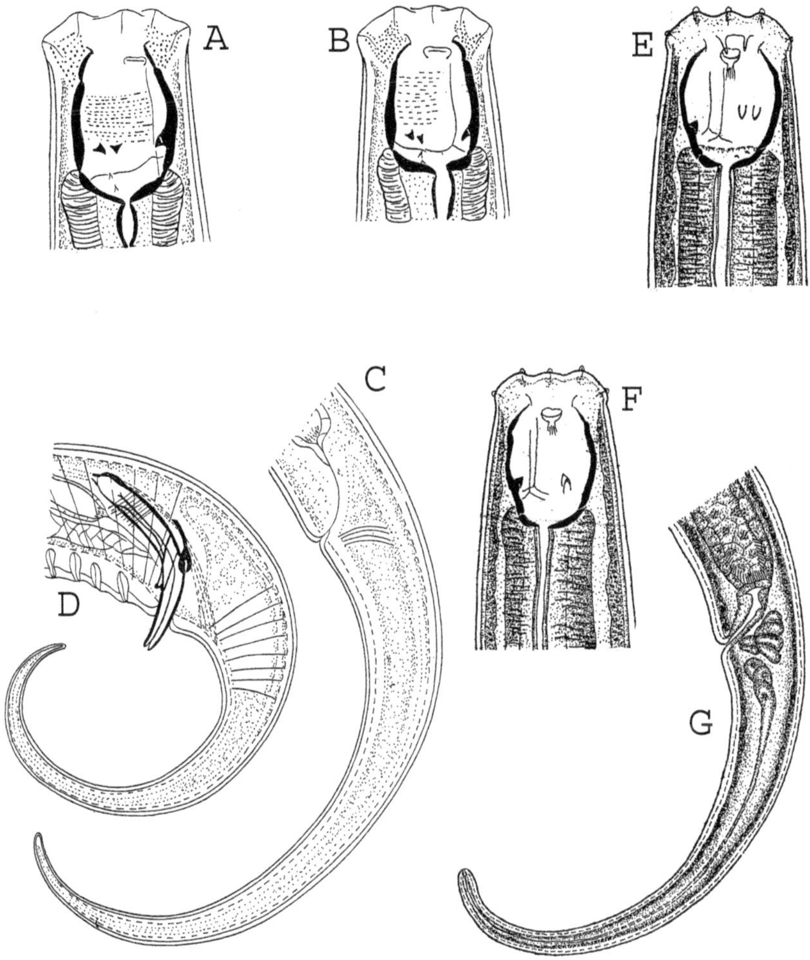

Fig. 43. *A-D:* Truxonchus subacutus *(Mulvey, 1961) Siddiqi, 1984. A, B: Anterior region; C: Female tail; D: Male tail; E-G:* Micatonchus reversus *Jairajpuri* et al., *2001. E, F: Anterior region; G: Female tail. A-D after Mulvey (1961); E-G after Jairajpuri* et al. *(2001).*

TYPE SPECIES

Truxonchus subacutus (Mulvey, 1961b) Siddiqi, 1984a
= *Anatonchus subacutus* Mulvey, 1961b

OTHER SPECIES

T. alleni (Mulvey, 1961b) Siddiqi, 1984a
=*Anatonchus alleni* Mulvey, 1961b
T. dolichurus (Ditlevsen, 1911) Siddiqi, 1984a
=*Mononchus dolichurus* Ditlevsen, 1911
=*Mononchus* (*Anatonchus*) *dolichurus* Ditlevsen, 1911 (Cobb, 1916)
=*Anatonchus dolichurus* (Ditlevsen, 1911) Cobb, 1916
T. genovi (Peneva, Neilson & Nedelchev, 1999) **comb. n.**
=*Anatonchus genovi* Peneva, Neilson & Nedelchev, 1999
T. gracilicaudatus (Cobb, 1917) Andrássy, 1993b
=*Mononchus* (*Anatonchus*) *gracilicaudatus* Cobb, 1917
=*Anatonchus gracilicaudatus* Cobb, 1917
T. mulveyi (Altherr, 1968) Siddiqi, 1984a
=*Anatonchus mulveyi* Altherr, 1968
T. paralleni (Jairajpuri & Khan, 1982) Siddiqi, 1984a
=*Anatonchus* sp. *apud* Mulvey, 1961b
=*Truxonchus confusus* Siddiqi, 1984a
T. penevae **sp. n.**
=*Anatonchus* sp. *apud* Peneva, Neilson & Nedelchev, 1999

***Truxonchus subacutus* (Mulvey, 1961b) Siddiqi, 1984a**
(Fig. 43A-D)

MEASUREMENTS

Female: L = 2.5-2.9 mm; a = 36-46; b = 4.4-5.0; c = 7.6-9.3; c′ = 7-8; V = 60-66.
Male: L = 2.1-3.0 mm; a = 33-50; b = 4.2-5.6; c = 8.1-10.8.

DESCRIPTION

Lip region offset by slight depression. Amphidial apertures at level of anterior end of buccal cavity. Buccal cavity 48-52 × 36-40 μm in female and 42-50 × 31-38 μm in male, walls moderately thick and distinctly striated. Three small, equal-sized, teeth located in posterior third of buccal cavity. Pharyngo-intestinal junction tuberculate. Female genital system amphidelphic. Male with 80-90 μm long spicules, bifid accessory pieces and gubernaculum. Supplements 12-13. Tail conoid, arcuate with minutely rounded terminus, 280-340 μm long in female and 230-290 μm in male. Caudal glands and spinneret absent.

DISTRIBUTION

USA-California; Oregon.

Key to species of *Truxonchus* Siddiqi, 1984a

1 – Tail cylindroid with bluntly rounded tip, 2-2.5 anal body diam. long; ♀ (juv.): L = 3.2 mm; a = 39; b = 3.4; c = 27; c′ = 2-2.5; V = 68; ♂: unknown . . . . *mulveyi*

– Tail elongate conoid with sharp or finely rounded tip, 7-14 anal body diam. long 2

2 – Buccal cavity oblong, twice as long as wide; ♀: L = 5.8-6.5 mm; a = 36-43; b = 4.6-5.0; c = 4.3-5.4; c′ = 12-14; V = 49-55; ♂: L = 5.9-6.3 mm; a = 40-50; b = 4.5-5.0; c = 5.5-5.9; supplements 20 (The Netherlands, Denmark, Germany, UK, Poland, Norway, Faeroe Islands, Switzerland, Austria, Bohemia, Slovakia, Spain, USA, Canada) . . . . *dolichurus*

– Buccal cavity nearly globular, as long as or barely longer than wide 3

3 – Large nematodes, 4-6 mm; tail 10-12 anal body diam. long 4

– Smaller nematodes, 2-3.5 mm; tail < 8 anal body diam. long 5

4 – Caudal spinneret subventral; body 4.8-6.1 mm; teeth large; ♀: L = 4.8-6.0 mm; a = 50-71; b = 4.3-6.5; c = 8-12; c′ = 11-12; V = 60-67; ♂: L = 4.8-6.1 mm; a = 56-78; b = 4.8-5.9; c = 6.6-10.8; supplements 17-19 (USA-California; Oregon; Virginia) . . . . *gracilicaudatus*

– Caudal spinneret terminal; body 3.8-4.0 mm long; teeth comparatively small; ♀: L = 4.0-4.8 mm; a = 42-51; b = 4.8-5.2; c = 7.3-11; c′ = 10; V = 53-67; ♂: L = 3.8-4.6 mm; a = 42-51; b = 4.7-5.2; c = 8.3-10; supplements 15-16 (USA-California) . . . . *paralleni*

5 – Tail 7-8 anal body diam. long 6

– Tail 3-4 anal body diam. long 7

6 – Tail tip sharp, subacute, without spinneret; ♀: L = 2.5-3.5 mm; a = 36-46; b = 4.2-5.6; c = 8-13; c′ = 7-8; V = 60-66; ♂:

L = 2.1-3.2 mm; a = 33-50; b = 4.2-5.6; c = 8-13; supplements 12-19 (USA-California; Oregon) . . . . *subacutus*

– Tail tip finely rounded, with spinneret; ♀: L = 2.5-3.5 mm; a = 33-50; b = 4.4-5.5; c = 6.7-11; c′ = 8; V = 60-66; ♂: L = 2.7-3.0 mm; a = 37-48; b = 4.7-5.5; c = 8.5-10; supplements 12-17 (Georgia, USA-California) . . . . *alleni*

7 – Advulval papillae present; vagina with sclerotised pieces; ♀: L = 2.4-3.2 mm; a = 26-39; b = 4.0-4.4; c = 13-17; c′ = 3.0-4.4; V = 65-71; ♂: unknown (Bulgaria) . . . . *genovi*

– Advulval papillae absent; vagina without sclerotised pieces; ♀: L = 3.6 mm; a = 30; b = 4.8; c = 13.6; c′ = 4.2; V = 65; ♂: unknown (Bulgaria) . . . . *penevae*

Genus *Micatonchus* Jairajpuri, Tahseen & Choi, 2001a
(Fig. 43E-G)

DIAGNOSIS

Body large. Labial papillae prominent. Buccal cavity barrel-shaped, slightly longer than wide. Dorsal tooth triangular, situated in posterior half of buccal cavity with apex directed anteriad. Subventral teeth slightly longer, claw-like, retrorse, placed a little anterior to dorsal tooth but tips of all three teeth almost at same level. Pharyngo-intestinal junction prominently tuberculate. Female genital system amphidelphic. Males known for only *M. koreanus*, having arcuate spicules, bifid lateral guiding pieces and 13-15 ventromedian supplements. Tail long, cylindroid-conoid with rounded terminus. Caudal glands and spinneret present.

TYPE SPECIES

Micatonchus reversus Jairajpuri, Tahseen & Choi, 2001a

OTHER SPECIES

M. koreanus Khan, Lee, Choi & Park, 2003
M. ventralis Khan, Lee, Choi & Park, 2003

Micatonchus reversus Jairajpuri, Tahseen & Choi, 2001a
(Fig. 43E-G)

MEASUREMENTS

Female: L = 2.18-2.20 mm; a = 26-27; b = 4.2; c = 9; c′ = 5.5-5.9; V = 65-67.

DESCRIPTION

Lip region slightly offset, 47-52 × 15 μm. Amphids cup-shaped, apertures near anterior end of buccal cavity. Buccal cavity fairly broad, barrel-shaped, 45-47 × 35-40 μm, strongly sclerotised. Dorsal wall bearing small teeth near its base with apices directed anteriad. Subventral teeth retrorse, claw-like, with tip lying at level of dorsal tooth apex. Pharyngo-intestinal junction tuberculate. Female genital system amphidelphic, anterior genital branch slightly longer than posterior but both functional with reflexed ovary. Vulva a small, transverse, slit, *pars refringens vaginae* present. Advulval papillae absent. Tail 237-245 μm long, ventrally arcuate, gradually tapering to a rounded terminus. Caudal glands well developed, with prominent terminal opening. Male not known.

DISTRIBUTION

Korea.

REMARKS

Khan *et al.* (2003) added two more species to this rather unusual genus from Korea. Both are very similar to *M. reversus* and may even be conspecific. Males are known for *M. koreanus*, having arcuate spicules measuring 86-90 μm, bifid lateral guiding pieces and 13-15 ventromedian supplements. Khan *et al.* (2003) differentiated *M. koreanus* from *M. reversus* in having vulval flaps (Figs 1, 2 of Khan *et al.* do not show any such flap); vulval papillae and presence of cuticularised pieces in vagina (*M. reversus* also has cuticularised pieces in vagina), the only difference being the presence of vulval papillae. Since Jairajpuri *et al.* (2001) described *M. reversus* based on a single mature female and a young female and, as it is also quite common in species with vulval papillae to sometimes find individuals lacking such structures, the differentiating characters between *M. reversus* and *M. koreanus* are no longer valid and *M. koreanus* is probably a synonym of *M. reversus*.

Key to species of *Micatonchus* Jairajpuri, Tahseen & Choi, 2001a

1 – Vulval papillae present; buccal cavity 52-55 × 35-39 μm; ♀: L = 2.4-2.7 mm; a = 36-39; b = 4.1-4.5; c = 7.9-8.6; c′ = 6.6-7.1; V = 62-66; ♂: L = 2.1-2.4 mm; a = 33-37; b = 4.1-4.3; c = 7.8-9.5; c′ = 4.4-6.0; supplements 13-15 (Korea) . . . . *koreanus*
– Vulval papillae absent 2

2 – Spinneret terminal; ♀: L = 2.18-2.20 mm; a = 26-27; b = 4.2; c = 9; c′ = 5.5-5.9; V = 65-67; ♂: unknown . . . . *reversus*
– Spinneret subterminal; ♀: L = 2.2-2.5 mm; a = 32-34; b = 4.1-4.3; c = 8.5-9.3; c′ = 5.4-5.7; V = 63-65; ♂: unknown . . . . . . *ventralis*

Subfamily Miconchinae Andrássy, 1976

DIAGNOSIS

Buccal cavity roomy. Dorsal tooth located in anterior or posterior part of buccal cavity, directed anteriad. Subventral walls also with teeth, equal in size to dorsal tooth or somewhat smaller, arranged in one pair or two pairs, located anteriorly and/or posteriorly in buccal cavity.

TYPE GENUS

Miconchus Andrássy, 1958
= *Miconchoides* Jairajpuri & Khan, 1982

OTHER GENERA

Crassibucca Mulvey & Jensen, 1967
Paracrassibucca Baqri & Jairajpuri, 1974
Promiconchus Jairajpuri & Khan, 1982
Doronchus Andrássy, 1993b

Key to genera of Miconchinae Andrássy, 1976

1 – Buccal cavity with one dorsal and two subventral teeth 2
– Buccal cavity with one dorsal and four subventral teeth 4

2 – Dorsal tooth in anterior, subventral teeth in posterior part of buccal cavity . . . . *Promiconchus*
– All teeth in posterior part of buccal cavity, level with each other . . 3

3 – Tail showing sexual dimorphism; in female elongate-conoid with sharp tip, in male much shorter and plumper, with bluntly rounded tip . . . . *Doronchus*
– Tail without sexual dimorphism, similar in both sexes . . *Miconchus*

4 – Dorsal tooth anterior, one pair of subventral teeth at level with it, second pair located more posterior in buccal cavity; female prodelphic . . . . *Crassibucca*
– Dorsal tooth anterior, both pairs of subventral teeth located more posterior in buccal cavity; female amphidelphic . . . *Paracrassibucca*

Genus *Miconchus* Andrássy, 1958

= *Miconchoides* Jairajpuri & Khan, 1982
(Figs 41F-J; 44A-E)

DIAGNOSIS

Body 1-7 mm, generally 2-3 mm, long. Cuticle smooth. Buccal cavity barrel-shaped, flattened at base. Dorsal and two subventral teeth approximately equal in size and position, directed anteriorly and located nearly basal or midway in buccal cavity. No denticles or ridges present. Pharyngo-intestinal junction tuberculate. Female genital system predominantly amphidelphic, rarely monodelphic (either prodelphic or pseudo-prodelphic). Male known in only two-thirds of species. Spicules arcuate with lateral guiding pieces and gubernaculum. Copulatory supplements 9-24. Tail of both sexes similar in shape, more or less conoid, occasionally filiform, with pointed or rounded tip, 2-25 anal body diam. long. Caudal glands and spinneret either well developed or reduced.

TYPE SPECIES

Miconchus digiturus (Cobb, 1893) Andrássy, 1958

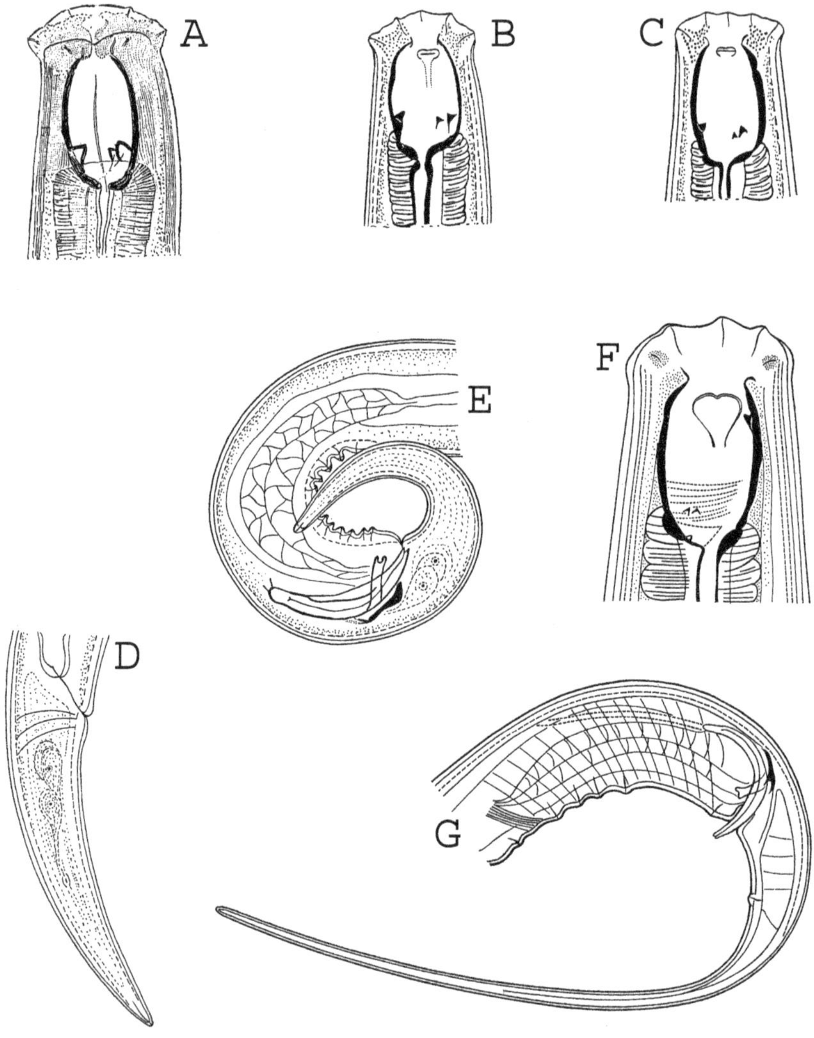

Fig. 44. *A-E:* Miconchus digiturus *(Cobb, 1893) Andrássy, 1958. A-C: Anterior region; D: Female tail; E: Male tail region; F, G:* Promiconchus microdonta *(Mulvey & Jensen, 1967) Jairajpuri & Khan, 1982. F: Anterior region; G: Male posterior region. A after Cobb (1917); B-G after Mulvey and Jensen (1967).*

= *Mononchus digiturus* Cobb, 1893
= *Mononchus* (*Iotonchus*) *digiturus* Cobb, 1893 (Cobb, 1916)
= *Iotonchus digiturus* (Cobb, 1893) Cobb, 1916

OTHER SPECIES

M. aquaticus Khan, Ahmad & Jairajpuri, 1978
M. araki Khan, Yoshiga, Okumura & Tanaka, 2008
M. baeticus Jiménez-Guirado, Alhama & Gutierrez, 1997
M. bulbicaudatus Jana, Chatterjee & Manna, 2008
M. californicus Mulvey, 1962
M. citri Khan, Ahmad & Jairajpuri, 1978
M. costaricense **sp. n.**
= *M. digiturus apud* Zullini, Loof & Bongers, 2002 *nec* Cobb, 1893
M. crenicaudatus Gagarin, 1984
M. dalhousiensis Jairajpuri, 1969
= *M. sardhanensis* Saha, Lal & Singh, 2006 **syn. n.**
M. effilatus (Schuurmans Stekhoven & Teunissen, 1938) Andrássy, 1958
= *Mononchus effilatus* Schuurmans Stekhoven & Teunissen, 1938
= *Mononchus (Iotonchus) effilatus* Schuurmans Stekhoven & Teunissen, 1938 (Goodey, 1951)
M. elegans Lal & Khan, 1988
M. eurinus Eroshenko, 1975
M. exilis (Cobb, 1917) Andrássy, 1958
= *Mononchus exilis* Cobb, 1917
M. fasciatus (Cobb, 1917) Andrássy, 1958
= *Mononchus (Iotonchus) fasciatus* Cobb, 1917
= *Iotonchus fasciatus* Cobb, 1917
M. gomezi Zullini, Loof & Bongers, 2002
M. hopperi Mulvey, 1962
M. japonicus Khan, Araki & Bilgrami, 2000
= *M. koreanus* Jairajpuri, Tahseen & Choi, 2001a
M. kansasensis Mulvey & Dickerson, 1970
M. longicaudatus Jiménez-Guirado, Peña-Santiago & Castillo-Castillo, 1993
M. oregensis Jensen & Mulvey, 1968
M. papillifer Andrássy, 2008
M. pararapax Mulvey & Jensen, 1967
M. parathornei Loof, 2006
M. rapax (Cobb, 1917) Andrássy, 1958
= *Mononchus (Iotonchus) rapax* Cobb, 1917
= *Iotonchus rapax* Cobb, 1917
M. rectangularis Jana, Chatterjee & Manna, 2008

M. regius (Cobb, 1917) Andrássy, 1958
=*Mononchus* (*Iotonchus*) *regius* Cobb, 1917
=*Iotonchus regius* Cobb, 1917
M. rex (Cobb, 1904) Andrássy, 1958
=*Mononchus rex* Cobb, 1904
=*Mononchus* (*Iotonchus*) *rex* Cobb, 1904 (Cobb, 1917)
=*Iotonchus rex* (Cobb, 1904) Cobb, 1917
M. schneideri (Meyl, 1955a) Andrássy, 1958
=*Iotonchus schneideri* Meyl, 1955a
M. studeri (Steiner, 1914) Andrássy, 1958
=*Mononchus studeri* Steiner, 1914
=*Mononchus* (*Iotonchus*) *studeri* Steiner, 1914 (Cobb, 1916)
=*Iotonchus studeri* (Steiner, 1914) Cobb, 1916
= *Mononchus* (*Anatonchus*) *studeri* Steiner, 1914 (Cobb, 1916)
M. thornei Mulvey & Jensen, 1967
M. triodontus Buangsuwon & Jensen, 1966
M. vulvapapillatus Choi, Khan & Lee, 1999

***Miconchus digiturus* (Cobb, 1893) Andrássy, 1958**
(Fig. 44A-E)

MEASUREMENTS

Female: L = 1.4-1.9 mm; a = 22-39; b = 3.7-5.2; c = 12-19; c′ = 3-4; V = 70-79.
Male: L = 1.5-1.7 mm; a = 36-37; b = 3.5-4.4; c = 13-18.

DESCRIPTION

Lip region offset by depression, labial papillae distinct. Amphidial apertures at level of anterior end of buccal cavity. Buccal cavity ovoid, 35-40 × 18-28 μm in female and 32-36 × 16-24 μm in male, provided with three equal-sized suprabasal teeth. Pharyngo-intestinal junction tuberculate. Female genital system mono-prodelphic, posterior branch rudimentary, *ca* as long as corresponding body diam. Tail elongate conoid, arcuate, similar in sexes, 90-100 μm long. Spinneret and caudal glands not observed. Male with 62-72 μm long spicules and 12-17 equally spaced supplements. Gubernaculum crescent-shaped, accessory pieces paired, bifurcate.

DISTRIBUTION

Colombia, Fiji Islands, India, Macedonia, St Lucia, USA, Venezuela, former Yugoslavia.

REMARKS

The population described as *M. digiturus* from Costa Rica by Zullini, Loof and Bongers (2002) is probably not *M. digiturus*. In this population the posterior genital branch is completely absent, whereas in *M. digiturus* a post-uterine sac is always present. Recently, Andrássy (2008) put a question mark on the conspecificity of Cobb's *digiturus* and the Costa Rican specimen. The Costa Rican specimens differ from *M. digiturus* in having comparatively larger buccal cavity (39-42 *vs* 35-36 μm), in the posterior position of the dorsal tooth apex (19-23 *vs* 46-50% of buccal cavity length from its base) and in the absence of a post-uterine sac (*vs* one vulval body diam. long sac present). Owing to these differences, the Costa Rican specimens are herein considered to represent a new species and are named as *Miconchus costaricense* **sp. n.**

Key to species of *Miconchus* Andrássy, 1958

1 – Female monodelphic or pseudo-monodelphic (posterior ovary present but strongly reduced) 2
– Female didelphic 6

2 – Post-uterine branch completely absent 3
– Post-uterine branch 1-2 corresponding body diam. long 4

3 – Apex of dorsal tooth at 41-46% of buccal cavity length from its base; advulval papillae present; ♀: L = 2.16 mm; a = 31; b = 3.9; c = 12.5; c′ = 4.5; V = 78; ♂: L = 1.98 mm; a = 32; b = 4.2; c = 18; c′ = 2.6; supplements 11 (Ecuador) . . . . *papillifer*
– Apex of dorsal tooth at 24-29% of buccal cavity length from its base; advulval papillae absent; ♀: L = 1.3-1.7 mm; a = 23-34; b = 3.5-3.9; c = 9-12; c′ = 3.5-4.2; V = 69-73; ♂: unknown (Japan, Korea) . . . . *japonicus*

4 – Apex of dorsal tooth midway in buccal cavity; ♀: L = 1.6-1.8 mm; a = 27-33; b = 3.6-4.3; c = 12-15; c′ = 3; V = 68-74; ♂: L = 1.8 mm; a = 35; b = 4; c = 16; supplements 15 (Thailand) . . . . *triodontus*

– Apex of dorsal tooth in posterior third of buccal cavity 5

5 – Post-uterine sac *ca* as long as corresponding body diam.; ♀: L = 1.4-1.9 mm; a = 22-39; b = 3.7-5.2; c = 12-19; c′ = 3-4; V = 70-79; ♂: L = 1.5-1.7 mm; a = 36-37; b = 3.5-4.4; c = 13-18; supplements 12-17 (former Yugoslavia, India, Fiji Islands, USA, St Lucia, Colombia, Venezuela, Macedonia) . . . . . . *digiturus*

– Posterior genital branch as long as two corresponding body diam.; ♀: L = 1.8-2.5 mm; a = 30-39; b = 3.7-4.5; c = 16-19; c′ = 3.3-3.8; V = 72-79; ♂: L = 1.99 mm; a = 38; b = 3.8; c = 17; c′ = 2.6; supplements 13 (Costa Rica) . . . . *gomezi*

6 – Teeth basal in position, originating at very end of interlateralia (lateral walls) 7

– Teeth suprabasal in position or further anterior, to midway in buccal cavity, conspicuously originating before proximal ends of interlateralia (lateral walls) 13

7 – Tail filiform, 15-25 anal body diam. long; ♀: L = 2.5-3.3 mm; a = 45-62; b = 5.2-5.9; c = 3.2-4.5; c′ = 15-25; V = 49-58; ♂: L = 2.3-2.9 mm; a = 30-49; b = 5.0-5.6; c = 3.4-4.4; supplements 10-14 (Spain) . . . . *longicaudatus*

– Tail conoid to elongate, not filiform, 2-8 anal body diam. long . . 8

8 – Tail 7-8 anal body diam. long 9

– Tail 2-5 anal body diam. long 10

9 – Body very large, almost 5 mm, slender; ♀: L = 4.8 mm; a = 59; b = 4.6; c = 8.9; c′ = 8; V = 68; ♂: unknown (Brazil) . . . . *schneideri*

– Body much smaller, *ca* 1 mm, plump; ♀ unknown; ♂: L = 1.1 mm; a = 18; b = 3; c = 4.8; supplements ten (Zaire) . . . . *effilatus*

10 – Caudal spinneret present; ♀: L = 1.4-2.2 mm; a = 24-38; b = 3.6-4.6; c = 12-20; c′ = 2-4; V = 60-70; ♂: L = 1.4-2.0 mm; a = 25-32; b = 3.7-4.2; c = 14-21; supplements 11-20 (The Netherlands, Germany, UK, Switzerland, Poland, Austria, Slovakia, Romania, Spain, France, Italy, Mauritius, USA, El Salvador) . . . . *studeri*

– Caudal spinneret absent 11

11 – Large species, *ca* 3 mm; ♀: L = 3.2 mm; a = 40; b = 5; c = 40; c′ = 4; V = 62; ♂: L = 2.8 mm; a = 43; b = 5; c = 23; supplements 11 (USA-Oregon) . . . . *oregensis*

– Smaller species, 1.5-2 mm 12

12 – Tail plump, 2.5-3 anal body diam. long; buccal teeth strong; ♀: L = 1.8-2.0 mm; a = 21-31; b = 3.6-4.0; c = 13-18; c′ = 2.5-3.5; V = 68-73; ♂: unknown (India) . . . . *citri*
– Tail slender, 5.0-5.5 anal body diam. long; buccal teeth small; ♀: L = 1.5-1.6 mm; a = 21-24; b = 3.8-4.2; c = 9-10; c′ = 5.0-5.5; V = 61-64; ♂: unknown (India) . . . . *elegans*

13 – Apices of teeth midway in buccal cavity; tail always short, 2-3 anal body diam. long 14
– Apices of teeth at 30-40% of buccal cavity length from its base; tail of varying length 18

14 – Body large, 3-4 mm; supplements 21-24 15
– Body smaller, 1.5-2 mm; supplements 14-15 16

15 – Tail sharply pointed at tip; ♀: L = 2.7-2.9 mm; a = 23; b = 3.6-4.2; c = 18-22; c′ = 2.2-2.5; V = 70-72; ♂: L = 2.8-3.1 mm; a = 23-25; b = 3.6-4.2; c = 18-23; supplements 21-23 (Russia) . . . . *crenicaudatus*
– Tail rounded at tip; ♀: L = 3.9 mm; a = 32; b = 4; c = 21; c′ = 2; V = 70; ♂: L = 3.4 mm; a = 28; b = 4.4; c = 26; supplements 24 (Russia-Far East) . . . . *eurinus*

16 – Tail short, c = 56; ♀: L = 2.0 mm; a = 43; b = 3.7; c = 56; c′ = 1.8; V = 78; ♂: L = 2.0 mm; a = 45; b = 4.2; c = 50; supplements 14 (Romania, Georgia, Australia) . . . . *exilis*
– Tail longer, c = 9-18 17

17 – Body small, L = 1.4-1.6 mm; tail 3 anal body diam. long; V = 70-72; ♀: L = 1.4-1.6 mm; a = 25-31; b = 3.6-3.9; c = 15-18; c′ = 3; V = 70-72; ♂: L = 1.5-1.6 mm; a = 29-31; b = 3.4-4.0; c = 18-20; supplements 14-15 (USA-Kansas) . . . . *kansasensis*
– Body comparatively long, L = 2.0-2.5 mm; tail 4.4-5.9 anal body diam. long; V = 63-68; ♀: L = 2.0-2.5 mm; a = 34-36; b = 3.7-4.0; c = 9.6-11.7; c′ = 4.4-5.9; V = 63-68; ♂: L = 2.0-2.1 mm; a = 33-37; b = 4.0-4.1; c = 12-13; c′ = 3.0-3.2; supplements 11-12 (Malaysia) . . . . *parathornei*

18 – Large species, 6-7 mm long 19
– Small species, 2.0-3.5 mm long 20

19 – Tail 1200-1300 μm or 13 anal body diam. long; ♀: L = 6.5-7.0 mm; a = 45; b = 5; c = 5.3; c′ = 13; V = 58; ♂: L = 6.5-

7.0 mm; a = 43; b = 5; c = 7.1; supplements 17 (New Zealand) .... *rex*

– Tail 600 μm or seven anal body diam. long; ♀: L = 6.2 mm; a = 43; b = 5.9; c = 10; c′ = 7; V = 63; ♂: unknown (USA-Virginia) .... *regius*

20 – Tail with spinneret........ 21

– Tail without spinneret........ 25

21 – Body *ca* 4 mm long; ♀: L = 3.7 mm; a = 38; b = 4.7; c = 10; c′ = 6; V = 61; ♂: unknown (Poland, USA-Virginia)...... *rapax*

– Body 2-3 mm long 22

22 – Advulval papillae present 23

– Advulval papillae absent........ 24

23 – Buccal cavity 51-60 × 33-39 μm; advulval papillae both anterior and posterior to vulva equally developed; spicules 80-105 μm; ♀: L = 2.68-3.75 mm; a = 38-51; b = 4.4-5.3; c = 14-18; c′ = 3.6-4.8; V = 61-67; ♂: L = 2.45-3.28 mm; a = 37-53; b = 4.6-5.5; c = 17-22; c′ = 2.5-3.2; supplements 13-16 (Spain)..... *baeticus*

– Buccal cavity 43-48 × 30-32 μm; only advulval papillae anterior to vulva well developed, posterior ones weak or absent; spicules 62-70 μm; ♀: L = 1.9-2.8 mm; a = 24-37; b = 3.7-4.6; c = 15-21; c′ = 2.5-3.0; V = 66-73; ♂: L = 1.4-2.5 mm; a = 26-40; b = 3.9-4.6; c = 17-20; supplements 17-21 (Romania, India)... *aquaticus*

24 – Tail slender, 300-500 μm or 8-10 anal body diam. long; ♀: L = 2.2-2.9 mm; a = 25-51; b = 3.6-5.1; c = 5.2-7.9; c′ = 8-10; V = 58-65; ♂: L = 2.7-2.8 mm; a = 49-54; b = 4.4-4.8; c = 6.7-8.4; supplements 12-13 (Ivory Coast, Nigeria, USA-California) .... *pararapax*

– Tail stout, 150-250 μm or 4-6 anal body diam. long; ♀: L = 2.0-2.6 mm; a = 25-37; b = 3.6-4.0; c = 9-19; c′ = 4-6; V = 63-68; ♂: L = 2.0-2.4 mm; a = 36-40; b = 4.3-4.5; c = 22-25; supplements 15 (Egypt, Nigeria, St Lucia, El Salvador, Cuba) .... *thornei*

25 – Buccal capsule comparatively slender, *ca* twice as long as wide; apices of teeth at *ca* 40% of buccal cavity length from its base 26

– Buccal capsule broad, *ca* 1.5 times as long as wide; apices of teeth at *ca* 30% of buccal cavity length from its base 27

26 – Teeth usually small; tail five anal body diam. long; ♀: L = 2.1-2.4 mm; a = 36-42; b = 5.5-6.0; c = 7.8-10; c′ = 5.0-5.5; V = 60-66; ♂: unknown (USA-Florida) . . . . *fasciatus*
– Teeth normally developed; tail 7-9 anal body diam. long; ♀: L = 2.2-3.3 mm; a = 34-57; b = 4.2-6.0; c = 6.9-9.7; c′ = 7-9; V = 60-66; ♂: L = 2.3-2.9 mm; a = 33-45; b = 4.4-5.7; c = 8-14; supplements 14-15 (Romania, USA-Florida) . . . . *hopperi*

27 – Teeth small, less projecting; tail 140-160 μm long; ♀: L = 1.9-2.3 mm; a = 28-31; b = 3.5-4.0; c = 12-14; c′ = 3.5-4.0; V = 63-68; ♂: unknown . . . . *dalhousiensis*
– Teeth large, well projecting; tail 190-350 μm long; ♀: L = 2.4-3.2 mm; a = 31-51; b = 4.5-5.9; c = 7.4-14; c′ = 4.5-6.0; V = 59-70; ♂: L = 2.2-3.1 mm; a = 43-49; b = 5.3-5.9; c = 15-21; supplements 15 (Egypt, USA-California; Oregon) . . . . *californicus*

Genus *Crassibucca* Mulvey & Jensen, 1967
(Fig. 45A-D)

DIAGNOSIS

Slender nematodes, 1-1.5 mm long. Cuticle smooth. Buccal cavity oblong, narrower than in other genera of family, provided with five teeth. Dorsal tooth largest, situated in anterior third of buccal cavity. Subventral teeth arranged in two pairs, smaller pair at level of dorsal tooth, larger pair located in posterior third/fourth of buccal cavity. Pharyngo-intestinal junction tuberculate. Female reproductive system prodelphic, without a post-uterine sac. Male known in one species only. Spicules arcuate, gubernaculum simple or bidentate, accessory pieces present, eight supplements. Tail of both sexes similar, conoid-arcuate to elongate, 3-14 anal body diam. long. Caudal glands poorly developed, spinneret present or absent.

TYPE SPECIES

Crassibucca penicula Mulvey & Jensen, 1967

OTHER SPECIES

C. asterocaudata Jairajpuri, Ahmad & Sturhan, 1998

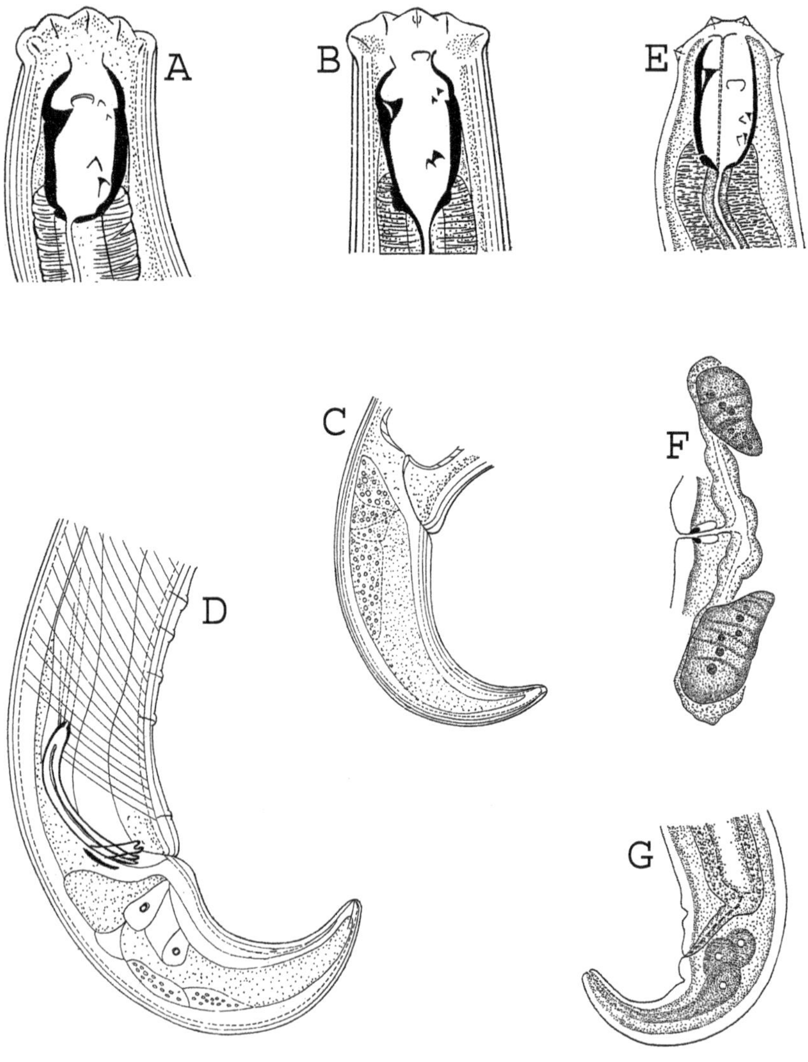

Fig. 45. *A-D:* Crassibucca penicula *Mulvey & Jensen, 1967. A, B: Anterior region; C: Female tail; D: Male posterior region; E-G:* Paracrassibucca paucidentatus *(Lordello, 1970) Jairajpuri & Khan, 1982. E: Anterior region; F: Female genital system; G: Female tail. A-D after Mulvey and Jensen (1967); E-G after Baqri and Jairajpuri (1974).*

C. colombica Siddiqi, 1984a
C. macrocauda Mulvey & Jensen, 1967

Crassibucca penicula Mulvey & Jensen, 1967
(Fig. 45A-D)

MEASUREMENTS

Female: L = 0.9-1.2 mm; a = 27-43; b = 3.4-3.9; c = 16-18; c′ = 2.5-3.0; V = 60-70.
Male: L = 1.1-1.2 mm; a = 30-37; b = 3.8-4.2; c = 20-29.

DESCRIPTION

Lip region offset. Buccal cavity *ca* twice as long as broad, thick-walled, 22-25 × 12-14 μm. Dorsal tooth relatively large, anterior, opposed by two small denticles and longitudinal rib on ventral walls. Subventral teeth nearly equal in size to dorsal tooth, situated in posterior third of buccal cavity. Female reproductive system mono-prodelphic. Vulval papillae absent. Tail 52-72 μm long, conoid, ventrally arcuate, without caudal glands or terminal opening. Male spicules 40-42 μm long, arcuate, lateral guiding pieces with bifid ends, gubernaculum short. Supplements eight, rather poorly developed.

DISTRIBUTION

Nigeria, St Lucia.

Key to species of *Crassibucca* Mulvey & Jensen, 1967

1 – Tail 12-14 anal body diam. long, very slender; ♀: L = 1.4-1.5 mm; a = 38-41; b = 3.8; c = 5.0-5.4; c′ = 12-14; V = 63; ♂: unknown (Nigeria) . . . . *macrocauda*
– Tail shorter, 2.5-4.5 anal body diam. long, fairly plump 2

2 – Tail with characteristic, usually star-like, ornamentation at tip; ♀: L = 1.16-1.29 mm; a = 34-41; b = 3.2-3.5; c = 12-13; c′ = 3.6-4.0; V = 70-75; ♂: unknown (Nicaragua) . . . . *asterocaudata*
– Tail lacking characteristic ornamentation at tip 3

3 – Tail 50-70 μm long; posterior subventral teeth nearly as strong as dorsal tooth; spinneret lacking; ♀: L = 0.9-1.2 mm; a = 27-43; b = 3.4-3.9; c = 16-18; c′ = 2.5-3.0; V = 60-70; ♂: L = 1.1-1.2 mm; a = 30-37; b = 3.8-4.2; c = 20-29; supplements eight (Nigeria, St Lucia) . . . . *penicula*

– Tail 110-120 μm long; posterior subventral teeth much smaller than dorsal tooth; spinneret present; ♀: L = 1.3 mm; a = 36-38; b = 3.6-3.7; c = 11-12; c′ = 4.5; V = 69-70; ♂: unknown (Colombia) .... *colombica*

Genus *Paracrassibucca* Baqri & Jairajpuri, 1974
(Fig. 45E-G)

DIAGNOSIS

Small nematodes, 0.7-1.38 mm long. Cuticle smooth. Buccal cavity oblong, comparatively narrow, armed with five teeth. Dorsal tooth large, in anterior third or fourth of buccal cavity, two pairs of subventral teeth, much smaller than dorsal tooth and lying in tandem in posterior half of buccal cavity. Pharyngo-intestinal junction tuberculate. Female genital system amphidelphic, vulva at 57-65% of body length. Male unknown. Tail short, ventrally arcuate, 2-3.5 anal body diam. long. Caudal gland and spinneret present.

TYPE AND ONLY SPECIES

Paracrassibucca paucidentatus (Lordello, 1970) Jairajpuri & Khan, 1982
= *Sporonchulus paucidentatus* Lordello, 1970
= *Paracrassibucca jenseni* Baqri & Jairajpuri, 1974

Paracrassibucca paucidentatus (Lordello, 1970) Jairajpuri & Khan, 1982
(Fig. 45E-G)

MEASUREMENTS

Female: L = 0.67-1.38 mm; a = 17-30; b = 3.1-4.0; c = 16-23; c′ = 2.1-3.5; V = 57-65.

DESCRIPTION

Lip region slightly offset. Amphids cup-shaped, aperture 5-6 μm wide, posterior to dorsal tooth apex. Buccal cavity 22-32 × 9-15 μm. Dorsal tooth at 73-81% of buccal cavity length from its base. Two pairs of subventral teeth slightly smaller than dorsal tooth and located in posterior half of buccal cavity, anterior pair slightly smaller than posterior and located at 40-54%, second pair located at 25-33% of buccal cavity length from its base. Pharyngo-intestinal junction tuberculate. Female genital system amphidelphic, vulva transverse, vagina short, surrounded by trapezoid cuticularised pieces. Tail short conoid, ventrally arcuate, 37-68 μm long. Caudal glands compact, with duct leading to a terminal opening. Male not known.

DISTRIBUTION

Brazil, Costa Rica, El Salvador, India.

Genus *Promiconchus* Jairajpuri & Khan, 1982
(Fig. 44F, G)

DIAGNOSIS

Body 1.3-1.9 mm long. Cuticle smooth. Buccal cavity barrel-shaped, spacious, armed with three teeth. Dorsal tooth located in anterior third/fourth of buccal cavity, rather weak, less developed than subventral teeth which are located in posterior third/fourth of buccal cavity. Pharyngo-intestinal junction tuberculate. Female mono-prodelphic, vulva at 63-70% of body length. Spicules arcuate, accessory pieces bifurcate. Supplements usually few in number, 5-6. Tail elongate, ventrally, or first ventrally then dorsally, bent, 4-10 anal body diam. long. Caudal glands and spinneret lacking.

TYPE SPECIES

Promiconchus microdontus (Mulvey & Jensen, 1967) Jairajpuri & Khan, 1982
= *Crassibucca microdonta* Mulvey & Jensen, 1967

OTHER SPECIES

P. conicaudatus (Altherr, 1971) Andrássy, 1993b
= *Crassibucca conicauda* Altherr, 1971
P. incultus (Carvalho, 1960a) Andrássy, 1993b
= *Mononchus* (*Cobbonchus*) *incultus* Carvalho, 1960a

= *Cobbonchus incultus* Carvalho, 1960a
= *Promiconchus siddiqii* (Ahmad & Jairajpuri, 1993) Andrássy, 1993b

Promiconchus microdontus (Mulvey & Jensen, 1967) Jairajpuri & Khan, 1982

(Fig. 44F, G)

MEASUREMENTS

Male: L = 1.5-1.7 mm; a = 34-38; b = 4.3-4.6; c = 5.3-5.7.

DESCRIPTION

Buccal cavity 29-30 × 20 μm or *ca* 1.5 times as long as broad; dorsal tooth small, situated in anterior third of buccal cavity. Subventral teeth slightly smaller than dorsal tooth, nearly basal in cavity. Amphidial aperture very wide, occupying nearly half cavity diam. Pharyngo-intestinal junction tuberculate. Female unknown. Male with 55-65 μm long spicules; lateral accessory pieces with bifurcate ends. Supplements 5 + 1, fairly well developed. Tail conoid, then abruptly cylindroid, 280-300 μm long, with acutely rounded terminus. Caudal glands and terminal opening absent.

DISTRIBUTION

Nigeria.

Key to species of *Promiconchus* Jairajpuri & Khan, 1982

1 – Tail *ca* ten anal body diam. long, simply bent ventrad; ♀: unknown; ♂: L = 1.5-1.7 mm; a = 34-38; b = 4.3-4.6; c = 5.3-5.7; supplements 5 + 1 (Nigeria) . . . . *microdontus*
– Tail 4-6 anal body diam. long, first bent ventrally, then dorsally . . 2

2 – Apex of dorsal tooth located in anterior third of buccal cavity; body 2 mm or longer; ♀ (juv.): L = 1.95 mm; a = 45; b = 3.8; c = 20; c′ = 4; ♂: unknown (Brazil) . . . . *conicaudatus*
– Apex of dorsal tooth located in anterior fourth of buccal cavity; body *ca* 1.5 mm long; ♀: L = 1.3-1.7 mm; a = 27-41; b = 3.8-4.2; c = 11-20; c′ = 5-6; V = 63-70; ♂: unknown (Dominica, Brazil) . . . . *incultus*

Genus *Doronchus* Andrássy, 1993b
(Fig. 46)

DIAGNOSIS

Body 1.8-3.6 mm long. Cuticle finely annulated. Labial region set off from body. Buccal cavity barrel-shaped, moderately roomy, slightly longer than cephalic diam., armed with three basal teeth lying at same level. Dorsal tooth distinctly larger than subventral teeth. Pharyngo-intestinal junction tuberculate. Female reproductive system amphidelphic, vulva located at 53-62% of body length. Spicules arcuate, with forked lateral pieces. Copulatory supplements 12-16. Tail sexually dimorphic: in female almost straight, elongate-conoid, with sharp terminus and 5-10 anal body diam. long; in male ventrally curved, broadly conoid with bluntly rounded terminus, only 1.5-2 anal body diam. long. Caudal glands and spinneret lacking.

TYPE SPECIES

Doronchus kirikiri (Yeates, 1967) Andrássy, 1993b
= *Miconchus kirikiri* Yeates, 1967

OTHER SPECIES

D. reflexus (Yeates, 1967) Andrássy, 1993b
= *Mononchus reflexus* Yeates, 1967

Doronchus kirikiri (Yeates, 1967) Andrássy, 1993b
(Fig. 46A-C)

MEASUREMENTS

Female: L = 3.20-3.64 mm; a = 38-41; b = 4.8-5.3; c = 6.9-7.6; c′ = 7.9-9.6; V = 54.7-59.8.
Male: L = 2.95-3.20 mm; a = 35-37; b = 4.5-5.6; c = 27-29; c′ = 1.5-1.6.

DESCRIPTION

Lip region slightly offset with usual two circles of papillae. Buccal cavity barrel-shaped, 54-57 × 32-36 μm in female and 57-60 × 32-37 μm

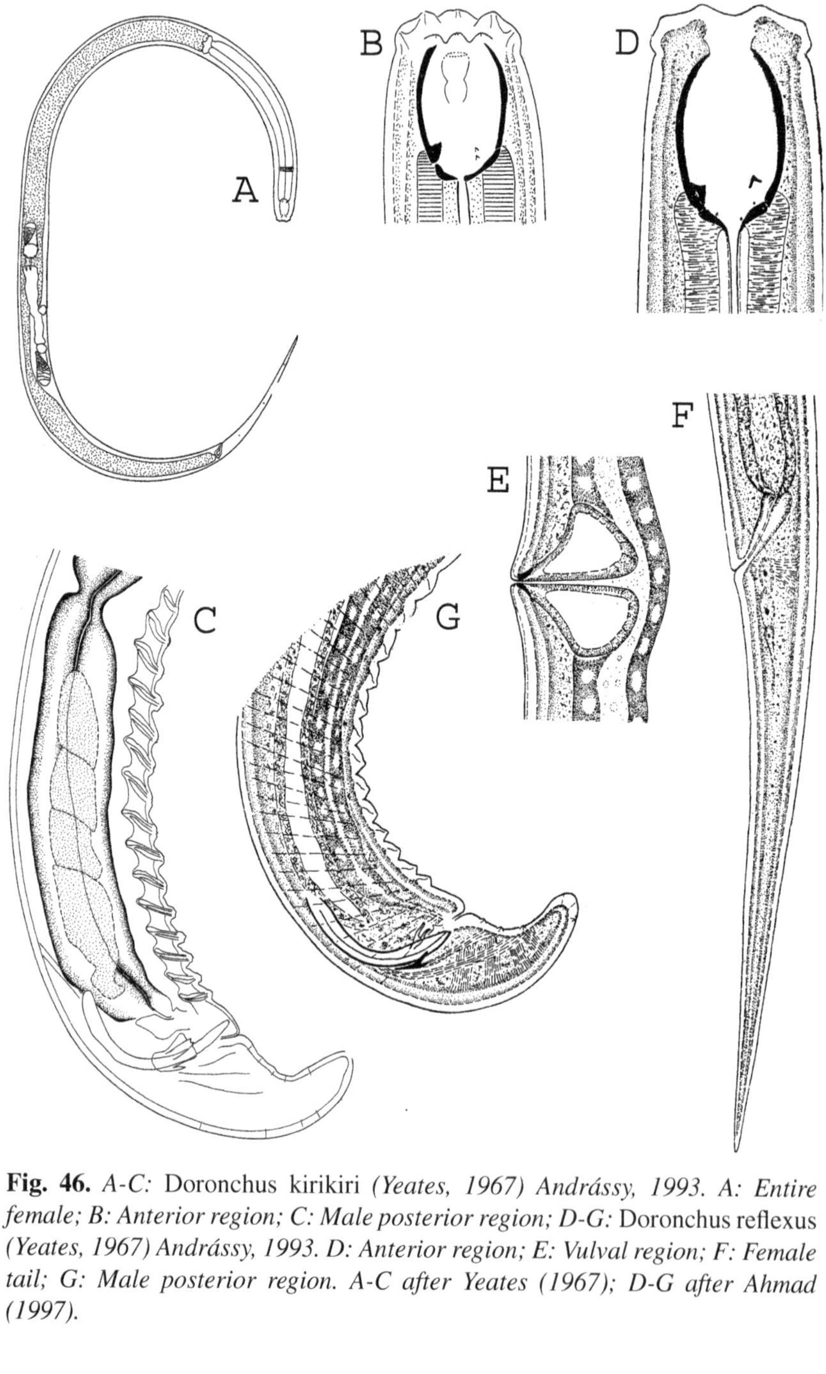

Fig. 46. *A-C:* Doronchus kirikiri *(Yeates, 1967) Andrássy, 1993. A: Entire female; B: Anterior region; C: Male posterior region; D-G:* Doronchus reflexus *(Yeates, 1967) Andrássy, 1993. D: Anterior region; E: Vulval region; F: Female tail; G: Male posterior region. A-C after Yeates (1967); D-G after Ahmad (1997).*

in male, moderately sclerotised with three anteriorly directed teeth, dorsal tooth largest, its apex at 75-80% of buccal cavity length from its base. Pharyngo-intestinal junction tuberculate. Female genital system amphidelphic. Male with 88-89 μm long, slender, strongly arcuate spicules; gubernaculum with broad lateral processes; lateral accessory pieces with simple bifid end; 14-15 ventromedian supplements. Tail elongate-conoid, slightly arcuate ventrally with moderately pointed terminus in female and short conoid with rounded terminus in male. Spinneret and caudal glands absent.

DISTRIBUTION

New Zealand.

Key to species of *Doronchus* Andrássy, 1993b

1 – Body short, 1.8-2.8 mm long; female tail 5-8 anal body diam. long; posterior fork of lateral guiding piece in male distally curved; ♀: L = 1.8-2.8 mm; a = 33-53; b = 4.6-6.0; c = 8-14; c′ = 4.5-8.3; V = 53-62; ♂: L = 2.1-2.3 mm; a = 39-52; b = 4.9-5.7; c = 26; supplements 12-13 (New Zealand). . . . . *reflexus*

– Body longer, 3.2-3.6 mm long; female tail 8-10 anal body diam. long; both forks of lateral guiding piece in male straight; ♀: L = 3.2-3.6 mm; a = 38-41; b = 4.8-5.3; c = 6.9-7.6; c′ = 8-10; V = 55-60; ♂: L = 2.9-3.2 mm; a = 35-37; b = 4.5-5.6; c = 27-29; supplements 14-16 (New Zealand) . . . . *kirikiri*

Nematology Monographs & Perspectives, 2010, Vol. 7, 193-238

Chapter 8

Family Iotonchidae Jairajpuri, 1969

Jairajpuri (1969) proposed the family Iotonchidae for the genera *Iotonchus*, *Hadronchus* and *Prionchulellus*. Mulvey (1978) added the genus *Parahadronchus* and Jairajpuri and Khan (1982) proposed a new genus *Jensenonchus* in this group. In their monograph on predatory nematodes, Jairajpuri and Khan (1982) accepted two subfamilies, the subfamily Iotonchinae for the genera in which the subventral teeth are totally absent and the subfamily Hadronchinae for the genera in which they are present. Siddiqi (1984a) erected the genera *Caputonchus* and *Mulveyellus* close to *Iotonchus* and also added a toothless mononch genus, *Nullonchus*, to this group. Ray and Das (1983) and Jairajpuri and Rahman (1984) added the genera *Hadronchulus* and *Hadronchoides*, respectively, from India. Andrássy (1993b) added another new genus, *Iotonchulus*. The classification of Jairajpuri and Khan (1982) is followed herein.

DIAGNOSIS

Buccal cavity roomy. Dorsal tooth located in either anterior or posterior half of buccal cavity (but totally lacking in *Nullonchus*), its apex directed anteriad. Subventral walls either completely unarmed or provided with smooth or serrate longitudinal ribs, or scattered denticles. Subventral teeth similar to dorsal tooth never occurring on subventral walls.

TYPE SUBFAMILY

Iotonchinae Jairajpuri, 1969

OTHER SUBFAMILY

Hadronchinae Khan & Jairajpuri, 1980

Key to subfamilies of Iotonchidae Jairajpuri, 1969

1 – Subventral teeth absent .... **Iotonchinae**
– Subventral teeth present .... **Hadronchinae**

Subfamily Iotonchinae Jairajpuri, 1969

DIAGNOSIS

Buccal cavity roomy. Dorsal tooth located in anterior or posterior half of buccal cavity, but lacking in *Nullonchus*. Subventral walls completely unarmed, no teeth or denticles.

TYPE GENUS

Iotonchus Cobb, 1916

OTHER GENERA

Jensenonchus Jairajpuri & Khan, 1982
Caputonchus Siddiqi, 1984a
Mulveyellus Siddiqi, 1984a
Nullonchus Siddiqi, 1984a
Iotonchulus Andrássy, 1993b

Key to genera of Iotonchidae Jairajpuri, 1969

1 – Buccal cavity without tooth, teeth or denticles .... *Nullonchus*
– Buccal cavity armed with a dorsal tooth 2
2 – Dorsal tooth located at or near base of buccal cavity ..... *Iotonchus*
– Dorsal tooth located in anterior fourth to somewhat posterior to middle of buccal cavity 3
3 – Thin longitudinal ridge facing dorsal tooth .... *Jensenonchus*
– No longitudinal ridge facing dorsal tooth 4
4 – Lip region sharply offset by deep constriction; tail straight .... *Caputonchus*
– Lip region not set off so sharply; tail more or less arcuate 5

5 – Tail long filiform, 10-20 anal body diam. long; terminal spinneret present . . . . *Iotonchulus*
– Tail conoid, 2-6 anal body diam. long; terminal spinneret lacking . . . . *Mulveyellus*

Genus *Iotonchus* Cobb, 1916

(Figs 47, 48)

DIAGNOSIS

Body 0.8-6.4 mm long. Cuticle smooth. Buccal cavity predominantly roomy, occasionally oblong, moderately broad (varying in length from 20-90 μm). Dorsal tooth always basal or suprabasal, not strong, often rather small. No other armature in buccal cavity. Pharyngo-intestinal junction tuberculate. Female genital system amphidelphic or monoprodelphic. Vulva located at mid-body to *ca* 80% of body length. Spicules more or less arcuate; bifurcate lateral accessory pieces present. Supplements 6-22 in number. Tail similar in both sexes, variable in shape and length, predominantly conoid or filiform, rarely very short and bluntly rounded, varying from 0.7-50 anal body diam. long. Caudal glands and spinneret mostly well developed, but in some species rudimentary or lacking, spinneret often subterminal.

TYPE SPECIES

Iotonchus gymnolaimus (Cobb, 1893) Cobb, 1916
= *Mononchus gymnolaimus* Cobb, 1893
= *Mononchus* (*Iotonchus*) *gymnolaimus* Cobb, 1893 (Cobb, 1916)

OTHER SPECIES

I. abrahami Loof, 2006
I. acuticaudatus Mulvey & Jensen, 1967
I. acutus Cobb, 1917
= *Mononchus* (*Iotonchus*) *acutus* Cobb, 1917
= *Iotonchus laticupulatus* Razzhivin, 1971
I. aequabilis Susulovsky, 1998
I. aequatorialis Vinciguerra & Orselli, 2006
I. anisostomus Buangsuwon & Jensen, 1966
I. apapillatus Dhanam & Jairajpuri, 1998

Fig. 47. Iotonchus trichurus *Cobb, 1917. After Cobb (1917).*

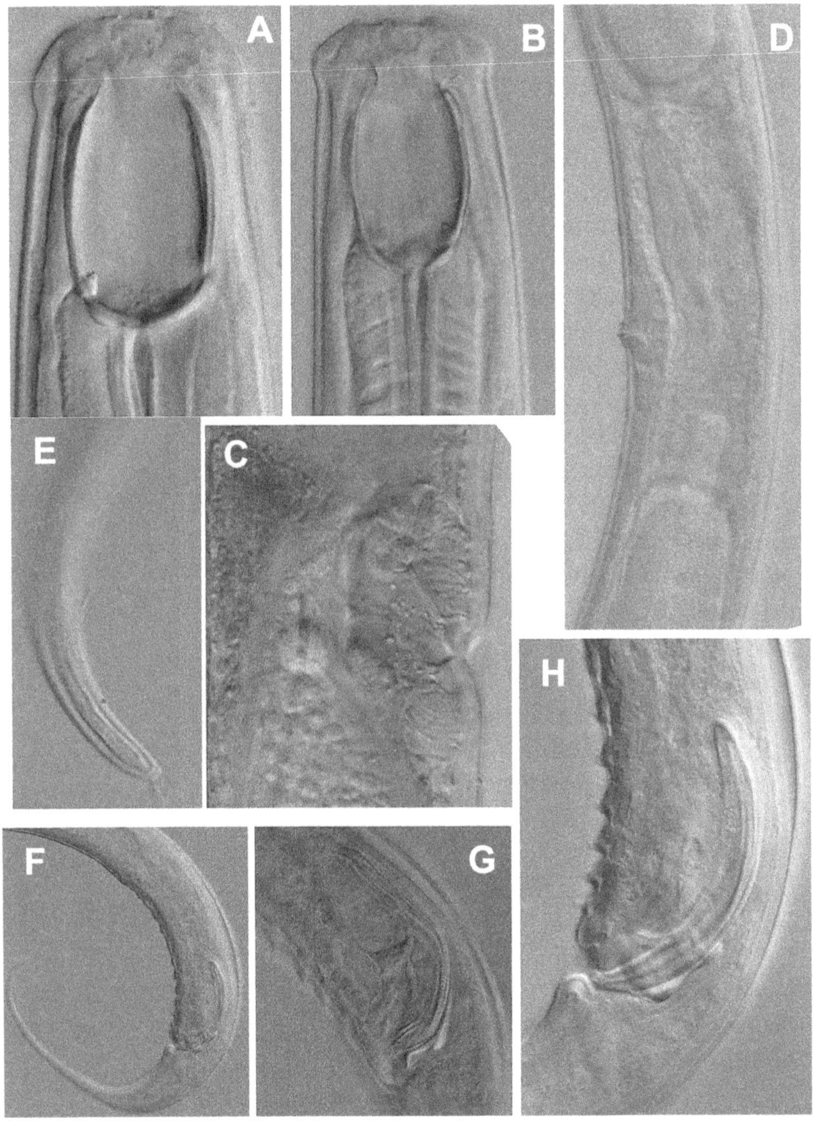

Fig. 48. Iotonchus parabasidontus *Mulvey & Jensen, 1967. A, B: Anterior region; C: Vulval region; D: Female genital system; E: Female tail terminus; F: Female posterior region; G, H: Male cloacal region.*

I. arcuatus Khan, Araki & Bilgrami, 2000
I. baqrii Jairajpuri, 1969
I. basidontus Clark, 1961a

= *Iotonchus prabhooi* Mohandas, 1979
I. brachylaimium Aboul-Eid & Ameen, 1982
I. brachylaimus Cobb, 1917
= *Mononchus* (*Iotonchus*) *brachylaimus* Cobb, 1917
I. caesar Alekseev, 2001
I. candelabri Yeates, 1992
I. carpathicus Popovici, 1990
I. chantaburensis Buangsuwon & Jensen, 1966
= *Iotonchus khani* Mohandas & Prabhoo, 1979a
I. clarki Mulvey & Jensen, 1967
I. consimilis Cobb, 1917
= *Mononchus* (*Iotonchus*) *consimilis* Cobb, 1917
I. cucumis Khan, So Deuk, Su Go & Yong, 2002
I. damsanensis Choi & Khan, 2000
I. devius Siddiqi, 2001
I. geminus Heyns & Lagerwey, 1965
I. globibucca Dhanam & Jairajpuri, 2002
I. goshiensis Khan, Yoshiga, Okumura & Tanaka, 2008
I. guineae Siddiqi, 2001
I. helicus Nguyen Vu Thanh, 2006
I. hinokumaensis Khan, Yoshiga, Okumura & Tanaka, 2008
I. indicus Jairajpuri, 1969
I. kherai Mohandas & Prabhoo, 1979a
I. kilumicus Siddiqi, 2001
I. kirbyi Siddiqi, 1984b
I. kirghistanicus Sultanalieva, 1983
I. koupensis Siddiqi, 2001
I. kvavadezi Eliava, Bagathuria & Chuchulashvili, 2005
I. lacuplanarum Yeates, 1992
I. lamottei Malcevschi, 1981
I. litoralis Coetzee, 1967b
I. longisaccatus Mohilal, Anandi & Dhanachand, 2000
I. longisaccus Siddiqi, 2001
I. loteniae de Bruin & Heyns, 1992a
I. magyar Andrássy, 1973
I. maragnus Clark, 1961a
I. mboticus Siddiqi, 2001
I. miamaensis Khan & Araki, 2002
I. microdontus Thong, 1971

I. montanus Yeates, 1992
I. muneris Siddiqi, 2001
I. nayari Mohandas & Prabhoo, 1979
= *Iotonchus shafii* Khan & Jairajpuri, 1980
I. ndu Siddiqi, 2001
I. nepotum Andrássy, 2008
I. nigeriensis Mulvey & Jensen, 1967
I. ogiensis Khan, Yoshiga, Okumura & Tanaka, 2008
I. onchus Jairajpuri, Tahseen & Choi, 2000
I. parabasidontus Mulvey & Jensen, 1967
I. paracutus Vinciguerra & Orselli, 2000
I. parageminus Jimenez-Guirado, 1994
I. paratrichurus Siddiqi, 2001
I. pauli Heyns & Lagerwey, 1965
I. percivali Clark, 1961a
I. pharoensis Aboul-Eid & Ameen, 1982
I. pseudodigonicus Ahmad & Jairajpuri, 1983
I. pusillus Loof, 2006
I. rapidulus Siddiqi, 1984b
I. recessus Yeates, 1992
I. rayongensis Buangsuwon & Jensen, 1966
I. rinae Coetzee, 1967b
I. risoceiae Carvalho, 1955
= *Mononchus* (*Iotonchus*) *risoceiae* Carvalho, 1955
= *Mononchus* (*Iotonchus*) sp. *apud* Carvalho, 1953
= *Iotonchus carvalhoi* Andrássy, 1958
I. rotundicaudatus Peña-Santiago & Jiménez-Guirado, 1991
I. sacculatus Siddiqi, 2001
I. sagaensis Khan, Araki & Bilgrami, 2000
I. silvallus Ahmad & Jairajpuri, 1983
I. singaporensis Ahmad, Baniyamuddin & Jairajpuri, 2005
I. southi Dhanam & Jairajpuri, 1998
I. spinicaudatus Coetzee, 1967b
I. stockdilli Yeates, 1988
I. tarjani Mulvey & Jensen, 1967
I. tenuidentatus (Kreis, 1924) Goodey, 1951
= *Mononchus tenuidentatus* Kreis, 1924
= *Mononchus* (*Iotonchus*) *tenuidentatus* Kreis, 1924 (Goodey, 1951)
I. terminus Mohilal, Anandi & Dhanachand, 2000

I. thui Nguyen Vu Thanh, 2006
I. togoensis Siddiqi, 2001
I. transkeiensis Heyns & Lagerwey, 1965
= *Iotonchus thailandensis* Buangsuwon & Jensen, 1966
I. trichurus Cobb, 1917
= *Mononchus* (*Iotonchus*) *trichurus* Cobb, 1917
= *Mononchus* (*Iotonchus*) *piracicabae* Lordello, 1953
= *Iotonchus piracicabae* Lordello, 1953 (Andrássy, 1958)
= *Mononchus* (*Iotonchus*) *piracicaboides* Carvalho, 1960b
= *Iotonchus piracicaboides* Carvalho, 1960b (Heyns & Lagerwey, 1965)
I. uisongensis Choi & Khan, 2000
I. vulvapapillatus Andrássy, 1964
I. zullinii Malcevschi, 1981

SPECIES INQUIRENDAE

I. manipurensis Gambhir & Dhanachand, 1990 (based on single female, probably *I. parabasidontus* with broken tail)
I. minutus Dhanam & Jairajpuri, 2002 (based on male only)
I. obtusus Choi, Khan & Lee, 1999 (based on single abnormal female)

Iotonchus gymnolaimus (Cobb, 1893) Cobb, 1916

MEASUREMENTS

Female: L = 2.9 mm; a = 40; b = 4.2; c = 7.1; c′ = 9; V = 67.

DESCRIPTION

It was briefly described by Cobb (1893) from Fiji Island, Rio de Janeiro, Brazil, and also from Virginia, USA. According to the description given by him, the amphids were located a little posterior to the lips. Buccal cavity thin-walled with a small dorsal tooth near its base and ventral walls showing projections near front of buccal cavity, of which Cobb said "probably representing the optical section of the junction of elements in the pharyngeal wall". Gonad single, anterior to a less prominent vulva. Tail long, tapering regularly towards terminus with a swollen tip and distinct spinneret. Cobb wrote, "this nemativorous species will probably prove cosmopolitan". Unfortunately, it was never redescribed. Because of the inadequate description of type species, we are providing here a description of *I. trichurus*, a monodelphic species of comparatively common occurrence. Among amphidelphic species, *I. parabasidontus* Mulvey & Jensen, 1967 is quite common and light microphotographs are provided (Fig. 48).

Iotonchus trichurus **Cobb, 1917**
(Fig. 47)

MEASUREMENTS

Female: L = 1.2-2.2 mm; a = 28-46; b = 3.5-5.4; c = 3-5; c′ = 15-20; V = 52-65.
Male: L = 1.7 mm; a = 28; b = 4.4; c = 3.6.

DESCRIPTION

Lip region offset, wider than adjoining body, 21-28 μm diam. Amphids prominent, near anterior end of buccal cavity. Buccal cavity 26-32 × 13-22 μm. Dorsal tooth at base of metarhabdion. Oblique subventral walls (telorhabdions) bearing two small denticles (foramina) each. Pharyngo-intestinal junction tuberculate. Female genital system mono-prodelphic. Posterior branch completely absent. *Pars refringens vaginae* present. Tail filiform, whip-like, 260-540 μm long. Small caudal glands and ventrally subterminal spinneret present.

DISTRIBUTION

Brazil, Campbell Islands, India, Malaysia, Mauritius, New Caledonia, New Zealand, Nigeria, Singapore, St Lucia.

Key to the species of *Iotonchus* Cobb, 1916

1 – Female genital organ unpaired, prodelphic or asymmetric, pseudo-prodelphic with rudimentary posterior ovary 2
– Female genital organ paired, amphidelphic 29

2 – Genital organ prodelphic, at most with short post-uterine sac 3
– Genital organ pseudo-prodelphic, with posterior rudimentary ovary 26

3 – Tail short, 2.2-2.8 anal body diam. long; ♀: L = 1.03-1.21 mm; a = 22-25; b = 3.1-3.8; c = 13-16; c′ = 2.2-2.8; V = 66-70; ♂: unknown (Papua New Guinea) . . . . *nepotum*
– Tail > 4 anal body diam. long 4

4 – Larger species, *ca* 3 mm long; buccal cavity *ca* 60 μm long. 5
– Smaller species, 2 mm or shorter; buccal cavity 50 μm or shorter 6

5 – Tail terminus swollen, bulbous; ♀: L = 2.9 mm; a = 40; b = 4.2; c = 7.1; c′ = 9; V = 67; ♂: unknown (Fiji Island, Brazil, USA-Virginia) . . . . *gymnolaimus*
– Tail terminus cylindrical, not swollen; ♀: L = 2.8-3.5 mm; a = 34-41; b = 4.5-4.9; c = 6.6-7.6; c′ = 8; V = 68-69; ♂: unknown (India, Brazil, USA-Hawaii) . . . . *consimilis*

6 – Post-uterine sac well developed, 1-3 corresponding body diam. long 7
– Post-uterine sac very short or absent 14

7 – Post-uterine sac two or more corresponding body diam. long 8
– Post-uterine sac 1-1.5 corresponding body diam. long 10

8 – Spinneret rudimentary; tail six anal body diam. long, uniformly tapering to tip; ♀: L = 1.6 mm; a = 32-35; b = 4.1-4.4; c = 7.2-7.9; c′ = 7; V = 65-66; ♂: L = 1.6 mm; a = 31; b = 4; c = 8.3; supplements nine (Ivory Coast) . . . . *lamottei*
– Spinneret well developed 9

9 – Spinneret subterminal, ventral; tail longer (c′ = 10), V = 63; ♀: L = 2.1 mm; a = 43; b = 4.6; c = 5.8; c′ = 10-11; V = 63; ♂: unknown (New Caledonia) . . . . *recessus*
– Spinneret terminal; tail shorter (c′ = 5), V = 71; ♀: L = 1.8-2.4 mm; a = 32-37; b = 4.1; c = 9.1-10.9; c′ = 5.2-5.4; V = 71; ♂: L = 2.35 mm; a = 47; b = 3.9; c = 7.3; c′ = 7.8; supplements four (Ivory Coast) . . . . *longisaccus*

10 – Spinneret absent 11
– Spinneret present 12

11 – Tail very thin with acute terminus, c′ = 9.5, post-uterine sac 1.4 corresponding body diam. long; ♀: L = 2.4 mm; a = 37; b = 4.1; c = 6.4; c′ = 9.4; V = 68.7; ♂: unknown (Ivory Coast) . . . *devius*
– Tail gradually tapering to rounded terminus, c′ = 5.5, post-vulval uterine sac as long as corresponding body diam.; ♀: L = 1.6-

1.9 mm; a = 25-35; b = 3.9-4.6; c = 10-13; c′ = 4; V = 68-72; ♂: L = 1.4-1.7 mm; a = 28-36; b = 3.9-4.5; c = 12-16; supplements 6-8 (Nigeria) . . . . *nigeriensis*

12 – Spinneret terminal; ♀: L = 2.0 mm; a = 33; b = 4.0; c = 7.1; c′ = 7.1; V = 70; ♂: L = 1.9-2.0 mm; a = 35-43; b = 3.9-4.7; c = 5.9-6.9; supplements 4-6 (Guinea) . . . . *guineae*

– Spinneret subterminal, dorsal 13

13 – Tail 5-6 anal body diam. long; gubernaculum quite thin; ♀: L = 1.9-2.1 mm; a = 29-34; b = 3.8-4.7; c = 9-12; c′ = 5-6; V = 71-75; ♂: L = 2.0-2.3 mm; a = 27-34; b = 4.0-4.4; c = 9-12; supplements 6-8 (Nigeria) . . . . *acuticaudatus*

– Tail four anal body diam. long; gubernaculum swollen; ♀: L = 1.3-2.0 mm; a = 26-36; b = 3.8-4.7; c = 9.6-13; c′ = 4; V = 69-76; ♂: L = 1.3-2.0 mm; a = 29-37; b = 3.9-4.6; c = 10-13; supplements 7-10 (Ivory Coast) . . . . *zullinii*

14 – Post-uterine sac a short diverticulum (<1 body diam. long) 15

– Post-uterine sac practically absent 19

15 – Tail 6-7 anal body diam. long; ♀: L = 1.68-2.05 mm; a = 28-35; b = 4.3-4.4; c = 7.6-8.0; c′ = 6.3-6.6; V = 69.7-72.1; ♂: L = 1.54-1.81 mm; a = 35-39; b = 3.9-4.5; c = 6.8-8.2; supplements 8-9 (Guinea) . . . . *sacculatus*

– Tail longer than eight anal body diam. long 16

16 – Body length *ca* 1.5 mm, V = 55-62 17

– Body length > 1.8 mm, V = 64-80 18

17 – Buccal cavity large (36-37×19-20 μm) with flat base; ♀: L = 1.49-1.52 mm; a = 33-36; b = 4.1-4.4; c = 4.1-4.7; c′ = 10.8-13.7; V = 57.6-62.4; ♂: unknown (Cameroon) . . . . *muneris*

– Buccal cavity smaller (30-32 × 15-19 μm) with conical base; ♀: L = 1.3-1.48 mm; a = 23-35; b = 4.2-4.6; c = 3.1-4.6; c′ = 10.3-15.0; V = 55.4-58.6; ♂: unknown (Cameroon) . . . . *paratrichurus*

18 – Buccal cavity shorter and relatively wider (32-38 × 23-29 μm); ♀: L = 1.9-2.0 mm; a = 36-39; b = 4.6; c = 6.1-6.3; c′ = 10; V = 63-64; ♂: L = 1.8-2.1 mm; a = 38-45; b = 4.2-4.9; c = 6.3-7.2; supplements 9-12 (New Caledonia) . . . . *lacuplanarum*

– Buccal cavity longer and relatively narrower (46-54 × 25-31 μm); ♀: L = 1.8-2.6 mm; a = 31-38; b = 3.7-4.5; c = 6-7; c′ = 8-11;

V = 64-80; ♂: L = 2.2 mm; a = 35; b = 4.3; c = 9; supplements 12 (India) .... *southi*

19 – Tail short (*ca* four anal body diam. long), conoid, with bluntly rounded terminus; ♀: L = 1.7 mm; a = 30; b = 3.7; c = 13; c′ = 4-5; V = 68; ♂: unknown (Thailand) .... *anisostomus*
– Tail > 6 anal body diam. long 20

20 – Body shorter (L = 0.8-1.3 mm) 21
– Body longer (L = 1.4-1.8 mm) 24

21 – Buccal cavity 22-25 × 9-14 μm 22
– Buccal cavity 31-36 × 18-20 μm 23

22 – Tail terminus subacute, without a distinct gland opening; ♀: L = 0.70-1.3 mm; a = 24-28; b = 3.6-4.0; c = 7.2-8.1; c′ = 4.7-5.9; V = 64-67; ♂: unknown (Malaysia) .... *pusillus*
– Tail terminus rounded, gland opening distinct; ♀: L = 0.8-1.0 mm; a = 24-32; b = 3.5-4.5; c = 3.8-6.0; c′ = 6.0-7.5; V = 59-65; ♂: unknown (Thailand, Singapore, Malaysia) .... *chantaburensis*

23 – Tail terminus bulb-like; ♀: L = 1.2-1.3 mm; a = 29-31; b = 3.7-4.2; c = 5.5-6.0; c′ = 6; V = 61-64; ♂: unknown (Singapore) .... *microdontus*
– Tail terminus not swollen; ♀: L = 1.03-1.23 mm; a = 26-31; b = 3.7-4.2; c = 5.1-6.7; c′ = 6.3-8.3; V = 59-65; ♂: unknown (Singapore) .... *singaporensis*

24 – Buccal cavity relatively larger (28-32 × 32-37 μm); ♀: L = 1.5-1.7 mm; a = 28-37; b = 4.0-4.6; c = 5.7-6.0; c′ = 8-11; V = 62-70; ♂: L = 1.4-1.7 mm; a = 31-35; b = 4.3-4.8; c = 5-7; supplements 9-10 (India) .... *baqrii*
– Buccal cavity smaller and narrower 25

25 – Buccal cavity 16-19 × 26-33 μm; ♀: L = 1.2-1.8 mm; a = 28-46; b = 3.5-5.4; c = 3-5; c′ = 15-20; V = 52-65; ♂: L = 1.7 mm; a = 28; b = 4.4; c = 3.6; supplements 8-10 (India, Singapore, Mauritius, Nigeria, St Lucia, Brazil, New Zealand, New Caledonia, Campbell Island, Malaysia, Costa Rica) .... *trichurus*
– Buccal cavity 24-25 × 41-42 μm; ♀: L = 1.4-1.8 mm; a = 27-36; b = 3.9-4.3; c = 4.8-5.6; c′ = 7-12; V = 62-65; ♂: unknown (India, Malaysia) .... *silvallus*

26 – Body *ca* 3 mm long 27

– Body shorter, *ca* 1.5-2.0 mm long 28

27 – Sphincter between uterus and oviduct; buccal cavity *ca* 60 μm long; ♀: L = 2.68-3.01 mm; a = 40-51; b = 4.6-5.0; c = 5.8-12.3; V = 64-72; ♂: L = 2.74-2.78 mm; a = 54; b = 4.7-4.8; c = 6.8; supplements 11 (Malaysia) . . . . *abrahmani*

– No sphincter between uterus and oviduct; buccal cavity 42-47 μm long; ♀: L = 2.9-3.3 mm; a = 40-46; b = 4.2-4.6; c = 6.8-7.8; c′ = 8-9; V = 67-72; ♂: unknown (Fiji Island) . . . . *kirbyi*

28 – Tail 11-12 anal body diam. long; posterior gonad two body diam. long; ♀: L = 1.4-1.6 mm; a = 31-34; b = 4.2-4.7; c = 4.1-4.7; c′ = 11-12; V = 60-63; ♂: L = 1.4-1.7 mm; a = 33-41; b = 4.4-4.7; c = 4.6-5.1; supplements 6-8 (India) . . . . *pseudodigonicus*

– Tail 13-18 anal body diam. long; posterior gonad three body diam. long; ♀: L = 1.7-2.3 mm; a = 41-51; b = 4.2-5.1; c = 3.9-4.4; c′ = 13-18; V = 55-66; ♂: unknown (Fiji Island) . . . . . . *rapidulus*

29 – Tail broadly rounded, hemispheroid, shorter to slightly longer than anal body diam. 30

– Tail conoid-arcuate or filiform, generally much longer than anal body diam. 31

30 – Lip region sharply set off by constriction; body 2-3 mm long, slender; ♀: L = 2.3-3.2 mm; a = 45-63; b = 5.9-7.7; c = 55-108; c′ = 0.8-1.4; V = 62-69; ♂: unknown (Spain) . . . *rotundicaudatus*

– Lip region slightly set off by depression; body 4 mm long, less slender; ♀: L = 4.2 mm; a = 38; b = 4.9; c = 86; c′ = 0.7; V = 69; ♂: unknown (Hungary) . . . . *magyar*

31 – Tail unusually long, whip-like, 40-50 anal body diam. long, vulva far anterior; ♀: L = 3.6-4.2 mm; a = 59-69; b = 6.3-7.1; c = 2.3-2.9; c′ = 40-50; V = 36-39; ♂: L = 3.3 mm; a = 53; b = 6.2; c = 27; supplements 12 (New Zealand) . . . . *maragnus*

– Tail never as long, at most 25 anal body diam. long; vulva at or posterior to mid-body 32

32 – Body large, 5.9-7.5 mm long 33

– Body at most 4.5 mm long, generally shorter 34

33 – Body slender; tail very long, *ca* 25 anal body diam. long; ♀: L = 5.9 mm; a = 71; b = 5.0; c = 3.8; c′ = 25; V = 51; ♂: L = 6.1 mm; a = 74; b = 5.1; c = 4.1; supplements 18 (New Zealand) . . . . *percivali*

– Body not as slender; tail short, *ca* five anal body diam. long; ♀: L = 7.3-7.5 mm; a = 35-37; b = 4-4.4; c = 11.2-12; c′ = 5.2-5.5; V = 55-58; ♂: unknown (Russia) . . . . *caesar*

34 – Caudal spinneret present 35
– Caudal spinneret absent 58

35 – Spinneret subterminal 36
– Spinneret terminal 46

36 – Spinneret subdorsal 37
– Spinneret subventral 39

37 – Tail 14-15 anal body diam. long; ♀: L = 2.21-2.25 mm; a = 42-44; b = 5.2-5.3; c = 4.2-4.3; c′ = 14-15; V = 52-53; ♂: L = 2.0-2.1 mm; a = 46-47; b = 5.2-5.6; c = 4.0-4.3; supplements ten (Cameroon) . . . . *mboticus*
– Tail 2.5-8 anal body diam. long 38

38 – Tail 2.5-4 anal body diam. long; ♀: L = 1.8-2.4 mm; a = 39-53; b = 5.1-6.2; c = 17-25; c′ = 2.5-4.3; V = 63-69; ♂: L = 1.7-2.2 mm; a = 35-52; b = 5.1-6.1; c = 24-32; supplements 9-13 (Spain) . . . . *parageminus*
– Tail 7-8 anal body diam. long; ♀: L = 2.0 mm; a = 37; b = 4.9; c = 9.5; c′ = 7-8; V = 57; ♂: unknown (Thailand) . . *rayongensis*

39 – Larger species, L = 3.5-4.3 mm 40
– Smaller species, L = 1.5-3.4 mm 41

40 – Advulval papillae absent; ♀: L = 3.5-3.8 mm; a = 38-40; b = 4.7-4.8; c = 6.4-6.9; c′ = 9-10; V = 57-60; ♂: L = 3.7 mm; a = 43; b = 4.7; c = 8.4; supplements 12 (New Caledonia) . . . . *montanus*
– Advulval papillae present; ♀: L = 3.8-4.3 mm; a = 50-53; b = 5.2-5.7; c = 4.8-5.5; c′ = 14-16; V = 50-53; ♂: L = 3.7-3.9 mm; a = 48-62; b = 4.9-5.1; c = 6.1-6.8; supplements 10-11 + 2-4 (Kenya, Uganda, Nigeria) . . . . *vulvapapillatus*

41 – Species small, L = 1.2-2 mm; advulval papillae absent; ♀: L = 1.2-2.0 mm; a = 21-39; b = 4.0-4.8; c = 5-8; c′ = 6-10; V = 56-65; ♂: unknown (India, St Lucia, El Salvador, Malaysia) . . *indicus*
– Species not as small, L = 2.2-3.4 mm; advulval papillae present or absent 42

42 – Advulval papillae absent 43
– Advulval papillae present 45

43 – Buccal cavity roomy, 49-52 × 35-41 μm; supplements 12-16; ♀: L = 2.3-2.9 mm; a = 38-42; b = 4.2-4.6; c = 6.4-7.5; c′ = 7.5-9; V = 61-63; ♂: L = 2.6-2.8 mm; a = 38-43; b = 4.3-4.5; c = 7.1-8.6 (New Caledonia) . . . . *candelabra*
– Buccal cavity not as wide; supplements 10-12 44

44 – Tail 17-18 anal body diam. long; ♀: L = 2.6-2.7 mm; a = 50-57; b = 4.9-5.2; c = 4.0-4.5; c′ = 17-18; V = 53-55; ♂: L = 2.4-2.5 mm; a = 49-57; b = 4.8-5.1; c = 4.1-4.6; supplements 10-11 (Cameroon) . . . . *kilumicus*
– Tail 9-15 anal body diam. long; ♀: L = 1.8-2.7 mm; a = 39-47; b = 4.2-4.8; c = 4.3-6.9; c′ = 9-15; V = 50-57; ♂: L = 1.8-2.2 mm; a = 42-45; b = 4.3-4.8; c = 4.4-6.9; supplements 10-12 (Cameroon) . . . . *ndu*

45 – Dorsal tooth basal; buccal cavity 57-60 × 29-30 μm; ♀: L = 2.1-2.5 mm; a = 30-36; b = 4.1-4.2; c = 4.1-5.5; c′ = 8-11; V = 54-58; ♂: L = 1.8 mm; a = 29; b = 4; c = 5.8; supplements 12 (Cameroon) . . . . *koupensis*
– Dorsal tooth suprabasal, 20-25 μm from base; buccal cavity 66-75 × 37-40 μm; ♀: L = 2.5-3.4 mm; a = 26-33; b = 4.2-4.6; c = 6.0-8.3; c′ = 7-8; V = 61-62; ♂: L = 2.1-2.9 mm; a = 32-38; b = 4.1-4.9; c = 7-9; supplements 9-11 (India) . . . . *kherai*

46 – Body > 3 mm long 47
– Body < 3 mm long 50

47 – Advulval papillae present; ♀: L = 3.2-3.5 mm; a = 44-50; b = 4.6-5.0; c = 7.6-8.4; c′ = 7.5-8.3; V = 60-61; ♂: L = 2.8-3.1 mm; a = 45-48; b = 4.5-4.7; c = 11.2-14; supplements 12-13 (Japan) . . . . *miamaensis*
– Advulval papillae absent 48

48 – Buccal cavity capacious, 64 × 53 μm; ♀: L = 3.2 mm; a = 33; b = 4.8; c = 17; c′ = 3-4; V = 65; ♂: L = 3.5 mm; a = 40; b = 5.3; c = 17; supplements 16 (USA-Virginia) . . . *brachylaimus*
– Buccal cavity longer, 67-80 μm long 49

49 – Buccal cavity narrower, 40-44 μm wide; vulva posterior; ♀: L = 2.7-3.5 mm; a = 35-39; b = 4.4-5.0; c = 10-13; c′ = 4.3-5.8; V = 62-74; ♂: L = 2.9 mm; a = 41; b = 4.6-4.7; c = 11-15; supplements 17-19 (India) . . . . *apapillatus*

– Buccal cavity broader, 56-57 μm wide; vulva anterior; ♀: L = 2.7-4.3 mm; a = 30-37; b = 4.2-4.9; c = 8-11; c′ = 6-8; V = 59-64; ♂: L = 3.1-3.7 mm; a = 36-39; b = 4.5-5.1; c = 11-11; supplements 14-18 (Hungary, India, Mauritius, South Africa, Brazil, USA-Hawaii) . . . . *risoceiae*

50 – Smaller species, never reaching 2 mm long; ♀: L = 1.4-1.9 mm; a = 27-38; b = 4-5; c = 10-17; c′ = 3.5-5.4; V = 61-72; ♂: L = 1.3-1.9 mm; a = 28-40; b = 4-5; c = 15-19; supplements 9-12 (South Africa) . . . . *rinae*

– Larger species (>2 mm long) 51

51 – Advulval papillae present 52

– Advulval papillae absent 55

52 – Buccal cavity narrowing posteriorly; ♀: L = 2.2-2.7 mm; a = 27-50; b = 4.0-4.9; c = 6.0-10.5; c′ = 6.5-9; V = 56-66; ♂: L = 2.0-2.4 mm; a = 27-50; b = 4.2-4.7; c = 9-15; supplements 12-13 + 2 (India, Nigeria, South Africa, Cameroon) . . . . *parabasidontus*

– Buccal cavity not narrowing posteriorly 53

53 – Buccal cavity with broad base; dorsal tooth pointing anteriad; ♀: L = 2.0-2.1 mm; a = 24; b = 4-5; c = 10-15; c′ = 3.5-4.0; V = 61-68; ♂: L = 2.6-2.8 mm; a = 34-37; b = 4-5; c = 16-21; supplements 13-17 (South Africa, Korea, Japan) . . . . *litoralis*

– Buccal cavity not as broad at its base 54

54 – Female tail longer (c′ = 4-6); male tail narrowing abruptly; ♀: L = 2.0-2.7 mm; a = 27-40; b = 4.0-4.6; c = 8-14; c′ = 4-6; V = 60-70; ♂: L = 2.0-2.6 mm; a = 30-37; b = 3.8-4.7; c = 8-16; supplements 12-16 (India) . . . . *nayari*

– Female tail shorter (c′ = 3-4); male tail narrowing gradually; ♀: L = 1.6-2.4 mm; a = 25-37; b = 3.9-4.7; c = 12-17; c′ = 3-4; V = 64-68; ♂: L = 1.7-2.0 mm; a = 28-44; b = 3.9-4.5; c = 12-21; supplements 11-14 (South Africa, Argentina) . . . . *geminus*

55 – Female tail very long (c′ = 13-17); ♀: L = 2.2-2.8 mm; a = 41-48; b = 4.8-5.2; c = 4.4-5.9; V = 51-56 (Togo) . . . . *togoensis*

– Female tail not as long (c′ < 10) 56

56 – Female tail longer (c′ = 8-10); vulva anterior; ♀: L = 2.2-2.4 mm; a = 20-30; b = 4.4-4.7; c = 4.3-4.8; V = 51-54 (Ecuador) . . . . *aequatorialis*

– Female tail shorter; V > 55 57

57 – Vulva anterior; supplements 14; ♀: L = 1.7-2.4 mm; a = 25-32; b = 3.7-4.3; c = 6.5-7.2; c′ = 5-7; V = 55-60; ♂: L = 1.9 mm; a = 30; b = 3.8; c = 8.6 (India, Thailand, Singapore, Ivory Coast, Colombia, New Zealand, Malaysia) .... *basidontus*
– Vulva posterior; supplements 10-12; ♀: L = 2.5-2.9 mm; a = 40-42; b = 4.1-4.6; c = 9.3-11.1; V = 60-65; c′ = 4.7-6.0; ♂: L = 2.2-2.6 mm; a = 39-46; b = 4.1-4.6; c = 11.5-14.1; supplements 10-12 (Japan) .... *sagaensis*

58 – Tail 2-5 anal body diam. long.............................59
– Tail 7-15 anal body diam. long.............................64

59 – Female tail very short, *ca* two anal body diam. long...........60
– Female tail longer, at least three anal body diam. long.........62

60 – Buccal cavity short and wide (42 × 30 μm); ♀: L = 1.4-1.8 mm; a = 20-23; b = 3.2-3.7; c = 23-28; c′ = 1.5-2; V = 72-75; ♂: L = 1.6 mm; a = 19-22; b = 3.3-3.8; c = 20-25; supplements 22 (Kirgizia) .... *kirghistanicus*
– Buccal cavity longer and relatively narrower (52-63 × 31-37 μm) 61

61 – Lip region very low, labial papillae weakly developed; buccal cavity almost rectangular; ♀: L = 1.5-1.6 mm; a = 23-26; b = 2.8-3.2; c = 17.1-19.4; c′ = 2.0-2.4; V = 67-73; ♂: L = 1.5-1.7 mm; a = 24-27; b = 3.2-3.3; c = 19.6-23; supplements 20-22 (Korea) .... *uisongensis*
– Lip region higher, labial papillae protruding; buccal cavity narrowing posteriorly; ♀: L = 1.6-1.7 mm; a = 20-22; b = 2.9-3.0; c = 17.2-21.7; c′ = 1.7-2.2; V = 71-74.6; ♂: unknown (Korea) .... *damsanensis*

62 – Vagina without sclerotised pieces; ♀: L = 1.3-1.5 mm; a = 35-42; b = 4.8-5.1; c = 8.4-9.3; c′ = 5; V = 59-63; ♂: L = 1.3-1.4 mm; a = 38-45; b = 4.6-5.1; c = 10-11; supplements 8-9 (South Africa).. *pauli*
– Vagina with sclerotised pieces 63

63 – Buccal cavity very roomy, almost spheroidal; female tail straight; ♀: L = 1.9-2.2 mm; a = 23-26; b = 5.0-5.1; c = 9.4-10.9; c′ = 3.5-4.6; V = 62-63; ♂: L = 1.6 mm; a = 23-24; b = 4.6; c = 10.6-11.3; supplements 10-11 (Italy) .... *paracutus*

– Buccal cavity not so wide; female tail ventrally curved; ♀: L = 1.6-2.4 mm; a = 28-40; b = 4.4-6.0; c = 11-16; c′ = 3-5; V = 61-66; ♂: L = 1.4-1.9 mm; a = 28-40; b = 4.4-5.5; c = 13-16; supplements 9-12 (Kazakhstan, South Africa, USA-California; Virginia) . . . . *acutus*

64 – Female tail very long, up to 15 anal body diam. long 65
– Female tail no longer than 12 anal body diam. long 66

65 – L = 2.5-3.5 mm; vulva posterior; ♀: L = 2.5-3.5 mm; a = 32-44; b = 4.2-5.1; c = 4.3-6.6; V = 55-60; c′ = 9-15; ♂: L = 2.1-3.7 mm; a = 33-47; b = 4.3-5.1; c = 4.9-6.7; supplements 11-13 (Romania) . . . . *carpathicus*
– L = 1.7-2.2 mm; vulva anterior; ♀: L = 1.7-2.2 mm; a = 27-41; b = 4.0-5.0; c = 3.3-4.0; c′ = 15; V = 47-51; ♂: L = 1.6-1.7 mm; a = 34-39; b = 4.3-4.4; c = 4; supplements 11 + 1 (Nigeria, Ivory Coast, Guinea, Cameroon) . . . . *tarjani*

66 – Body > 3 mm long; ♀: L = 3.2-4.2 mm; a = 41-49; b = 5.2-5.7; c = 7-9; c′ = 7-10; V = 58-73; ♂: L = 3.2-3.7 mm; a = 45-54; b = 5-6; c = 8-9; supplements 11-12 (New Zealand) . . . *stockdilli*
– Body < 3 mm long 67

67 – Advulval papillae present 68
– Advulval papillae absent 70

68 – Vagina with sclerotised pieces 69
– Vagina without sclerotised pieces; ♀: L = 2.4-2.5 mm; a = 35-36; b = 4.6-4.8; c = 7-8; c′ = 7; V = 61; ♂: L = 2.1 mm; a = 34; b = 4.6; c = 8; supplements one (Korea) . . . . *onchus*

69 – Buccal cavity 30-37 × 25 μm; ♀: L = 1.9-2.4 mm; a = 34-42; b = 4.3-5.1; c = 7.3-8.6; c′ = 7-9.5; V = 61-68; ♂: unknown (South Africa) . . . . *loteniae*
– Buccal cavity 54 × 37 μm; ♀: L = 2.4 mm; a = 31; b = 5.0; c = 7.1; c′ = 8.2; V = 59; ♂: unknown (India) . . . . *globibucca*

70 – Body > 2 mm long 71
– Body < 2 mm long 72

71 – V = 62-66; tail longer, c′ = 10-12; ♀: L = 2.2-3.0 mm; a = 33-42; b = 5-6; c = 6-10; c′ = 10-12; V = 62-66; ♂: L = 1.3-2.5 mm; a = 36-42; b = 5-6; c = 9-13; supplements 13-14 (South Africa) . . . . *spinicaudatus*

– V = 53-61; tail shorter, c′ = 6-9; ♀: L = 1.9-2.7 mm; a = 28-46; b = 4-5; c = 5-9; c′ = 6-9; V = 53-61; ♂: L = 2.1-2.5 mm; a = 30-42; b = 4.4-5.0; c = 8-10; supplements 12-17 (Nigeria, Kenya, Suriname, Costa Rica, USA-Florida) ...... *tenuidentatus*

72 – Lateral organs present; ♀: L = 1.5-1.7 mm; a = 28-48; b = 3.7-5.7; c = 6-8; c′ = 7-8; V = 58-66; ♂: unknown (India, Thailand, South Africa)................................. *transkeiensis*

– Lateral organs absent 73

73 – Tail strongly curved, clearly narrowing near terminus; ♀: L = 1.7-1.8 mm; a = 28-46; b = 4-5; c = 5-9; c′ = 6-9; V = 53-61; ♂: unknown (Japan) .... *arcuatus*

– Tail only slightly ventrally curved, with thicker terminal part; ♀: L = 1.7-2.0 mm; a = 28-34; b = 3.9-5.2; c = 4.6-5.9; c′ = 9-10; V = 52-59; ♂: unknown (Nigeria, Guinea) .... *clarki*

Note

Not included in key because of unavailability of literature: *I. brachylaimium*, *I. haroensis*, *I. kvavadezi*, *I. helicus* and *I. thui*.

Genus *Jensenonchus* Jairajpuri & Khan, 1982
(Fig. 49A-C)

DIAGNOSIS

Body 0.5-2.3 mm long. Cuticle smooth. Buccal cavity roomy, armed with a dorsal tooth situated in anterior third to mid-section of buccal cavity. Facing tooth, a small longitudinal ridge and/or a fine transverse rib on each subventral wall present. Pharyngo-intestinal junction tuberculate. Female genital system amphidelphic or prodelphic, in latter case without post-uterine sac. Vulva located at 60-80% of body length. Spicules arcuate with bifurcate accessory pieces. Supplements 9-15 in number. Tail similar in both sexes; either very short and rounded or 2-5 anal body diam. long and conoid-arcuate. Caudal glands reduced, terminal opening absent.

TYPE SPECIES

Jensenonchus ovatus (Jensen & Mulvey, 1968) Jairajpuri & Khan, 1982
= *Iotonchus ovatus* Jensen & Mulvey, 1968

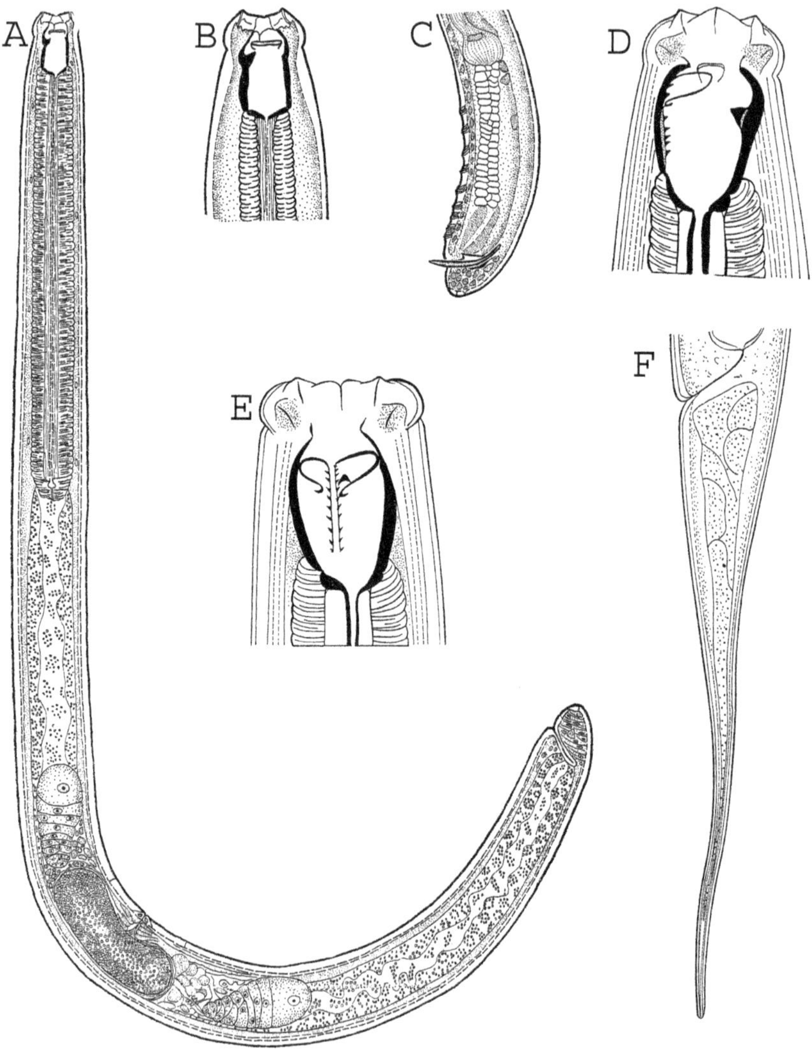

Fig. 49. *A-C:* Jensenonchus ovatus *(Jensen & Mulvey, 1968) Jairajpuri & Khan, 1982. A: Entire female; B: Anterior region; C: Male posterior region; D-F:* Prionchulellus cavenessi *Mulvey & Jensen, 1967. D, E: Anterior region; F: Female tail. A-C after Jensen and Mulvey (1968); D-F after Mulvey and Jensen (1967).*

OTHER SPECIES

J. alter Andrássy, 1993b

= *Iotonchus amphigonicus apud* Jensen & Mulvey, 1968 *nec* Thorne, 1924
J. amphigonicus (Thorne, 1924) Andrássy, 1993b
= *Mononchus amphigonicus* Thorne, 1924
= *Iotonchus amphigonicus* (Thorne, 1924) Andrássy, 1958
J. antedontoides (Coetzee, 1967b) Andrássy, 1993b
= *Iotonchus antedontoides* Coetzee, 1967b
= *Mulveyellus antedontoides* (Coetzee, 1967b) Siddiqi, 1984a
= *Iotonchus monhystera apud* Jairajpuri, 1970a; Khan & Jairajpuri, 1980 and Jairajpuri & Khan, 1982 *nec* Cobb, 1917
J. antedontus (Mulvey, 1963a) Andrássy, 1993b
= *Iotonchus antedontus* Mulvey, 1963a
= *Mulveyellus antedontus* (Mulvey, 1963a) Siddiqi, 1984a
J. vorax (Cobb, 1917) Andrássy, 1993b
= *Mononchus vorax* Cobb, 1917
= *Mononchus papillatus vorax* Cobb, 1917 (Micoletzky, 1922)
= *Iotonchus vorax* (Cobb, 1917) Mulvey, 1963a
= *Mulveyellus vorax* (Cobb, 1917) Siddiqi, 1984a
= *Mononchus sphagni* Brzeski, 1960
= *Clarkus sphagni* (Brzeski, 1960) Jairajpuri, 1970a
= *Iotonchus sphagni* (Brzeski, 1960) Loof & Winiszewska-Slipińska, 1993
= *Iotonchus celer* Susulovsky, 1988

Jensenonchus ovatus **(Jensen & Mulvey, 1968) Jairajpuri & Khan, 1982**

(Fig. 49A-C)

MEASUREMENTS

Female: L = 0.52-0.83 mm; a = 19-27; b = 3.5-3.7; c = 31-58; V = 59-63.
Male: L = 0.6-0.9 mm; a = 20-26; b = 3.5-4.7; c = 39-46.

DESCRIPTION

Lip region distinctly offset by depression with prominent angular lips. Buccal cavity rectangular, thick-walled, 16-19 × 9-12 μm in female and 18-22 × 8-12 μm in male. Dorsal tooth medium-sized, located in anterior third of buccal cavity. Distinct longitudinal rib and prominent transverse ridge on subventral walls at level with dorsal tooth. Amphidial aperture large, nearly 75% diam. of buccal cavity at level of dorsal tooth. Female genital system amphidelphic. Vaginal sclerotisation present. Usually one pair of pre- and one pair of post-vulval papillae present. Tail short, hemispheroid, less than anal body diam. long. Male with 9-10 irregularly spaced ventromedian supplements beginning within range of spicules. Spicules slender-arcuate, 44-53 μm long. Gubernaculum crescent-shaped, 18-24 μm long. Lateral accessory pieces prominent with bifurcate terminus. Caudal glands and spinneret absent.

DISTRIBUTION

USA-Oregon.

Key to species of *Jensenonchus* Jairajpuri & Khan, 1982

1 – Female genital system monodelphic-prodelphic; ♀: L = 0.7-1.5 mm; a = 21-35; b = 3.1-4.7; c = 10-19; c′ = 2-3; V = 72-81; ♂: L = 1.1-1.2 mm; a = 31-35; b = 3.6-4.0; c = 13-14; supplements 11 (India, South Africa, New Caledonia) . . . . *antedontoides*
– Female genital system didelphic-amphidelphic 2

2 – Tail in both sexes broadly rounded, shorter than anal body diam.; body very small, <1 mm long; ♀: L = 0.5-0.8 mm; a = 19-27; b = 3.5-3.7; c = 31-58; c′ = 0.6-0.7; V = 59-63; ♂: L = 0.6-0.9 mm; a = 20-26; b = 3.5-4.7; c = 39-46; supplements 9-10 (USA) . . . . *ovatus*
– Tail in both sexes conoid, ventrally bent, 2-5 anal body diam. long; body large, 2.3 mm long 3

3 – Apex of dorsal tooth situated in anterior-third of buccal cavity . . . 4
– Apex of dorsal tooth situated midway in buccal cavity 5

4 – Tail 4-5 anal body diam. long, sharply ventrally arcuate; ♀: L = 1.4-2.0 mm; a = 24-38; b = 3.4-4.3; c = 9-14; c′ = 4-5; V = 58-67; ♂: unknown (UK, Poland, Bohemia, Ukraine, Canada, USA) . . . . *vorax*
– Tail shorter, 2-3 anal body diam., not so sharply arcuate; ♀: L = 1.2-1.5 mm; a = 22-24; b = 3.1-4.0; c = 13-19; c′ = 2-3; V = 65-

69; ♂: L = 1.1-1.4 mm; a = 24-30; b = 3.1-3.8; c = 12-20; supplements 14-15 (India, USA) . . . . *antedontus*

5 – Transverse ribs on subventral walls at level of dorsal tooth; tail tip rounded; body *ca* 2 mm long; ♀: L = 2.3 mm; a = 29; b = 5; c = 22; c′ = 2.5; V = 68; ♂: L = 2.3 mm; a = 35; b = 4.9; c = 25; supplements 14-12 (USA) . . . . *amphigonicus*

– Transverse ribs on subventral walls situated more anteriad than dorsal tooth; tail tip pointed; body *ca* 1 mm long; ♀: L = 1.2-1.3 mm; a = 26; b = 3.2; c = 17-23; c′ = 2.5; V = 67-72; ♂: unknown (USA) . . . . *alter*

Genus *Caputonchus* Siddiqi, 1984a
(Fig. 50D-I)

DIAGNOSIS

Very small nematodes, 0.5-0.6 mm. Cuticle smooth. Lip region sharply set off by a deep constriction. Buccal cavity spacious but small, shorter than labial diam., dorsal tooth located in anterior quarter, pointing anteriad. No subventral ridges or denticles. Pharyngo-intestinal junction tuberculate. Female genital system amphidelphic, vulva somewhat posterior to mid-body. Tail straight, conical with minute rounded tip, 2.5-3 times as long as anal body diam. Caudal glands and terminal opening lacking. Male not known.

TYPE AND ONLY SPECIES

Caputonchus capitatus Siddiqi, 1984a

Caputonchus capitatus Siddiqi, 1984a
(Fig. 50D-I)

MEASUREMENTS

Female: L = 0.52-0.62 mm; a = 19-21; b = 3.4-3.7; c = 15-16; c′ = 2.4-2.7; V = 54-59.

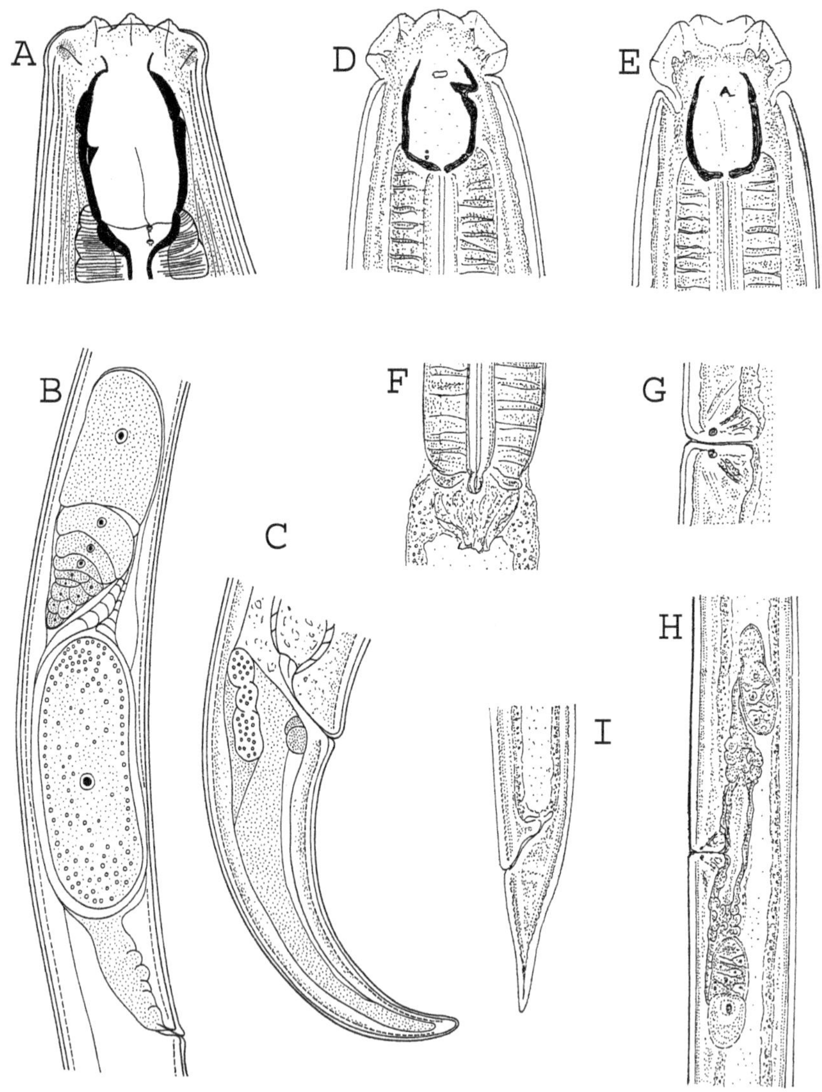

Fig. 50. *A-C:* Mulveyellus jairi *(Lordello, 1959) Siddiqi, 1984. A: Anterior region; B: Female genital system; C: Female tail; D-I:* Caputonchus capitatus *Siddiqi, 1984. D, E: Anterior region; F: Pharyngo-intestinal junction; G: Vulval region; H: Female genital system; I: Female tail. A-C after Mulvey and Jensen (1967); D-I after Siddiqi (1984).*

DESCRIPTION

Body almost straight, markedly tapering posteriorly. Lip region trapezoid, offset by a sharp constriction, 19-20 × 7-8 μm; papillae distinct. Amphidial apertures oval, 3-5 μm wide, at base of lip region. Buccal cavity 13-14 × 8-9 μm. Apex of dorsal tooth 8-9 μm from anterior end of body and 9.0-9.5 μm from base of buccal cavity. Female genital system amphidelphic. Vulva a small pore; vagina with small dot-like *pars refringens*. Tail conical with tip minutely rounded. Caudal glands and spinneret absent. Male not known.

DISTRIBUTION

West Indies.

Genus *Mulveyellus* Siddiqi, 1984a
(Figs 50A-C; 51A-E)

DIAGNOSIS

Body 0.8-3.1 mm long. Cuticle smooth. Buccal cavity barrel-shaped with a single dorsal tooth varying in location from anterior third to somewhat posterior to middle of buccal capsule. Subventral wall devoid of ribs or denticles. Pharyngo-intestinal junction tuberculate. Female genital system prodelphic or amphidelphic. Vulva at 52-81% of body length. Spicules with lateral accessory pieces. Supplements 11-15. Tail of both sexes similar in shape, conoid-arcuate, 2.5-6 anal body diam. long. Caudal glands reduced, spinneret absent.

TYPE SPECIES

Mulveyellus jairi (Lordello, 1959) Siddiqi, 1984a
= *Mononchus jairi* Lordello, 1959
= *Iotonchus jairi* (Lordello, 1959) Clark, 1961a

OTHER SPECIES

M. arenicola (Altherr, 1963) Andrássy, 1993b
= *Iotonchus arenicola* Altherr, 1963
M. monhystera (Cobb, 1917) Siddiqi, 1984a

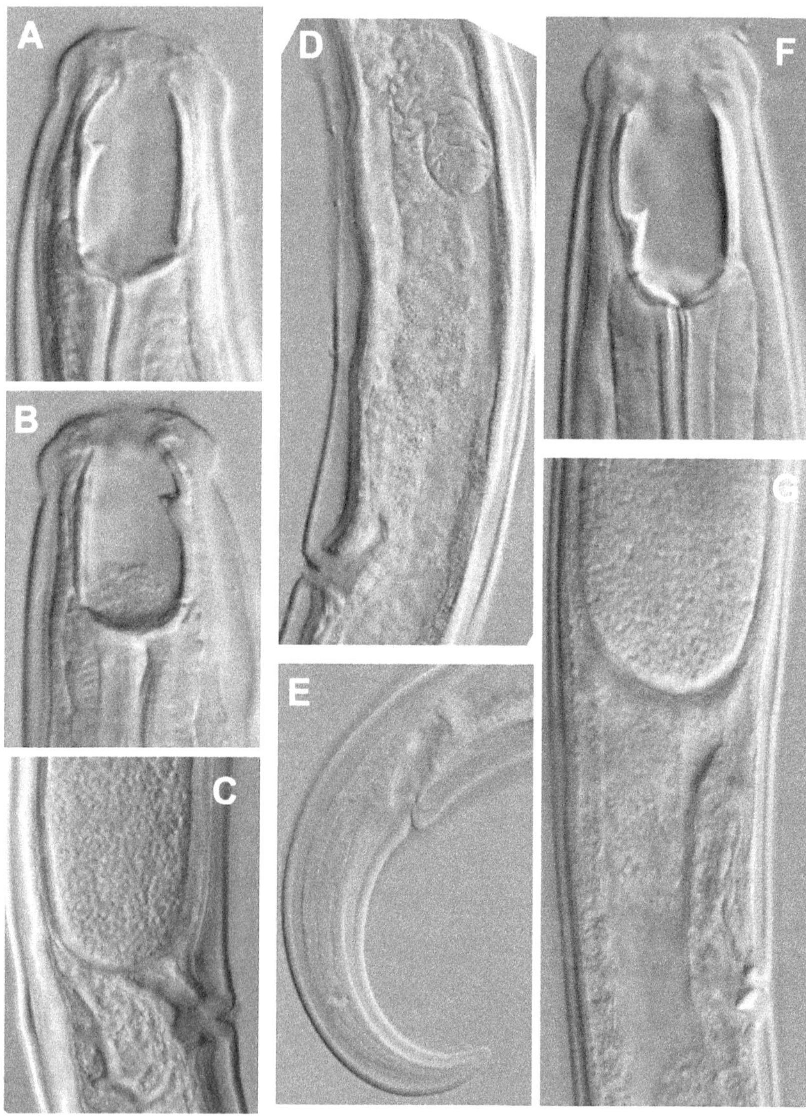

Fig. 51. *A-E:* Mulveyellus jairi *(Lordello, 1959) Siddiqi, 1984. A, B: Anterior region; C: Vulval region; D: Female genital system; E: Female tail; F, G:* Iotonchulus darreni *(Ahmad* et al., *2005) Vinciguerra & Orselli, 2006. F: Anterior region; G: Female genital system.*

= *Mononchus monhystera* Cobb, 1917
= *Iotonchus monhystera* (Cobb, 1917) Jairajpuri, 1970a
M. parazschokkei (Allgén, 1929) Andrássy, 1993b
= *Mononchus parazschokkei* Allgén, 1929
= *Mononchus* (*Iotonchus*) *parazschokkei* Allgén, 1929 (Goodey, 1951)
= *Iotonchus parazschokkei* (Allgén, 1929) Goodey, 1951
M. shamimi (Patil & Khan, 1982) Andrássy, 1993b
= *Iotonchus shamimi* Patil & Khan, 1982

Mulveyellus jairi (Lordello, 1959) Siddiqi, 1984a
(Figs 50A-C; 51A-E)

MEASUREMENTS

Female: L = 0.8-1.6 mm; a = 23-30; b = 3.4-4.3; c = 12-16; c′ = 2.5-3.0; V = 71-76.

DESCRIPTION

Lip region set off. Buccal cavity with fairly thick walls, 35-40 × 21-25 μm. Dorsal tooth relatively small, situated about midway in buccal cavity. Female reproductive system monodelphic, prodelphic, vulval lips raised above level of cuticle. Vulval papillae absent. Tail conoid, ventrally arcuate, 75-115 μm long, without glands or terminal opening. Male absent.

REMARKS

Malcevschi (1981) described a male of this species which seems to belong to *M. monhystera* (Cobb, 1917) by virtue of the anterior position of the dorsal tooth (Andrássy, 1993b).

DISTRIBUTION

Brazil, India, Nigeria, St Lucia.

Key to species of *Mulveyellus* Siddiqi, 1984a

1 – Female genital system mono-prodelphic, no post-uterine sac 2

– Female genital system amphidelphic 3

2 – Dorsal tooth in anterior third of buccal cavity; ♀: L = 1.0-1.3 mm; a = 23-30; b = 3.6-4.3; c = 12-17; c′ = 2.6-3.0; V = 72-81; ♂: L = 1.1-1.2 mm; a = 31-35; b = 3.6-4.0; c = 13-16; supplements 11-12 (Ivory Coast, South Africa, Argentina, Brazil, Colombia).. *monhystera*

– Dorsal tooth midway in buccal cavity; ♀: L = 0.8-1.6 mm; a = 23-30; b = 3.4-4.3; c = 12-16; c′ = 2.5-3; V = 71-76; ♂: unknown (India, Nigeria, St Lucia, Brazil) .... *jairi*

3 – Apex of dorsal tooth situated in anterior third of buccal cavity; ♀: L = 1.7 mm; a = 24; b = 3.6; c = 14; c′ = 3; V = 64; ♂: unknown (India) .... *shamimi*

– Apex of dorsal tooth situated posterior to middle of buccal cavity.. 4

4 – Tail 5-6 anal body diam. long, curled up; ♀: L = 3.1 mm; a = 41; b = 5.4; c = 13; c′ = 5-6; V = 52; ♂: unknown (Argentina)...... .... *arenicola*

– Tail 3.5-4 anal body diam. long, simply arcuate, not curled up; ♀: L = 1.1-2.8 mm; a = 23-27; b = 3.4-4.6; c = 8-15; c′ = 3.5-4; V = 57-65; ♂: L = 2.9 mm; a = 33-36; b = 3.7-4.1; c = 25-27; supplements 14-15 (Chile, New Zealand, Campbell Islands) .... *parazschokkei*

Genus *Nullonchus* Siddiqi, 1984d

(Fig. 52A-C)

DIAGNOSIS

Body 1-2 mm long. Cuticle smooth. Lip region slightly set off. Buccal cavity roomy, broadly ovoid, completely devoid of any teeth or denticles or ribs. Pharyngo-intestinal junction tuberculate; tubercles, however, in general smaller than in family. Female genital system mono-prodelphic or amphidelphic. Male not known. Tail conoid-arcuate, 2-5 anal body diam. long. Caudal glands and spinneret present.

TYPE SPECIES

Nullonchus levistomus Siddiqi, 1984d

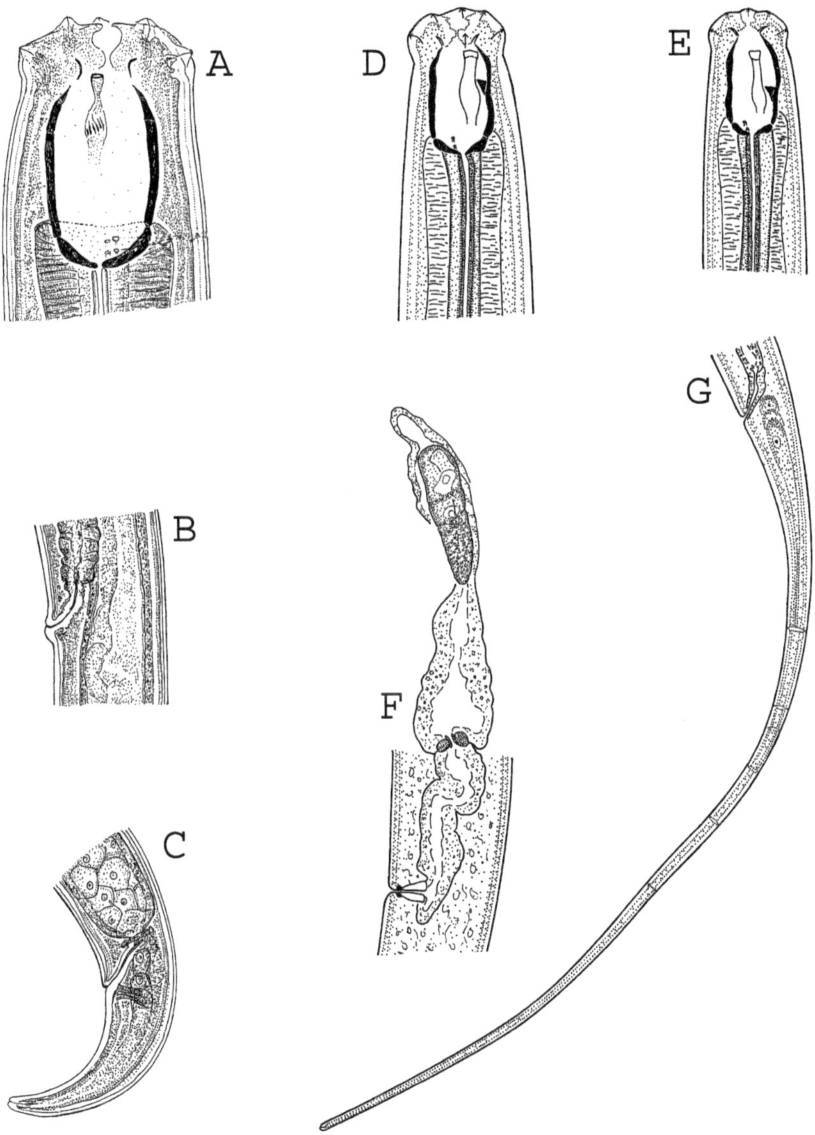

Fig. 52. *A-C:* Nullonchus levistomus *Siddiqi, 1984. A: Anterior region; B: Vulval region; C: female tail; D-G:* Iotonchulus longicaudatus *(Baqri* et al.*, 1978) Andrássy, 1993. D, E: Anterior region; F: Female genital system; G: Female tail. A-C after Siddiqi (1984); D-G after Baqri and Jairajpuri (1978).*

OTHER SPECIES

N. rafiqi Jana, Chatterjee & Manna, 2006
N. rapax Siddiqi, 1984d
N. valens Siddiqi, 1984d

Nullonchus levistomus Siddiqi, 1984d
(Fig. 52A-C)

MEASUREMENTS

Female: L = 1.0-1.5 mm; a = 23-28; b = 3.3-4.1; c = 12-18; c′ = 2.3-2.9; V = 74-76.

DESCRIPTION

Lip region broadly rounded, 29-38 μm diam., wider than adjoining body with slightly raised lips and distinct papillae. Amphidial aperture small, oval, at base of lip region, 12-15 μm from anterior end. Buccal cavity barrel-shaped, with smooth walls devoid of any teeth, denticles or ridges. Pharyngo-intestinal junction tuberculate. Genital system monodelphic-prodelphic. Vagina directed anteriad, without *pars refringens*. Post-uterine sac completely absent. Tail conoid, ventrally arcuate, 60-90 μm long. Caudal glands and terminal opening present. Male not known.

DISTRIBUTION

Colombia-Amazonas.

Key to species of *Nullonchus* Siddiqi, 1984d

1 – Female genital system amphidelphic; ♀: L = 2.7-2.9 mm; a = 32-35; b = 5.5-5.8; c = 8.4-8.7; c′ = 5.5-5.9; V = 57.6-58.7; ♂: unknown . . . . *rafiqi*
– Female genital system mono-prodelphic 2

2 – Post-uterine sac present, 1.5 corresponding body diam. long; buccal cavity *ca* 50 μm long; ♀: L = 1.9 mm; a = 31; b = 4; c = 10; c′ = 5; V = 72; ♂: unknown (Colombia) . . . . *valens*

– Post-uterine sac absent; buccal cavity 30-35 μm long 3

3 – Tail short, 2-3 anal body diam. long; vagina strongly arcuate, directed anteriad; ♀: L = 1.0-1.5 mm; a = 23-28; b = 3.3-4.1; c = 12-18; c′ = 2.3-3.1; V = 74-76; ♂: unknown (Colombia) . . . . *levistomus*

– Tail longer, 4-5 anal body diam.; vagina slightly oblique; ♀: L = 1.4-1.7 mm; a = 28-36; b = 4.0-4.1; c = 9.8-11.4; c′ = 4.3-5.1; V = 68-73; ♂: unknown (Colombia) . . . . *rapax*

Genus *Iotonchulus* Andrássy, 1993b

(Figs 51F, G; 52D-G)

Diagnosis

Body either small, near 1 mm, or very large close to 4 mm. Cuticle smooth. Buccal cavity barrel-shaped, dorsal tooth rather small, located close to midway in buccal cavity. Subventral armature lacking. Pharyngo-intestinal junction tuberculate. Female reproductive system amphidelphic or prodelphic, in latter case with a very short post-uterine sac. Spicules with small lateral accessory pieces. Supplements 11-13. Female tail filiform, 12-19 anal body diam. long, male tail comparatively shorter and more plump. Caudal glands and spinneret present.

Type species

Iotonchulus longicaudatus (Baqri, Baqri & Jairajpuri, 1978) Andrássy, 1993b
= *Iotonchus longicaudatus* Baqri, Baqri & Jairajpuri, 1978
= *Mulveyellus longicaudatus* (Baqri, Baqri & Jairajpuri, 1978) Siddiqi, 1984a
= *Iotonchus heynsi* Mohandas & Prabhoo, 1979a

Other species

I. bangkokensis (Buangsuwon & Jensen, 1966) Andrássy, 1993b
= *Iotonchus bangkokensis* Buangsuwon & Jensen, 1966
I. darreni (Ahmad, Baniyamuddin & Jairajpuri, 2005) Vinciguerra & Orselli, 2006
= *Iotonchus darreni* Ahmad, Baniyamuddin & Jairajpuri, 2005

I. ophiocercus (Clark, 1961a) Andrássy, 1993b
= *Iotonchus ophiocercus* Clark, 1961a

Iotonchulus longicaudatus (Baqri, Baqri & Jairajpuri, 1978) Andrássy, 1993b
(Fig. 52D-G)

MEASUREMENTS

Female: L = 1.08-1.32 mm; a = 39-45; b = 3.5-4.2; c = 3.4-4.0; c′ = 14-19; V = 59-61.

DESCRIPTION

Lip region offset, 17-21 × 8-11 μm, wider than adjoining body. Amphidial aperture near anterior end of buccal cavity, 8-11 μm from anterior extremity. Buccal cavity cylindrical, 21-24 × 12-15 μm. Apex of dorsal tooth 11-17 μm or 50-66% of buccal cavity length from its base. Pharyngo-intestinal junction tuberculate. Vulva transverse; vagina sclerotised distally. Genital system mono-prodelphic; posterior branch represented by a very small sac. Tail 283-334 μm long, filiform. Caudal glands and subterminal opening present. Male not known.

DISTRIBUTION

India, Kazakhstan.

Key to species of *Iotonchulus* Andrássy, 1993b

1 – Large species, 3.7-4.0 mm long; female amphidelphic; ♀: L = 3.8-4.0 mm; a = 41-47; b = 4.7-4.9; c = 4.7-5.2; c′ = 12-14; V = 54-62; ♂: L = 3.7-3.8 mm; a = 38-40; b = 4.6-5; c = 5.8-10; supplements 11-13 (New Zealand) .... *ophiocercus*
– Much smaller species, *ca* 1 mm long; female mono-prodelphic . . . 2

2 – Buccal cavity 26-28 × 15-16 μm; post-uterine sac completely absent; ♀: L = 1.09-1.15 mm; a = 32-34; b = 4.2-4.3; c = 4.4-5.1; c′ = 10-11; V = 60-63; ♂: unknown (Singapore) .... *darreni*
– Buccal cavity < 25 μm long; small post-uterine sac present 3

3 – Tail terminus bulbous; buccal cavity 22-23 × 13-15 μm; ♀: L = 08-0.9 mm; a = 33-43; b = 4.1-4.7; c = 2.9-4.0; c′ = 12-15; V = 52-60; ♂: unknown (Thailand)........................*bangkokensis*
 – Tail terminus cylindrical; buccal cavity 21-24 × 12-15 μm; ♀: L = 1.0-1.3 mm; a = 32-45; b = 3.4-4.6; c = 3.4-4.1; c′ = 13-19; V = 59-62; ♂: unknown (India, Kazakhstan).......*longicaudatus*

Subfamily Hadronchinae Khan & Jairajpuri, 1980

DIAGNOSIS

Dorsal tooth large; subventral teeth in two longitudinal rows of few to numerous denticles.

TYPE GENUS

Hadronchus Mulvey & Jensen, 1967

OTHER GENERA

Prionchulellus Mulvey & Jensen, 1967
Parahadronchus Mulvey, 1978
Hadronchulus Ray & Das, 1983
= *Morenchus* Dhanachand, Renubala & Mohilal, 1991
Hadronchoides Jairajpuri & Rahman, 1984

Key to genera of Hadronchinae Khan & Jairajpuri, 1980

1 – Dorsal tooth opposed by two longitudinal, serrate ridges.........2
 – Dorsal tooth opposed by fine scattered denticles................4

2 – Dorsal tooth located in posterior half of buccal cavity, mostly sub-basal, serrate ridges similarly posterior in location; caudal spinneret present.................................... *Parahadronchus*
 – Dorsal tooth located in anterior half of buccal cavity, serrate ridges occupying almost entire length of buccal cavity; caudal spinneret absent...3

3 – Subventral walls only with two ribs; female mono-prodelphic..... ....*Hadronchus*

– Subventral walls with serrate ridges forming hook-like loop anteriorly . . . . *Prionchulellus*

4 – Dorsal tooth in anterior part of buccal cavity, denticles almost along entire length of subventral walls; female mono-prodelphic . . . . *Hadronchoides*

– Dorsal tooth in posterior part of buccal cavity; denticles also restricted to posterior walls; female amphidelphic . . . *Hadronchulus*

Genus *Hadronchus* Mulvey & Jensen, 1967
(Fig. 53A-D)

DIAGNOSIS

Body 1.5-2.2 mm long. Cuticle smooth. Buccal cavity roomy, broad at both ends, as long as, or somewhat longer than labial diam. Dorsal tooth well developed, located in anterior half of buccal cavity at 60-70% of buccal cavity length from its base. Opposed by two denticulate or serrate ridges occupying almost entire length of subventral walls. Pharyngo-intestinal junction tuberculate. Female genital system mono-prodelphic, with or without post-uterine branch. Vulva at 61-69% of body length. Spicules arched. Supplements 8-10 in number. Tail in both sexes similar, conoid-arcuate, 4-6 anal body diam. long. Caudal glands and spinneret reduced.

TYPE SPECIES

Hadronchus bisexualis Mulvey & Jensen, 1967

OTHER SPECIES

H. monohystera Mulvey & Jensen, 1967

Hadronchus bisexualis Mulvey & Jensen, 1967
(Fig. 53A-D)

MEASUREMENTS

Female: L = 1.6-2.2 mm; a = 27-35; b = 3.7-4.7; c = 11.9-16.8; V = 61-65.

Male: L = 1.5-1.9 mm; a = 33-40; b = 3.7-4.1; c = 15.8-24.

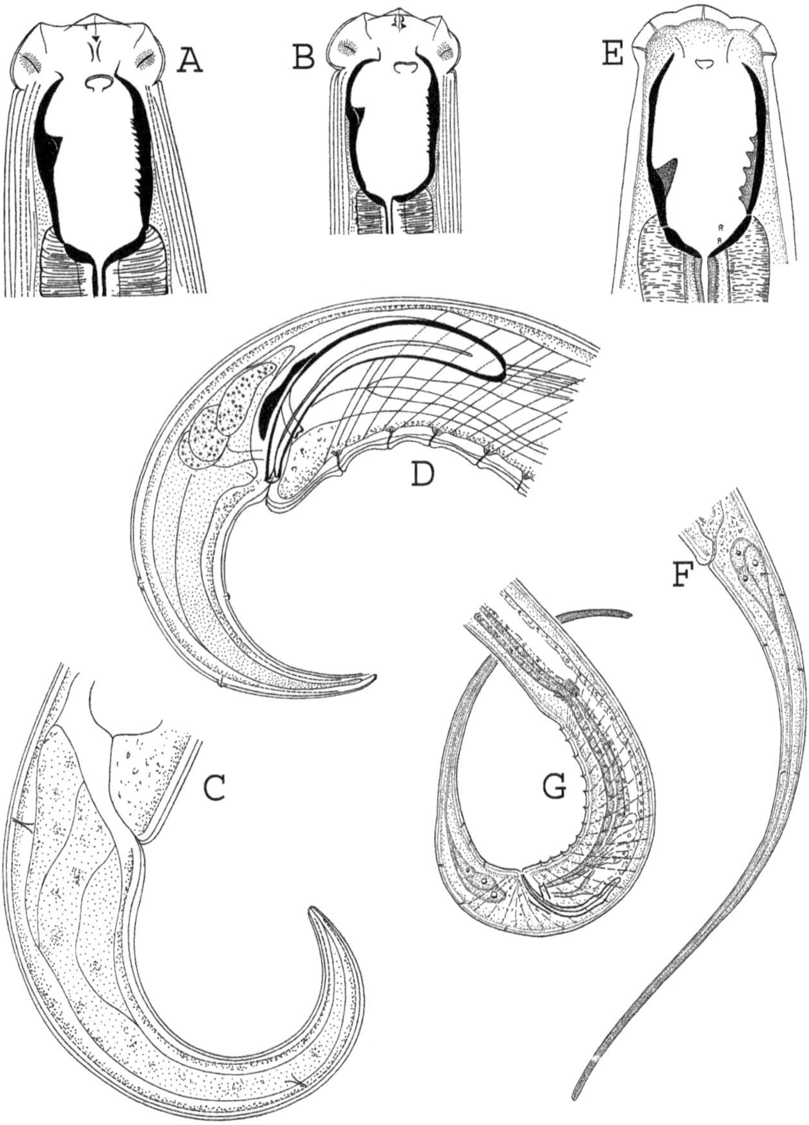

Fig. 53. *A-D:* Hadronchus bisexualis *Mulvey & Jensen, 1967. A, B: Anterior region; C: Female tail; D: Male tail region; E-G:* Parahadronchus andamanicus *(Jairajpuri, 1969) Mulvey, 1978. E: Anterior region; F: Female tail; G: Male posterior region. A-D after Mulvey and Jensen (1967); E-G after Jairajpuri (1969).*

DESCRIPTION

Labial region distinctly offset from body. Buccal cavity roomy with thick walls, 40-47 × 26-30 μm. Dorsal tooth medium-sized, located slightly anterior to middle of buccal cavity, its apex at 64-72% of buccal cavity length from its base, opposed by two longitudinal rows of small denticles. Pharyngo-intestinal junction tuberculate. Female genital system mono-prodelphic. Post-uterine sac with walls composed of large cells. Uterine egg 95-110 × 70-80 μm. Vulval papillae absent. Tail conoid, ventrally arcuate, 110-170 μm long, without caudal glands or terminal opening. Male with buccal cavity 36-42 × 22-25 μm. Spicules massive, arcuate, 75-90 μm long; gubernaculum short. Lateral accessory pieces with bifid ends. Supplements 8-10, fairly well developed, regularly arranged. Tail conoid, ventrally arcuate, 75-95 μm long, without glands or terminal opening.

DISTRIBUTION

Nigeria.

Key to species of *Hadronchus* Mulvey & Jensen, 1967

1 – Post-uterine sac present; tail as long as four anal body diam.; subventral denticles small; ♀: L = 1.6-2.2 mm; a = 27-35; b = 3.7-4.1; c = 12-17; c′ = 4; V = 61-65; ♂: L = 1.5-1.9 mm; a = 33-40; b = 3.7-4.1; c = 16-24; supplements 10-12 . . . . *bisexualis*
– Post-uterine sac absent; tail as long as six anal body diam.; subventral denticles comparatively large; ♀: L = 1.6-1.9 mm; a = 34-40; b = 3.6-3.9; c′ = 6; V = 64-69; ♂: unknown . . . . *monohystera*

Genus *Prionchulellus* Mulvey & Jensen, 1967
(Fig. 49D-F)

DIAGNOSIS

Small nematodes, 0.9-1.4 mm long. Cuticle smooth. Buccal cavity barrel shaped, *ca* 1.5 times as long as wide. Dorsal tooth in anterior third of buccal cavity, opposed by two rows of evenly spaced denticles along each longitudinal rib. Both longitudinal ribs extended anteriorly to form loops within anterior area of buccal cavity. Free ends of both loops curved to form small tooth-like structures. Pharyngo-intestinal junction tuberculate. Female genital system mono-prodelphic. Vulva at mid-body. Tail elongate, as long as six anal body diam., without caudal glands and spinneret. Male unknown.

TYPE AND ONLY SPECIES

Prionchulellus cavenessi Mulvey & Jensen, 1967

Prionchulellus cavenessi Mulvey & Jensen, 1967
(Fig. 49D-F)

MEASUREMENTS

Female: L = 0.9-1.4 mm; a = 18-27; b = 3.7-4.6; c = 6.1-7.7; V = 49-56.

DESCRIPTION

Labial region distinctly offset by a constriction. Amphidial apertures 5-6 μm wide, at level of anterior end of buccal cavity. Buccal cavity barrel-shaped, with thick walls, 27-31 $\times$ 19-21 μm. Dorsal tooth medium-sized, opposed by evenly spaced denticles along both longitudinal ribs. Both ribs with looped extensions, free ends forming hook-like structure. Pharyngo-intestinal junction tuberculate. Female genital system mono-prodelphic, with ovary generally reflexed to level of vulva. No vulval papillae or sperm in genital tract. Tail conoid in anterior third then abruptly cylindroid to terminus. Caudal glands and terminal pore absent. Male unknown.

REMARKS

Mulvey and Jensen (1967) recorded two aberrant forms of this species that varied in tail shape and length only.

DISTRIBUTION

Nigeria-Oyo Province.

Genus *Parahadronchus* Mulvey, 1978
(Figs 53E-G; 54)

DIAGNOSIS

Body 1.9-3.3 mm long. Cuticle smooth. Buccal cavity large, barrel-shaped, as long as, or longer than labial diam. Dorsal tooth large, located in posterior half of buccal cavity; opposed by two or four short longitudinal ridges, each bearing 2-8 small denticles, also located in posterior part of buccal cavity. Pharyngo-intestinal junction tuberculate. Female genital system amphidelphic or mono-prodelphic. Vulva at 54-75% of body length. Spicules comparatively slender and arcuate. Male with 10-13 supplements. Tail of both sexes similar, elongate to filiform, as long as 6-18 anal body diam. Caudal gland and spinneret present or absent.

TYPE SPECIES

Parahadronchus andamanicus (Jairajpuri, 1969) Mulvey, 1978
= *Hadronchus andamanicus* Jairajpuri, 1969

OTHER SPECIES

P. diphuensis (Phukan & Sanwal, 1981) Andrássy, 1993b
= *Hadronchus diphuensis* Phukan & Sanwal, 1981
P. egregius Andrássy, 1993b
P. magnus Dhanam & Jairajpuri, 1998
P. marami Renubala & Dhanachand, 1992
P. selangorensis Loof, 2006
P. shakili (Jairajpuri, 1969) Mulvey, 1978
= *Hadronchus shakili* Jairajpuri, 1969
= *Hadronchus karangensis* Phukan & Sanwal, 1981
= *Parahadronchus mangiferi* Saha, Lal & Singh, 2006 **syn. n.**
P. siroii Renubala & Dhanachand, 1992
P. subhonicus Dhanachand, Renubala & Mohilal, 1991
P. yuenae (Thong, 1971) Mulvey, 1978
= *Hadronchus yuenae* Thong, 1971

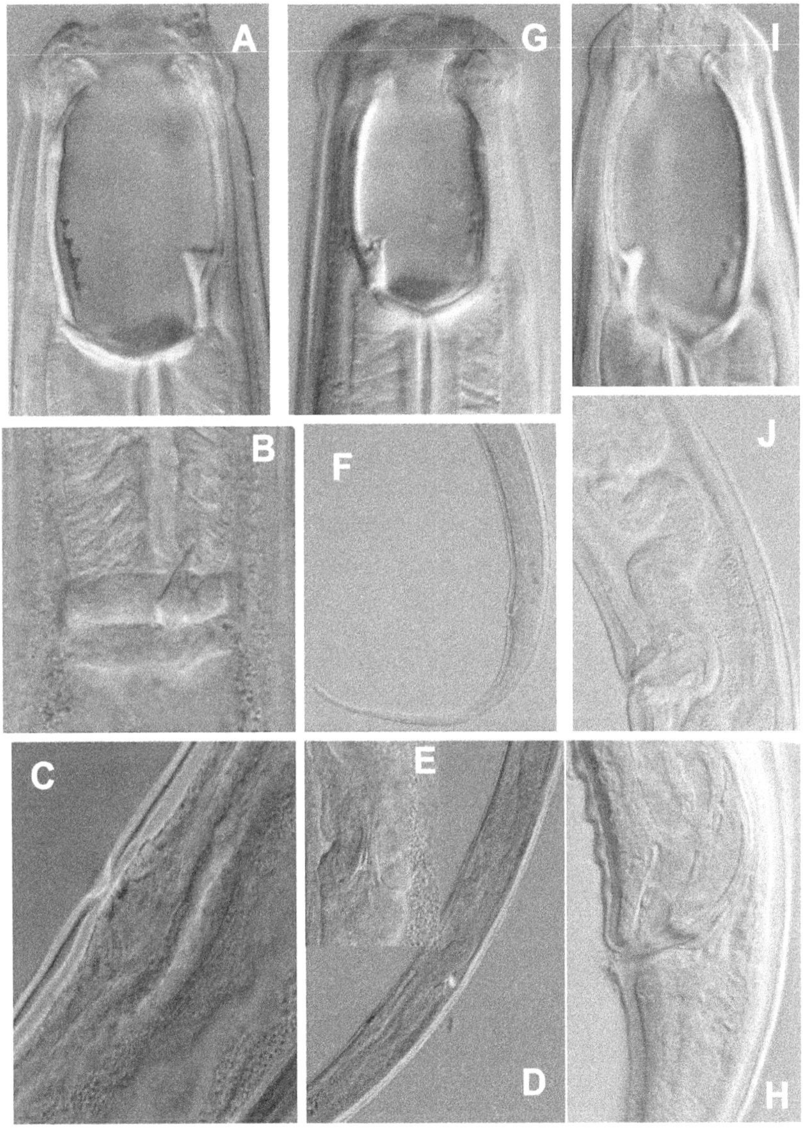

Fig. 54. *A-F:* Parahadronchus andamanicus *(Jairajpuri, 1969) Mulvey, 1978. A: Anterior region; B: Pharyngo-intestinal junction; C: Vulval region; D, E: Female genital system; F: Female tail; G, H:* Parahadronchus shakili *(Jairajpuri, 1969) Mulvey, 1978. G: Anterior region; H: Male cloacal region; I, J:* Parahadronchus yuenae *(Thong, 1971) Mulvey, 1978. I: Anterior region; J: Vulval region.*

Parahadronchus andamanicus (Jairajpuri, 1969) Mulvey, 1978
(Figs 53E-G; 54A-F)

MEASUREMENTS

Female: L = 2.24-3.22 mm; a = 30-41; b = 3.7-4.4; c = 5-13; c′ = 8-10; V = 64-75.
Male: L = 2.21-2.78 mm; a = 32-42; b = 4.0-4.4; c = 6-8; c′ = 6-7.

DESCRIPTION

Lip region offset, 46-65 × 18-26 μm, distinctly wider than adjoining body; labial papillae not very prominent. Amphidial apertures at 14-21 μm from anterior end. Buccal cavity 58-77 × 35-44 μm. Dorsal tooth large, near base of buccal cavity, its apex 26-35 μm from base of buccal cavity. Each subventral wall bearing 4-8 small denticles opposite to dorsal tooth. Pharyngo-intestinal junction tuberculate. Vulva transverse, with cuticularised lips. Pre- and post-vulval papillae present, their number variable; 0-3 pre- and 0-2 post-vulval. Anterior genital branch well developed, functional; posterior branch degenerate to form a uterine sac, 158-221 μm or 2-3 corresponding body diam. long. Tail 410-520 μm long, filiform. Caudal glands well developed, opening terminal. Male with buccal cavity 53-58 × 32-37 μm. Spicules 82-100 μm long medially; gubernaculum 23-28 μm and lateral accessory pieces 18-23 μm long. Supplements 12-13, regularly spaced. Tail 320-400 μm long. Caudal glands and spinneret well developed.

DISTRIBUTION

India-Andamans; Manipur.

Key to species of *Parahadronchus* Mulvey, 1978

1 – Female genital system mono-prodelphic 2
– Female genital system amphidelphic 4

2 – Post-uterine sac absent; tail 590-660 μm long; ♀: L = 2.3-2.9 mm; a = 35-41; b = 4.2-4.7; c = 3.9-4.7; c′ = 12-13; V = 56-62; ♂: unknown (Singapore, Malaysia) . . . . *yuenae*
– Post-uterine sac present, 2-3 body diam. long; tail 410-500 μm long. 3

3 – Buccal cavity 60-80 μm long, roomy; spinneret terminal; ♀: L = 2.2-3.2 mm; a = 30-41; b = 3.7-4.4; c = 5-13; c′ = 8-10; V = 64-75; ♂: L = 2.2-2.8 mm; a = 32-42; b = 4.0-4.4; c = 6-8; supplements 12-13 (India) . . . . *andamanicus*

– Buccal cavity 50 μm long, less roomy; spinneret subdorsal; ♀: L = 2.8 mm; a = 38; b = 4.7; c = 5.7; c′ = 8.6; V = 67; ♂: L = 2.2-2.5 mm; a = 36-39; b = 4.2-4.9; c = 6.6-7.1; c′ = 6.6-7.1; supplements 12-13 (India) . . . . *subhonicus*

4 – Subventral denticles arranged along four ridges; spinneret lacking; ♂: L = 2.5 mm; a = 42; b = 4.5; c = 4.2; c′ = 18; V = 54; ♂: unknown (Vietnam) . . . . *egregius*

– Subventral denticles arranged along two ridges; spinneret present 5

5 – Buccal cavity large, 69-84 μm long; dorsal tooth apex at 38-41% of buccal cavity length from its base; advulval papillae absent; ♀: L = 3.6-4.7 mm; a = 48-51; b = 4.4-4.9; c = 8-9; c′ = 7.9-9.9; V = 58-60; ♂: L = 4.0-4.1 mm; a = 50-57; b = 4.1-4.4; c = 10-11; c′ = 4.6; supplements 15-16 (India) . . . . *magnus*

– Buccal cavity medium-sized, <60 μm long; dorsal tooth apex > 40% of buccal cavity length from its base; vulval papillae usually present 6

6 – Tail > 500 μm long (c′ = 13-18) 7

– Tail < 430 μm (c′ = 6-11) 8

7 – Buccal cavity 46-50 × 29-37 μm; dorsal tooth apex at 41-46% of buccal cavity length from its base; ♀: L = 2.34-3.11 mm; a = 41-61; b = 5.0-6.3; c = 4.1-5.2; c′ = 5.2-18.6; V = 52-59; ♂: L = 2.18 mm; a = 42.5; b = 4.9; c = 4.5; c′ = 11.1; spicules 65 μm; supplements 11 (India) . . . . *siroii*

– Buccal cavity 55-57 × 32-33 μm; dorsal tooth apex at 45-49% of buccal cavity length from its base; ♀: L = 3.05-3.21 mm; a = 44-50; b = 3.9-4.9; c = 4.9-5.3; c′ = 13-16; V = 58-60; ♂: unknown (Malaysia) . . . . *selangorensis*

8 – Apex of dorsal tooth < 30% of buccal cavity length from its base; ♀: L = 2.47-2.58 mm; a = 24-28; b = 4.2-4.3; c = 6.9-7.3; c′ = 6.8-7.4; V = 60-62; ♂: L = 2.48 mm; a = 33; b = 4.7; c = 11; supplements 12 (India) . . . . *mangiferi*

– Apex of dorsal tooth > 30% of buccal cavity length from its base 9

9 – Spinneret subterminal; ♀: L = 2.38 mm; a = 46.6; b = 4.37; c = 6.3; c′ = 10.6; V = 57.7; ♂: L = 1.66-2.16 mm; a = 31.5-45.1; b = 3.7-4.6; c = 5.9-6.7; c′ = 7.6-8.3; supplements 8-9 (India).. *marami*

– Spinneret terminal.. 10

10 – Apex of dorsal tooth 40-60% of buccal cavity length from its base; subventral ridges with 3-6 denticles each; spicules 80-95 μm long; ♀: L = 2.3-3.3 mm; a = 32-47; b = 4.2-4.9; c = 5.5-11; c′ = 6-11; V = 56-70; ♂: L = 2.0-2.8 mm; a = 28-44; b = 3.4-4.6; c = 6.6-13; supplements 11-13 (India).................. *shakili*

– Apex of dorsal tooth at 60-70% of buccal cavity length from its base; subventral ridges with 2-3 denticles each; spicules 65-75 μm long; ♀: L = 1.9-2.2 mm; a = 34-38; b = 4.2-4.7; c = 4.9-5.3; c′ = 10-12; V = 56-59; ♂: L = 1.9-2.2 mm; a = 36-40; b = 4.4-4.8; c = 5-6; supplements 10-11 (India)............. *diphuensis*

Genus *Hadronchulus* Ray & Das, 1983

= *Morenchus* Dhanachand, Renubala & Mohilal, 1991

(Fig. 55A-D)

DIAGNOSIS

Body 1.8-3.2 mm long. Cuticle smooth. Buccal cavity moderately large. Dorsal tooth massive, in posterior part of buccal cavity. Subventral walls opposite dorsal tooth bearing minute denticles arranged mostly in posterior half. Pharyngo-intestinal junction tuberculate. Female genital system amphidelphic, posterior gonad occasionally less developed than anterior. Vulva located at 50-70% of body length. Spicules arcuate, supplements 10-13 in number. Tail of both sexes similar, elongate-conoid to filiform, as long as 4-20 anal body diam. Caudal gland and spinneret present.

TYPE SPECIES

Hadronchulus shamimi Ray & Das, 1983

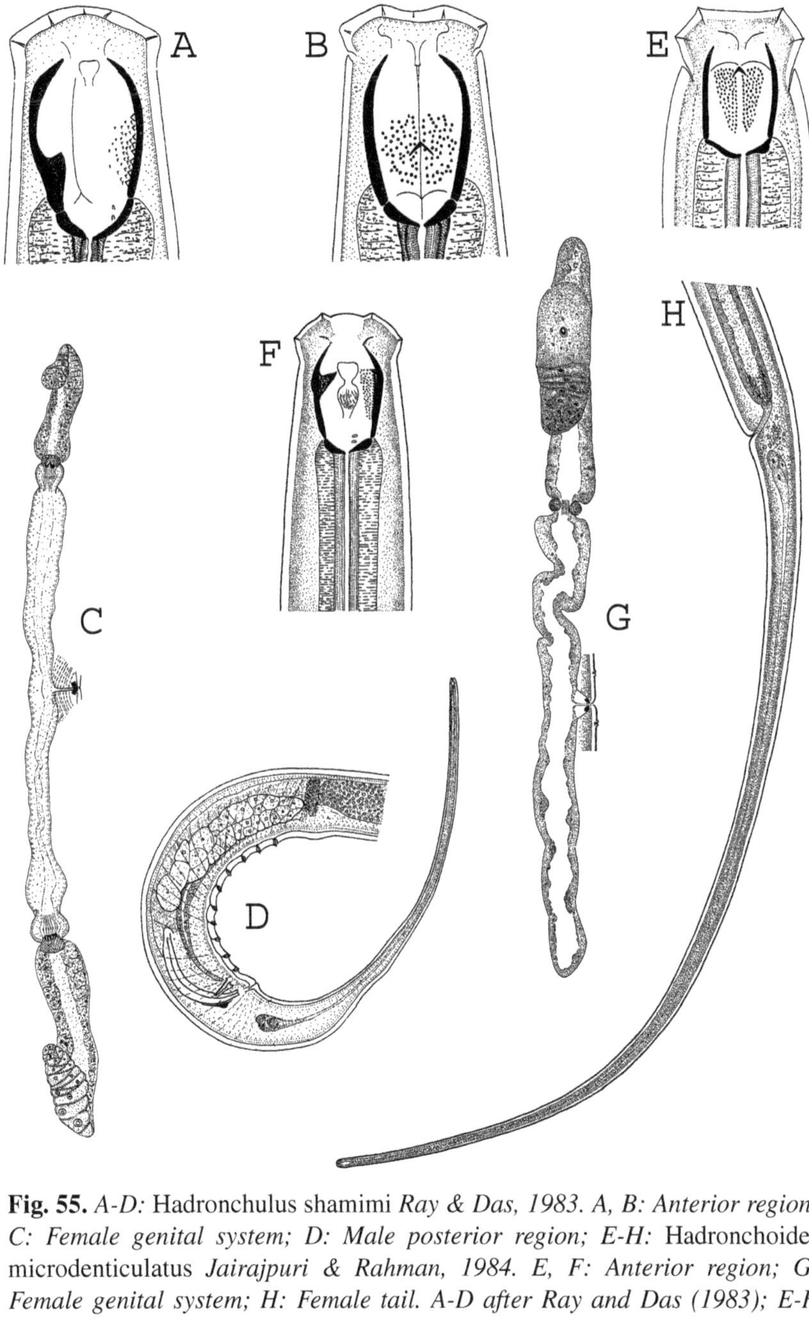

Fig. 55. *A-D:* Hadronchulus shamimi *Ray & Das, 1983. A, B: Anterior region; C: Female genital system; D: Male posterior region; E-H:* Hadronchoides microdenticulatus *Jairajpuri & Rahman, 1984. E, F: Anterior region; G: Female genital system; H: Female tail. A-D after Ray and Das (1983); E-H after Jairajpuri and Rahman (1984).*

OTHER SPECIES

H. denticulus (Dhanachand, Renubala & Mohilal, 1991) Andrássy, 1993b
=*Morenchus denticulus* Dhanachand, Renubala & Mohilal, 1991
H. grandis (Patil & Khan, 1982) Andrássy, 1993b
=*Sporonchulus grandis* Patil & Khan, 1982

***Hadronchulus shamimi* Ray & Das, 1983**
(Fig. 55A-D)

MEASUREMENTS

Female: L = 2.67-2.83 mm; a = 39-44; b = 4.5-4.7; c = 5.8-6.3; c′ = 14-15; V = 62-68.
Male: L = 2.37-2.75 mm; a = 42-46; b = 4.5-4.7; c = 6.6-7.6.

DESCRIPTION

Lip region offset, 53 × 20 μm. Amphidial apertures 18 μm from anterior end. Buccal cavity 55-63 × 35 μm. Dorsal tooth large, pointing anteriad, located in posterior half of buccal cavity, its apex 25-30 μm from base of buccal cavity, opposing dorsal tooth are scattered many small denticles in several irregular rows occupying little more than posterior half of buccal cavity. Pharyngo-intestinal junction tuberculate. Vulva transverse with cuticularised lips; vulval papillae absent. Genital system amphidelphic, both branches well developed. Tail 483 μm long. Caudal glands and spinneret poorly developed. Male with buccal cavity 50 × 27-30 μm. Spicules 90-100 μm; gubernaculum well developed; lateral accessory pieces 17-22 μm. Supplements 10-13, regularly spaced.

DISTRIBUTION

India-Orissa.

Key to species of *Hadronchulus* Ray & Das, 1983

1 – Subventral denticles few in number, forming a wavy pattern; ♀: L = 2.7-3.1 mm; a = 48-61; b = 4.9-5.3; c = 3.5-5.0; c′ = 15-21;

V = 51-53; ♂: L = 3.0-3.2 mm; a = 46; b = 4.7-5.0; c = 4.4-4.5; supplements 10-13 (India) . . . . *denticulus*

– Subventral denticles numerous, rasp-like 2

2 – Tail four anal body diam. or 110-155 μm long; both ovaries equally developed; ♀: L = 2.0-2.2 mm; a = 31-37; b = 3.6-4.0; c = 14-18; c′ = 3.5-4.5; V = 67-70; ♂: L = 1.85 mm; a = 39; b = 3.6; c = 18; supplements 12 (India) . . . . *grandis*

– Tail 14-15 anal body diam. or 480 μm long; posterior ovary less developed than anterior; ♀: L = 2.6-2.8 mm; a = 39-44; b = 4.5-4.7; c = 5.8-6.3; c′ = 14-15; V = 62-68; ♂: L = 2.4-2.7 mm; a = 42-46; b = 4.5-4.7; c = 6.6-7.6; supplements 10-13 (India) . . . . *shamimi*

Genus *Hadronchoides* Jairajpuri & Rahman, 1984

(Fig. 55E-H)

DIAGNOSIS

Small to medium-sized nematodes. Cuticle smooth. Lips large, lip region set off from body. Buccal cavity moderately roomy. Dorsal tooth large, sharply pointed, located in anterior half of buccal cavity. Subventral walls armed with numerous small denticles gradually decreasing in number posteriorly. Pharyngo-intestinal junction tuberculate. Female genital system mono-prodelphic with short to long post-uterine sac. Vulva post-equatorial. Male not known. Tail filiform, 15-20 anal body diam. long. Caudal glands present, spinneret terminal.

TYPE SPECIES

Hadronchoides microdenticulatus Jairajpuri & Rahman, 1984

OTHER SPECIES

H. maximus Loof, 2006

Hadronchoides microdenticulatus Jairajpuri & Rahman, 1984
(Fig. 55E-H)

MEASUREMENTS

Female: L = 1.12-1.123 mm; a = 39-41; b = 4.8-4.9; c = 2.8-3.8; c′ = 15-21; V = 52-60.

DESCRIPTION

Lip region offset, 18-19 × 9-10 μm, distinctly wider than adjoining body. Amphidial apertures slightly anterior to dorsal tooth apex. Dorsal tooth moderately developed, in anterior third of buccal cavity, its apex at 15-16 μm or 75-80% of length of buccal cavity from its base, opposed by minute scattered subventral denticles, their number gradually decreasing posteriorly. Pharyngo-intestinal junction tuberculate. Vulva transverse, small cuticularised pieces present at vulva-vagina junction. One pre- and one post-vulval papillae present. Genital system pseudo-monoprodelphic. Anterior gonad well developed with distinct sphincter at oviduct-uterus junction. Posterior branch represented by large sac measuring 73-102 μm or 2.5-3.4 vulval body diam. Tail filiform, 323-410 μm long. Caudal glands small, opening terminal. Male not known.

DISTRIBUTION

India-Arunachal Pradesh.

Key to species of *Hadronchoides* Jairajpuri & Rahman, 1984

1 – Buccal cavity 20-21 × 9-10 μm; apex of dorsal tooth at 75-80% of buccal cavity length from its base; post-uterine sac long; ♀: L = 1.1-1.2 mm; a = 39-41; b = 4.8-4.9; c = 2.8-3.8; c′ = 15-21; V = 52-60; ♂: unknown . . . . *microdenticulatus*

– Buccal cavity 42-44 × 24-25 μm; apex of dorsal tooth at 59-64% of buccal cavity length from its base; post-uterine sac very small; ♀: L = 2.22-2.68 mm; a = 47-56; b = 4.6-4.9; c = 3.8-4.3; c′ = 18-21; V = 58-64; ♂: unknown . . . . *maximus*

Nematology Monographs & Perspectives, 2010, Vol. 7, 239-257

Chapter 9

Suborder Bathyodontina Coomans & Loof, 1970

The genus *Bathyodontus* Fielding, 1950 was described by Fielding (1950), who doubtfully placed it under the subfamily Nygolaiminae Thorne, 1935 of the family Dorylaimidae de Man, 1876. Fielding mentioned that the type species, *B. cylindricus* Fielding, 1950, was predatory in its feeding behaviour. Clark (1961d), while revising the classification of the Order Enoplida, removed this genus from Dorylaimidae and placed it into the new family Bathyodontidae along with two other genera, *Oionchus* Cobb, 1913 and *Mirolaimus* Andrássy, 1956, within the new superfamily Mononchoidea. Previously, *Oionchus* and *Mirolaimus* had been placed by Thorne (1939) and Andrássy (1956), respectively, under the subfamily Nygolaiminae. Clark (1961d), when proposing the superfamily Mononchoidea and the family Bathyodontidae, remarked that the latter represented a transitional kinship between Mononchoidea and Dorylaimoidea. s'Jacob and Loof (1962) synonymised *Mirolaimus* with *Bathyodontus*. De Coninck (1965), and also Jairajpuri and Loof (1966), transferred the genus *Mononchulus* Cobb, 1918 from the family Oncholaiminae, superfamily Enoploidea, to the family Bathyodontidae. The former author also proposed the subfamilies Bathyodontinae and Mononchulinae. Jairajpuri and Loof (1966) had also recognised, besides *Bathyodontus*, *Oionchus* and *Mononchulus*, the genus *Stephanium* Rahm, 1938 in this group. They accepted *Mirolaimus* as a synonym of *Bathyodontus* and *Enoplochilus* as a synonym of *Oionchus*.

Jairajpuri (1969), when proposing the Order Mononchida, raised Bathyodontidae to the rank of a superfamily and the two subfamilies of De Coninck to the rank of families. Coomans and Loof (1970) did some excellent work on this group of nematodes and raised it to the status of an independent suborder, the Bathyodontina, with the superfamilies Bathyodontoidea and Mononchuloidea, each with a single family. Under Bathyodontidae they placed, along with *Bathyodontus*, the genus *Cryptonchus* Cobb, 1913, and under Mononchulidae the following three genera were included: *Mononchulus*, *Oionchus* and *Stephanium*.

Andrássy (1976) did not agree with the proposal of the suborder Bathyodontina and regarded this group of nematodes as representing only a superfamily under Mononchina with the families Bathyodontidae and Mononchulidae. Under the former he placed *Bathyodontus* and *Mirolaimus*, thereby indicating that the synonymy of the two genera was not acceptable to him. The latter view was in conformity with Khan (1972) who also considered these two genera as separate. Under Mononchulidae Andrássy recognised three genera, *viz.*, *Mononchulus*, *Oionchus* (syn. *Enoplochilus*) and *Rahmium* Andrássy, 1973 (= *Stephanium*). He did not agree with the placing of *Cryptonchus* under Bathyodontidae and also regarded *Cryptonchulus* Sukul, 1969 as a possible synonym of the former.

Lorenzen (1981) pointed out that Cryptonchinae Chitwood, 1937 has priority over Bathyodontidae Clark, 1961d, and hence Bathyodontidae and Bathyodontoidea Clark, 1961d were synonymised with Cryptonchidae and Cryptonchoidea, respectively, by Siddiqi (1983). He also raised this group to the rank of an Order, Bathyodontida Siddiqi, 1983, and recognised two superfamilies, *viz.*, Cryptonchoidea Chitwood, 1937 (Siddiqi, 1983) and Mononchuloidea De Coninck, 1965 (Coomans & Loof, 1970).

De Ley and Blaxter (2002, 2004) accepted this group as the suborder Bathyodontina with two superfamilies, Cryptonchoidea and Mononchuloidea, the former with two families, Cryptonchidae and Bathyodontidae, and the latter with a single family, the Mononchulidae. Zullini and Peneva (2006) followed the same classification and included the genera *Cryptonchus* and *Bathyodontus* under Cryptonchoidea and the genera *Mononchulus* and *Oionchus* under Mononchuloidea. The genus *Rahmium* Andrássy, 1973 was considered *genus inquirenda* because of the poor description. All four genera are of rare occurrence.

Suborder Bathyodontina Coomans & Loof, 1970

Diagnosis

Body length 0.7-2.2 mm. Cephalic papillae arranged in two or three circles. Buccal cavity rather narrow, composed of several sections, embedded almost entirely in pharyngeal tissue, weakly to strongly sclerotised. Distinct subventral tooth of varying size present, denticles present or absent. Pharynx with five gland nuclei with anterior one dorsal, located about middle of pharynx far posterior to gland orifice, followed by a ventrosublateral pair and, near posterior pharyngeal end, another ventrosublateral pair.

TYPE SUPERFAMILY

Cryptonchoidea Chitwood, 1937

OTHER SUPERFAMILY

Mononchuloidea De Coninck, 1965

Key to superfamilies of Bathyodontina Coomans & Loof, 1970

1 – Buccal cavity thin, narrow, tooth very small; polymyarian .... **Cryptonchoidea**
– Buccal cavity comparatively wide, tooth grooved, large; meromyarian .... **Mononchuloidea**

Superfamily Cryptonchoidea Chitwood, 1937

DIAGNOSIS

Cephalic papillae arranged in two circles. Buccal cavity long, cylindrical, sometimes tubular in its posterior portion, rhabdions weakly sclerotised, tooth very small, DN anterior to S_1N; S_2N far posterior to S_2O. Cardia large.

TYPE FAMILY

Cryptonchidae Chitwood, 1937

OTHER FAMILY

Bathyodontidae Clark, 1961d

Key to families of Cryptonchoidea Chitwood, 1937

1 – Buccal cavity long and cylindrical . . . . **Cryptonchidae**
– Buccal cavity anteriorly wide, cylindrical, posteriorly narrow, tubular . . . . **Bathyodontidae**

Family Cryptonchidae Chitwood, 1937

DIAGNOSIS

Buccal cavity long, cylindrical.

TYPE AND ONLY GENUS

Cryptonchus Cobb, 1913

Genus *Cryptonchus* Cobb, 1913

= *Gymnolaimus* Cobb, 1913
= *Ditlevsenia* Micoletzky, 1925
= *Cryptonchulus* Sukul, 1969
(Figs 56, 57)

DIAGNOSIS

Body 0.8-2.4 mm long, slender. Six lips, 12 labial papillae. Amphids stirrup-shaped, sub-labial, up to 1.6 lip region diam. from anterior end. Buccal cavity encircled by pharyngeal tissue, comprising long cylinder with small denticle in left ventrosublateral base. Nerve ring at 20-30% of pharyngeal length from anterior end. Pharynx cylindrical, composed of a long muscular corpus and a short constricted basal region. Sequence (from anterior to posterior direction) of pharyngeal gland nuclei and orifices is: (nerve ring), DO, DN, S_1O, S_1N, S_2O, S_2N. Cardia bulboid, multinucleated. Cardiac glands absent. Tail elongate conoid to filiform, with glands, spinneret not always discernible. Female monoprodelphic; ovary reflexed. Vulva near mid-body or slightly posterior. Eggs (as those of related taxa) much longer than broad. Testes two, opposed, outstretched. Spicules arcuate, without gubernaculum. Row of ventromedian supplements present.

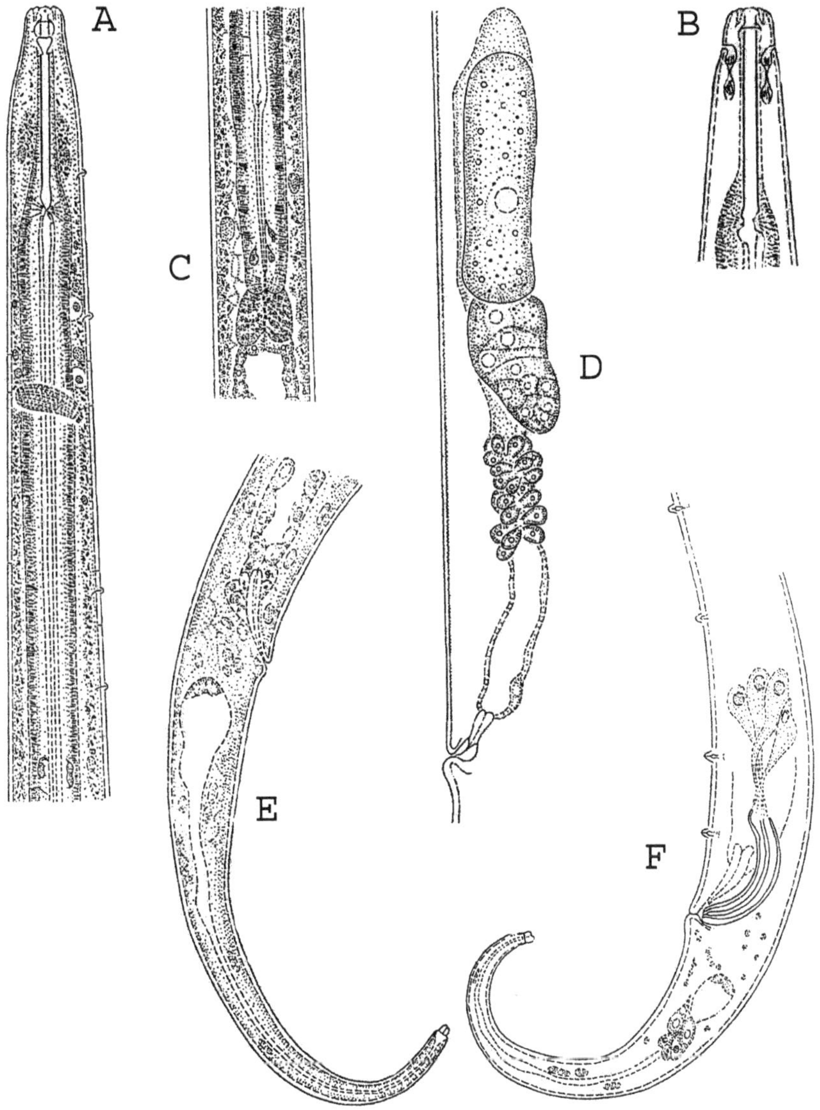

Fig. 56. Cryptonchus tristis *(Ditlevsen, 1911) Filipjev, 1934. A: Pharyngeal region; B: Anterior region; C: Pharyngo-intestinal junction; D: Female genital system; E: Female tail; F: Male posterior region. After Anderson (1968).*

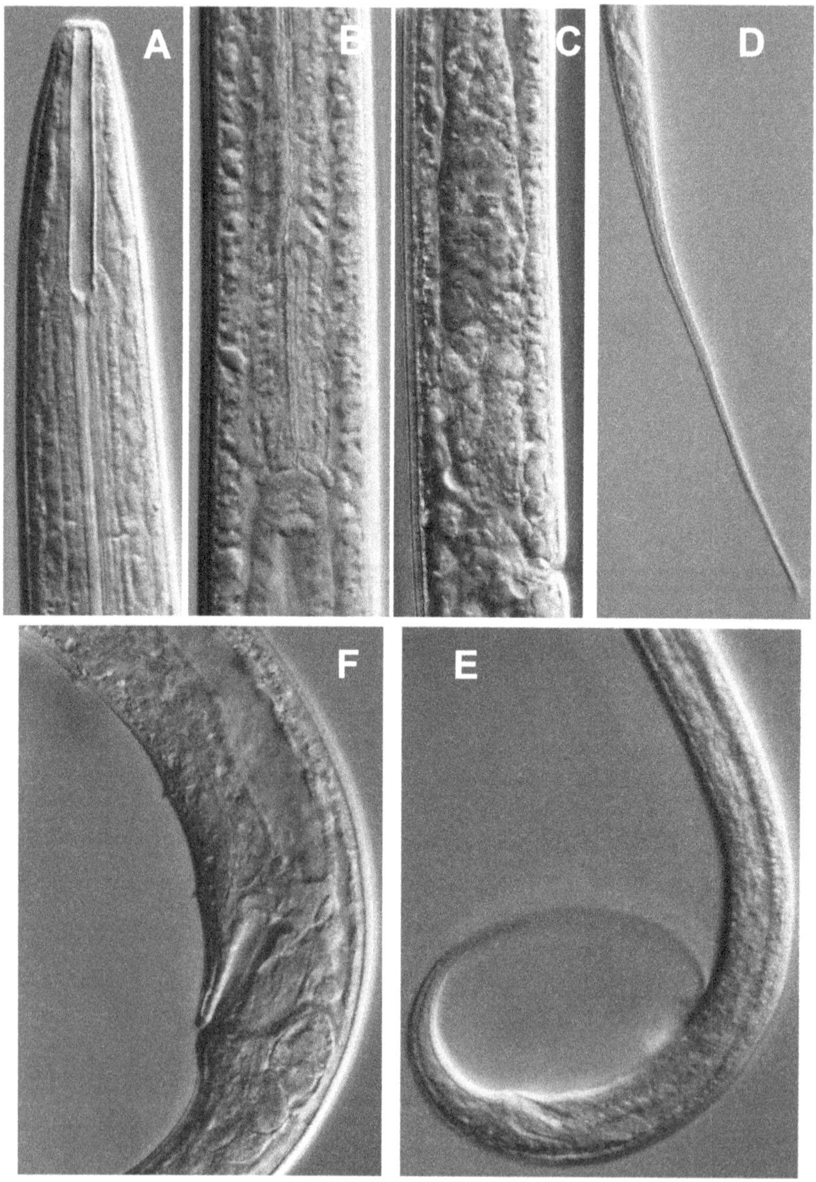

Fig. 57. Cryptonchus abnormis *(Allgén, 1933) Schuurmans Stekhoven, 1951. A: Anterior region; B: Posterior pharynx; C: Female genital system; D: Female tail; E, F: Male posterior region.*

TYPE SPECIES

C. tristis (Ditlevsen, 1911) Filipjev, 1934
= *Cylindrolaimus tristis* Ditlevsen, 1911
= *Cryptonchus nudus* Cobb, 1913
= *Ditlevsenia tristis* (Ditlevsen, 1911) Micoletzky, 1925
= *Gymnolaimus nudus* (Cobb, 1913) Timm, 1957 (= *C. abnormis*)

OTHER SPECIES

C. abnormis (Allgén, 1933) Schuurmans Stekhoven, 1951
= *Cylindrolaimus abnormis* Allgén, 1933
C. exilis (Ditlevsen, 1911) Filipjev, 1934
= *Cryptonchulus exilis* Sukul, 1969

***Cryptonchus tristis* (Ditlevsen, 1911) Filipjev, 1934**
(Fig. 56)

MEASUREMENTS

Female: L = 1.8-2.4 mm; a = 45-63; b = 3.5-4.2; c = 10-13; V = 50-56.
Male: L = 1.8-2.0 mm; a = 48-52; b = 3.6-3.8; c = 11-12.

DESCRIPTION

Lip region continuous; lips six, apices slightly elevated. Each lip bearing two median papillae, one perioral, one on outer lip contour. Amphids stirrup-shaped with small slit-like aperture. Buccal cavity triquetrous, cylindrical, 56-72 μm long; walls thickened at base of head and forming a refractive ring, base of buccal cavity anisomorphic. Left subventral wall of buccal cavity base recurved inward, adjoining left subventral wall of telostom concave, bearing a single median denticle. Opposing right subventral and dorsal walls convex, protruding into lumen. Pharynx cylindrical, 506-602 μm long, composed of a corpus and a constricted basal region 63-94 μm long. Cardia 18-33 μm long, slightly overlapping intestine. Female genital system monodelphic-prodelphic, ovary reflexed. Oviduct obscure; vagina short, directed obliquely anterior. Post-uterine branch absent. Male with arcuate spicules, gubernaculum absent. Ventromedian supplements 3-4.

Tail conoid, ventrally arcuate with blunt terminus, 163-208 μm long in female and 158-170 μm in male. Spinneret terminal.

DISTRIBUTION

Cosmopolitan.

Cryptonchus abnormis (Allgén, 1933) Schuurmans Stekhoven, 1951
(Fig. 57)

MEASUREMENTS

Female: L = 1.0-1.15 mm; a = 38-46; b = 4.5-4.8; c = 2.9-3.5; V = 41-45.

DESCRIPTION

Differs from *C. tristis* in having a long, filiform, tail.

DISTRIBUTION

Bangladesh, Congo Republic, India, Ivory Coast, Java, New Zealand, Paraguay, Surinam, USA-Florida.

Family Bathyodontidae Clark, 1961b

DIAGNOSIS

Buccal cavity anteriorly wider, cylindrical, posteriorly narrow, tubular.

TYPE AND ONLY GENUS

Bathyodontus Fielding, 1950
= *Mirolaimus* Andrássy, 1956

Genus *Bathyodontus* Fielding, 1950

=*Mirolaimus* Andrássy, 1956
(Figs 58; 59A-E)

DIAGNOSIS

Body 0.8-1.2 mm long. Oral opening surrounded by six lips bearing two circles of papillae, inner one consisting of six, outer one ten. Buccal cavity long, composed of three sections: a cylindrical anterior part with cuticularised walls comprising three sectors (one dorsal, two subventral); a shorter and much narrower portion with much thicker walls; and a long tubular part. Buccal cavity completely (or almost so) surrounded by pharyngeal tissue. Pharynx stout, cylindrical, muscular and glandular. Cardiac glands absent. Female genital system amphidelphic. Tail short, rounded, with glands and spinneret. Male with two equal spicules and a series of ventromedian supplements.

TYPE SPECIES

B. cylindricus Fielding, 1950

OTHER SPECIES

B. indicus (Khan, 1972) Jairajpuri & Khan, 1982
=*Mirolaimus indicus* Khan, 1972
B. mirus (Andrássy, 1956) Andrássy *in* Hopper & Cairns, 1959
=*Mirolaimus mirus* Andrássy, 1956

Bathyodontus cylindricus Fielding, 1950

(Fig. 58A-E)

MEASUREMENTS

Female: L = 0.79-1.02 mm; a = 13-22; b = 3.0-3.6; c = 33-54; c′ = 0.6-0.8; V = 52-60.
Male: L = 0.89 mm; a = 19; b = 3.5; c = 42; c′ = 0.6.

DESCRIPTION

Lip region almost continuous with body contour, 13-16 μm diam., 6-9 μm high, slightly offset. Lips conoid with amalgamated bases. Buccal

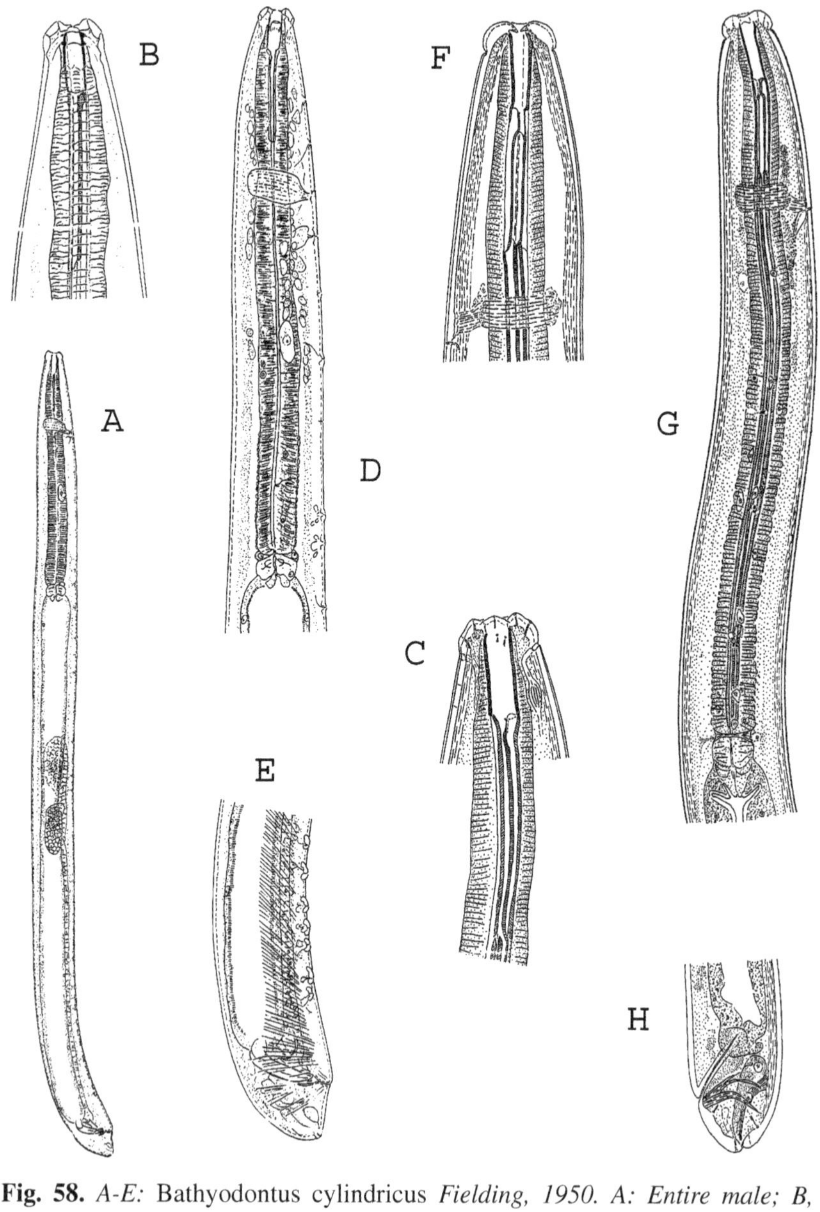

Fig. 58. *A-E:* Bathyodontus cylindricus *Fielding, 1950. A: Entire male; B, C: Anterior region; D: Pharyngeal region; E: Male posterior region; F-H:* Bathyodontus mirus *(Andrássy, 1956) Andrássy* in *Hopper & Cairns, 1959. F: Anterior region; G: Pharyngeal region; H: Female tail. A, B, D, E after De Ley and Coomans (1989); C, F-H after Coomans and Loof (1970).*

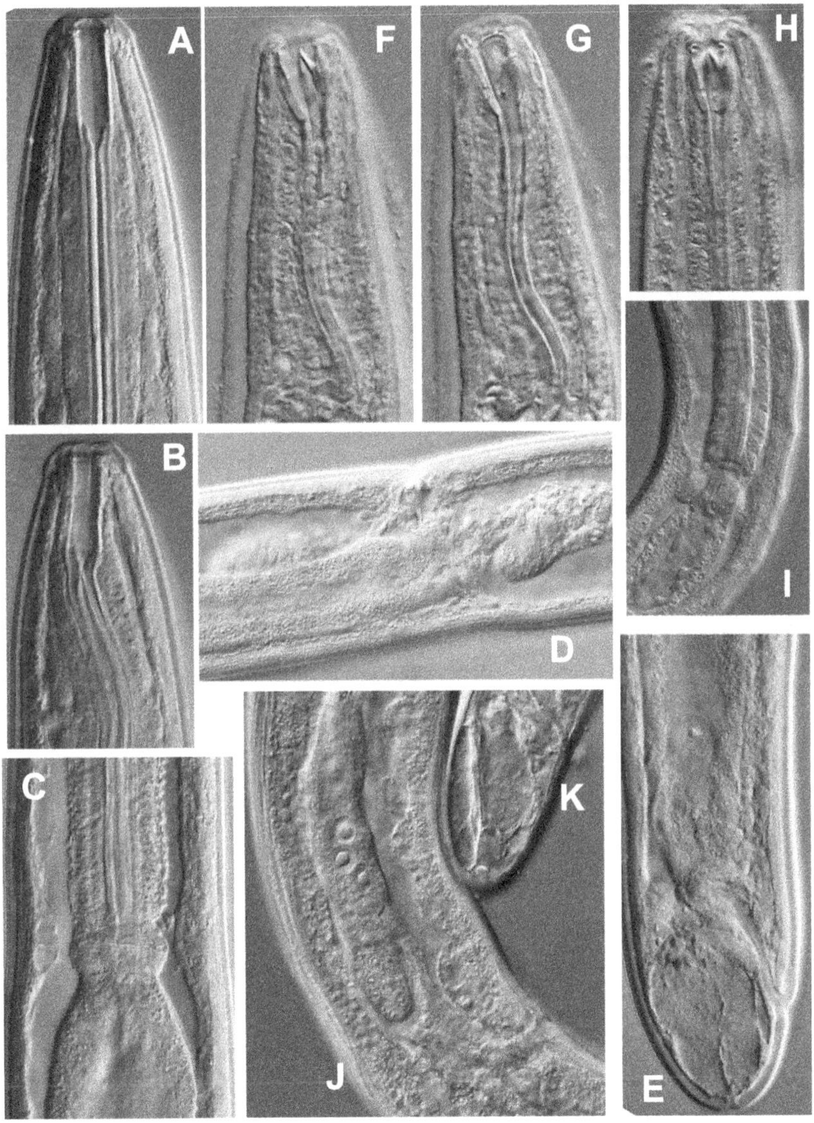

Fig. 59. *A-E:* Bathyodontus indicus *(Khan, 1972) Jairajpuri & Khan, 1982. A, B: Anterior region; C: Pharyngo-intestinal junction; D: Vulval region; E: Female tail region; F-K:* Oionchus obtusus *Cobb, 1913. F-H: Anterior region; I: Pharyngo-intestinal junction; J: Vulval region; K: Female tail.*

cavity 52-69 μm long, anterior, wider, part of buccal cavity 16-19 μm long, 5-6 μm diam., posterior subsection 31-42 μm long, much longer than anterior part. Ventrosublateral denticle small but distinctly visible. Vulva a transverse slit. Female genital organs amphidelphic. Tail 14-16 μm, rounded, less than one anal body diam. long. Spinneret terminal. Male with 37 pre-cloacal supplements. Spicules plump, slightly curved with very large manubrium, 34 μm long. Gubernaculum 8 μm long, lateral guiding pieces 11 μm long with bifurcate indented tip.

DISTRIBUTION

Colombia, Egypt, India, Nigeria, Senegal, Sudan, USA-California, Venezuela.

Key to species of *Bathyodontus* Fielding, 1950

1 – Vulva pre-equatorial; spinneret subterminal; ♀: L = 0.63-0.87 mm; a = 15-25; b = 3.2-5.1; c = 38.7-45.5; c′ = 0.64-0.74; V = 42.2-45.5; ♂: unknown (India) . . . . *indicus*
 – Vulva post-equatorial; spinneret terminal 2

2 – Lip region offset by marked constriction, lips clearly demarcated; body diam. at base of pharynx twice lip region diam.; ventrosublateral denticle minute; ♀: L = 1.20-1.45 mm; a = 20-23; b = 3.2-3.6; c = 52-59; V = 54-57; ♂: unknown (India, The Netherlands) . . . . *mirus*
 – Lip region slightly offset, lips with amalgamated bases, body diam. at base of pharynx thrice lip region diam.; ventrosublateral denticle distinct; ♀: L = 0.79-0.92 mm; a = 13-19; b = 3.0-3.3; c = 50-54; V = 58-60; ♂: unknown (Colombia, Egypt, India, Nigeria, Venezuela, USA-California) . . . . *cylindricus*

Superfamily Mononchuloidea De Coninck, 1965

DIAGNOSIS

Cephalic papillae arranged in three circlets. Buccal cavity strongly sclerotised, with large anterior subventral tooth and 2-6 rows of transverse denticles. DN at level with S_1N; S_2N at level with S_2O. Spinneret conspicuous.

TYPE AND ONLY FAMILY

Mononchulidae De Coninck, 1965

Family Mononchulidae De Coninck, 1965

DIAGNOSIS

Same as above.

TYPE GENUS

Mononchulus Cobb, 1918

OTHER GENUS

Oionchus Cobb, 1913

Key to genera of Mononchulidae De Coninck, 1965

1 – Tail long, cylindrical .... *Mononchulus*
– Tail short, hemispherical .... *Oionchus*

Genus *Mononchulus* Cobb, 1918
(Figs 60, 61)

DIAGNOSIS

Body 0.7-1.3 mm long, relatively plump. Buccal cavity encircled by pharyngeal tissue. Cheilostom with thick, longitudinally striated walls. Prostom short, sclerotised, wider than cheilostom. Mesostom with sclerotised walls, a ventrosublateral tooth-like process and *ca* six rows of transverse denticles. Metastom with heavily sclerotised walls and with a rather large, grooved, tooth on right or left ventrosublateral wall. Telostom elongate with tri-radial walls. Amphidial apertures inconspicuous slits located near tip of large tooth. Pharynx almost cylindrical, only slightly thickening posteriad. Sequence of pharyngeal gland nuclei and orifices is: DO, S_1O, $DN+S_1N$, S_2N+S_2O. Three large cardiac gland cells present. Female genital system monoprodelphic,

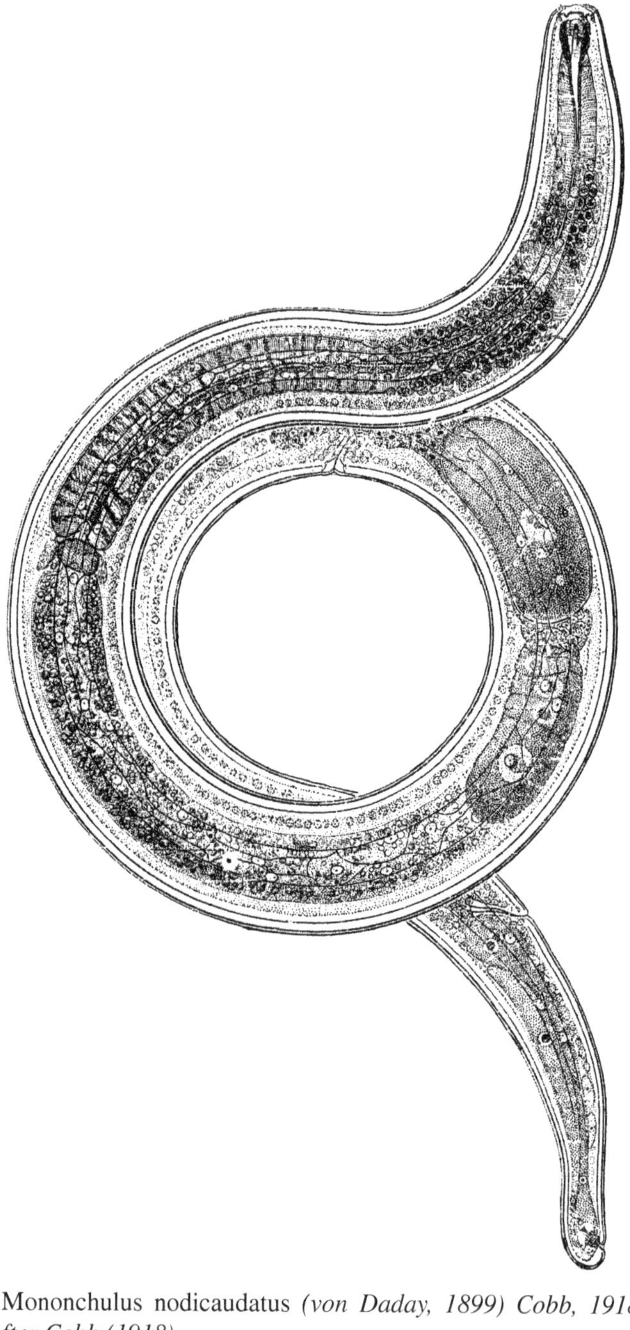

Fig. 60. Mononchulus nodicaudatus *(von Daday, 1899) Cobb, 1918, entire female. After Cobb (1918).*

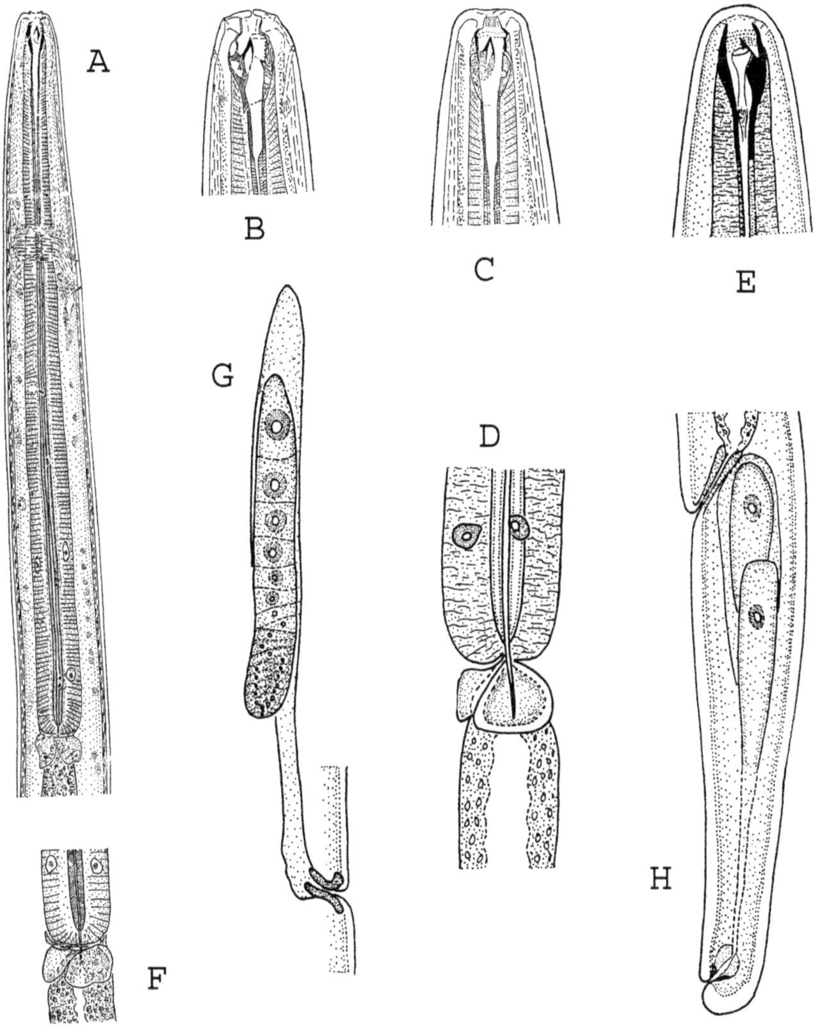

Fig. 61. Mononchulus nodicaudatus *(von Daday, 1899) Cobb, 1918. A: Pharyngeal region; B, C, E: Anterior region; D, F: Pharyngo-intestinal junction; G: Female genital system; H: Female tail. A-D after Coomans and Loof (1970); E-H after Jairajpuri and Khan (1982).*

ovary reflexed, posterior genital branch rudimentary. Male with paired spicules, gubernaculum and row of ventromedian supplements. Tail elongate, cylindrical, with glands and a large, ventrally subterminal, spinneret.

TYPE AND ONLY SPECIES

M. nodicaudatus (von Daday, 1899) Cobb, 1918
= *Prismatolaimus nodicaudatus* von Daday, 1899

Mononchulus nodicaudatus (von Daday, 1899) Cobb, 1918
(Figs 60, 61)

MEASUREMENTS

Female: L = 1.01-1.21 mm; a = 25-34; b = 4.0-4.2; c = 10-11; V = 61-64.

DESCRIPTION

Lip region 12-14 μm diam., 6-8 μm high. Amphidial aperture 3 μm wide, located at 6-8 μm from anterior end of body. Buccal cavity 22-24 μm long, 8 μm diam., embedded in pharyngeal tissue up to base of main tooth. Opposite base of main tooth there are two small teeth, one slightly posterior to other, six rows of denticles present at level of subventral tooth. Nerve ring located at 82-93 μm from anterior end of body. Pharyngo-intestinal junction with distinct glands. Female genital system monoprodelphic. Weakly developed sphincter at oviduct-uterus junction. Tail long, cylindrical, 92-113 μm or 3-4 anal body diam. long. Caudal glands present, spinneret subterminal.

REMARKS

The above description is based on Jairajpuri and Khan (1982). Jairajpuri and Loof (1966) provided a description of this species based on specimens collected by Dr R.W. Timm from East Pakistan (now Bangladesh). Both Cobb and Schneider said that the tooth was located on the right subventral side but Jairajpuri and Loof (1966) have shown it to be on the left subventral side. Coomans and Loof (1970) found that, of the eight specimens they examined, the tooth was on the right subventral metarhabdion in four specimens and on the left subventral metarhabdion in the others.

DISTRIBUTION

Bangladesh, India, Suriname.

Genus *Oionchus* Cobb, 1913
= *Enoplochilus* Kreis, 1932
(Figs 59F-K; 62)

DIAGNOSIS

Body 0.7-1.2 mm long, rather plump. Buccal cavity encircled by pharyngeal tissues. Prostom short, sclerotised, wider than cheilostom. Mesostom sclerotised with a ventrosublateral, posteriorly placed small tooth (near base of large metastomal tooth) and 2-4 rows of transverse denticles. Metastom sclerotised with a very large grooved tooth on right or left ventrosublateral side. Telostom elongate with triradiate walls. Amphids near tip of large tooth or slightly posterior. Pharynx nearly cylindrical, slightly thickening posteriorly. The sequence of pharyngeal gland nuclei and orifices is: DO, S_1O, $DN+S_1N$, S_2N+S_2O. Three cardiac glands present. Vulva somewhat posterior to mid-body. Tail short, rounded, with glands and a terminal (or slightly dorsally subterminal) spinneret.

TYPE SPECIES

O. obtusus Cobb, 1913
= *O. acutangulus* Ali & Joshi, 1970
= *O. setosus* Sukul, 1971

OTHER SPECIES

O. obtusicaudatus (Kreis, 1932) Thorne, 1935
= *Enoplochilus obtusicaudatus* Kreis, 1932
O. paraobtusus Jairajpuri & Khan, 1982

SPECIES INQUIRENDAE

O. didelphus Sukul, 1971
O. rotundicaudatus (Kannan, 1961b)
= *Mononchulus rotundicaudatus* Kannan, 1961b

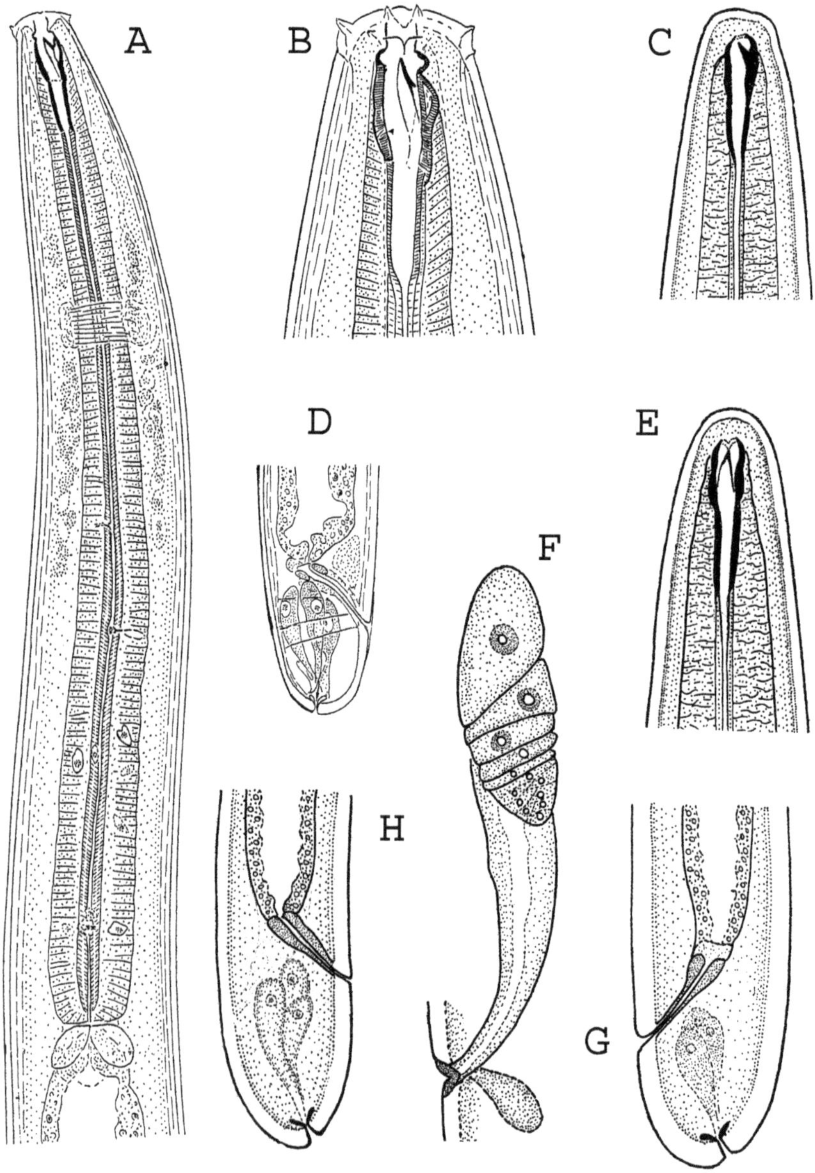

Fig. 62. *A-G:* Oionchus obtusus *Cobb, 1913. A: Pharyngeal region; B, C, E: Anterior region; D: Female tail; F: Female genital system; G: Female tail; H:* Oionchus paraobtusus *Jairajpuri & Khan, 1982, female tail. A-C after Coomans and Loof (1970); D-H after Jairajpuri and Khan (1982).*

Oionchus obtusus Cobb, 1913
(Figs 59F-K; 62A-G)

MEASUREMENTS

Female: L = 0.67-1.2 mm; a = 19-33; b = 3.0-3.9; c = 41-55; c′ = 0.8-0.9; V = 58-67.

DESCRIPTION

Lip region 12-14 μm diam., 4-6 μm high. Amphidial apertures 2 μm wide, located 5-7 μm from anterior end of body. Anterior wider part of buccal cavity 11-13 μm long, 6-8 μm diam. Tooth large, subventral walls bearing two transverse rows of denticles. Nerve ring located at 70-80 μm from anterior end of body. Female genital system mono-prodelphic, posterior branch represented by a small sac, measuring 6-8 μm or less than one vulval body diam. long. Sphincter indistinct, at oviduct-uterus junction. Tail hemispheroid, 18-22 μm or less than one anal body diam. long. Spinneret subterminal dorsally. Male not known.

DISTRIBUTION

Africa, Central and South America, India.

Oionchus paraobtusus Jairajpuri & Khan, 1982
(Fig. 62H)

MEASUREMENTS

Female: L = 0.84-1.04 mm; a = 23-30; b = 3.2-3.7; c = 26-30; c′ = 1.2-1.6; V = 60-64.

DESCRIPTION

Closely related to *O. obtusus* but differing in having comparatively longer tail and subventral position of spinneret.

DISTRIBUTION

India.

Nematology Monographs & Perspectives, 2010, Vol. 7, 259-264

Addendum

The following species, arranged alphabetically by genus, are not included in the keys, either because they were originally overlooked or because of subsequent publication. For the sake of completeness, a brief diagnostic synopsis of each of these species is given below.

Clarkus patricius Susulovsky, 1994

Female: L = 2.93-3.58 mm; a = 28-34; b = 4.3-4.8; c = 19-26; c′ = 2.3-3.1; V = 62-69.
Male: L = 2.77-3.37 mm; a = 30-35; b = 4.0-4.8; c = 26-29; c′ = 1.5-18.

Buccal cavity 41-46 × 20-23 μm in female and 40-43 × 20-23 μm in male, dorsal tooth apex at 78-88% from base of buccal cavity, vulval papillae present both anterior and posterior to vulva, tail conoid, ventrally arcuate, 139-172 μm long with rudimentary caudal glands, terminal opening absent. Male spicules 99-116 μm long, gubernaculum 20-23 μm long, lateral accessory pieces 20-26 μm long, 18-23 ventromedian supplements.

Distribution: Ukraine.

Coomansus ulsani Choi, Eoun, Khan & Lee, 1999

Female: L = 1.2-1.5 mm; a = 23.5-26; b = 3.4-3.8; c = 13.6-14.8; c′ = 2.7-3.2; V = 65-68.

Buccal cavity 36-39 × 21-23 μm; dorsal tooth apex at 36-42% of buccal cavity from anterior end, a thin longitudinal ridge opposite dorsal tooth present.

Distribution: Korea.

Iotonchus aequabilis Susulovsky, 1998

Female: L = 2.21-2.53 mm; a = 36-43; b = 5.4-5.6; c = 34-41; c′ = 1.3-1.8; V = 70-72.

Buccal cavity 44-48 × 28-33 μm; genital system amphidelphic, vulval papillae absent; tail 54-71 μm long, conoid, gradually narrowing to an obtusely rounded terminus, caudal glands and spinneret absent.

Distribution: Ukraine.

Iotonchus cucumis Khan, Park, Bae & Shin, 2002

Female: L = 2.9-3.2 mm; a = 39-43; b = 4.2-4.3; c = 10.7-12.2; c′ = 4.7-5.2; V = 64-66.

Buccal cavity 57-60 × 34-39 μm, genital system amphidelphic, three pre- and three post-vulval papillae present. Tail 238-285 μm long, caudal glands distinct, spinneret terminal.

Note

As per the figure provided by the authors, the dimensions of the buccal cavity are 45 × 30 μm. If correct, this species might be a junior synonym of *I. parabasidontus* Mulvey & Jensen, 1967.

Distribution: Korea.

Iotonchus hinokumaensis Khan, Yoshiga, Okumura & Tanaka, 2008

Female: L = 2.0-2.2 mm; a = 33-44; b = 4.7-5.4; c = 9.4-10.8; c′ = 5.3-6.1; V = 60-63.
Male: L = 2.1-2.3 mm; a = 41-44; b = 4.4-5.4; c = 8.7-9.3; c′ = 5.9-6.3.

Buccal cavity 37-40 × 24-26 μm, female genital system amphidelphic, spicules 68-75 μm, gubernaculum 14-16 μm, ventromedian supplements 9-11. Tail elongate, 205-228 μm in female and 225-254 μm in male. Caudal glands and spinneret absent.

Note

The absence of caudal glands and a spinneret is very unusual for this species. In *Iotonchus* species with a long tail, the caudal glands and spinneret are generally present.

Distribution: Japan.

Iotonchus goshiensis Khan, Yoshiga, Okumura & Tanaka, 2008

Female: L = 1.5-1.6 mm; a = 28-30; b = 3.7-4.1; c = 24-26; c′ = 1.7-1.9; V = 63-65.

Buccal cavity 40-42 × 26-27 μm, genital system amphidelphic with both branches equally developed. Tail short, 60-66 μm long, conoid with rounded terminus.

Distribution: Japan.

Iotonchus ogiensis Khan, Yoshiga, Okumura & Tanaka, 2008

Female: L = 2.0-2.3 mm; a = 33-36; b = 4.1-4.4; c = 18-20; c′ = 2.7-3.3; V = 62-63.
Male: L = 2.1-2.3 mm; a = 32-34; b = 4.4-4.6; c = 14-15; c′ = 2.5-2.7.

Buccal cavity 38-41 × 24-26 μm, genital system amphidelphic with both branches equally developed, spicules 90-95 μm, gubernaculum 21-24 μm, lateral guiding pieces 12-14 μm, 12-14 ventromedian supplements. Tail elongate conoid with rounded terminus, 105-118 μm long in female and 145-162 μm in male. Caudal glands and spinneret absent.

Note

It is very unusual for the male tail to be longer than that of the female tail, as is recorded for this species.

Distribution: Japan.

Miconchus araki Khan, Yoshiga, Okumura & Tanaka, 2008

Female: L = 2.8-3.1 mm; a = 27-32; b = 4.2-4.5; c = 12.2-13.5; c′ = 4.3-4.5; V = 66-68.

Buccal cavity 60-64 × 48-52 μm, dorsal tooth and two subventral teeth in posterior half of buccal cavity, genital system amphidelphic, both branches equally developed. Tail elongate conoid, 225-245 μm long. Caudal glands distinct, spinneret terminal.

Distribution: Japan.

Miconchus bulbicaudatus Jana, Chatterjee & Manna, 2008

Female: L = 2.1-2.2 mm; a = 38-40; b = 4.1-4.3; c = 14.2-14.7; c′ = 3.1-3.3; V = 63.5-66.5.

Buccal cavity 48-49 × 25-26 μm, apex of dorsal tooth and two subventral teeth at 31% from base of buccal cavity and at same level; genital system amphidelphic. Tail conoid arcuate, gradually tapering at terminus, tail tip rounded with bulbous structure at tip, caudal glands distinct, spinneret terminal.

Distribution: India.

Miconchus rectangularis Jana, Chatterjee & Manna, 2008

Female: L = 1.77-1.82 mm; a = 26-26.5; b = 3.5-3.9; c = 24.6-24.7; c′ = 2.0-2.1; V = 71-72.

Buccal cavity very narrow, rectangular in shape, 51 × 17 μm, dorsal and two subventral teeth at 31% from base of buccal cavity; genital system amphidelphic. Tail conoid arcuate, gradually tapering at terminus. Caudal glands distinct, spinneret terminal.

Distribution: India.

Miconchus vulvapapillatus Choi, Euon, Khan & Lee, 1999

Female: L = 2.7-3.6 mm; a = 29-36; b = 4.1-4.5; c = 18.4-21; c′ = 2.7-3.2; V = 65-69.

Male: L = 3.4-3.7 mm; a = 33-34; b = 4.2-4.6; c = 22-24; c′ = 1.8.

Buccal cavity 53-60 × 33-25 μm, apex of dorsal tooth at 56-61% and subventral teeth at 53-56% from base of buccal cavity, female genital system amphidelphic, 5-8 prominent vulval papillae present in contiguous series, spicules 1.5-1.7 times anal body diam. long, gubernaculum 36-43 μm, lateral guiding pieces 22-24 μm with bifid terminus, 28-31 ventromedian supplements. Caudal glands and spinneret absent.

Distribution: Korea.

Mylonchulus jinjuensis Khan, Choi, Lee & Choi, 2000

Female: L = 1.9-2.2 mm; a = 42-54; b = 3.6-3.9; c = 31-39; c′ = 1.5-1.8; V = 66-70.
Male: L = 1.8-2.0 mm; a = 45-54; b = 3.6-3.9; c′ = 1.4-1.6.

Buccal cavity 26-34 × 16-20 μm, dorsal tooth typical of genus, subventral teeth nearly basal, relatively large, subventral denticles fairly large, arranged in 6-7 rows, spicules 58-67 μm, gubernaculum 15-22 μm, lateral guiding pieces 15-18 μm, 10-11 ventromedian supplements. Caudal glands in tandem, spinneret slightly dorsally subterminal.

Note

The nearly basal position of the subventral teeth is very unusual for this genus.

Distribution: Korea.

Mylonchulus taeguensis Khan, Choi, Lee & Choi, 2000

Female: L = 1.48-1.54 mm; a = 37-41; b = 3.2-3.5; c = 28-36; c′ = 1.2-1.4; V = 67-70.
Male: L = 1.47-1.77 mm; a = 38-44; b = 3.2-3.6; c = 28-36; c′ = 1.1-1.5.

Buccal cavity 29-31 × 18-23 μm, dorsal tooth typical of genus, subventral teeth present, subventral denticles in 6-7 rows, spicules 52-64 μm, gubernaculum 20-26 μm, lateral guiding pieces 15-18 μm,

12-13 ventromedian supplements. Caudal glands in tandem, spinneret terminal.

Distribution: Korea.

References

ABOUL-EID, H.Z. & AMEEN, H.H. (1982). Species of the genera *Iotonchus*, *Mononchus* and *Mylonchulus* (Nematoda: Mononchida) from Egypt. *Zoological Society of Egypt Bulletin* 31, 51-65.

AHMAD, W. (2000). New and known species of Mononchida (Nematoda) from Mbalmayo reserve forest, Cameroon. *Annales Zoologici* 50 (*Special issue: M.W. Brzeski (1937-1999) in memoriam*), 145-149.

AHMAD, W. & JAIRAJPURI, M.S. (1983). Two new species of *Iotonchus* (Nematoda: Mononchida) from the Silent Valley, Kerala, India. *Systematic Parasitology* 5, 83-87.

AHMAD, W. & JAIRAJPURI, M.S. (1993). Description of *Promiconchus siddiqii* sp. n. from Dominica (Nematoda: Mononchida). *Afro-Asian Journal of Nematology* 3, 96-98.

AHMAD, W., BANIYAMUDDIN, M. & JAIRAJPURI, M.S. (2005). Three new and a known species of Mononchida (Nematoda) from Singapore. *Journal of Nematode Morphology and Systematics* 7, 97-107.

ALEKSEEV, V.M. (2001). A new species of the predatory nematodes of the genus *Iotonchus* (Nematoda: Mononchida) from Khanka lake. *Zoologichesky Zhurnal* 80, 749-753.

ALI, S.M. & JOSHI, P.M. (1970). On new species of the genus *Oionchus* Cobb, 1913 (Nematoda: Bathyodontidae) from Marathawada, India. *Marathawada University Journal of Science* 9, 53-55.

ALLGÉN, C. (1929). Über einige freilebende Moosenematoden. *Nytt Magasin for Natuurvidenskapene* 67, 211-230.

ALLGÉN, C. (1933). Freilebende Nematoden aus dem Trondhjemsfjord. *Capita Zoologica* 4, 1-162.

ALTHERR, E. (1950). Les nématodes du Parc national Suisse. (Nématodes libres du sol). *Ergebnisse der Wissenschaftlichen Untersuchung des Schweizerischen Nationalparks* 3, 3-46.

ALTHERR, E. (1953). Nématodes du sol du Jura vaudois et français. I. *Bulletin Société Vaudoise des Sciences Naturelles* 65, 429-460.

ALTHERR, E. (1958). Nématodes du bassin inférieur de la Wesser et des dunes d'Héligoland. Espèces nouvelles ou incomplètemente décrites. *Memorias de la Sociedad de Ciencias Naturales* 12, 45-63.

ALTHERR, E. (1960). Results from the Danish expeditions to the French Cameroons (1949-1950). XXVIII – Nématodes limnicoles. *Bulletin de l'Institut Français d'Afrique Noire* 22, 770-787.

ALTHERR, E. (1963). Contributions à la connaissance de la faune des sables submergés en Lorraine. Nématodes. *Annales de Spéléologie* 18, 53-98.

ALTHERR, E. (1968). Nématodes de la nappe phréatique du réseau fluvial de la Saale (Thuringe) et psammiques du Lac Stechlin (Brandebourg du nord). *Limnologica* 6, 247-320.

ALTHERR, E. (1971). Contribution à la connaissance des nématodes cavernicoles de Roumanie. *Bulletin de l'Institut Français d'Afrique Noire* 71, 23-46.

ALTHERR, E. (1972). Contribution à la connaissance des nématodes rithrostygopsammiques et rithrostygopséphiques de Suède. *Revue Suisse de Zoologie* 79, 881-902.

ALTHERR, E. (1974). Les nématodes du sous-sol graveleux-sablonneux des rives de la Fulda (Hesse) et de la Weser inférieur. *Bulletin de l'Institut Français d'Afrique Noire* 72, 19-35.

ALTHERR, E. (1976). La faune des eaux profondes interstitielles de la région de Wiesbaden. *Bulletin de la Société Vaudoise des Sciences Naturelles* 73, 97-116.

ALTHERR, E. (1977). Contribution à la connaissance des nématodes de l'estuaire de l'Amazone. *Amazoniana* 6, 145-159.

ALTHERR, E. & DELAMARE DEBOUTTEVILLE, C.L. (1972). Nématodes interstitiels des eaux douces des États-Unis d'Amérique (États de Washington, du Colorado et du Massachusetts) récoltés par Cl. Delamare Deboutteville. *Annales de Spéléologie* 27, 683-760.

ANANDI, Y., MOHILAL, N. & DHANACHAND, CH. (1997). Study on soil nematodes of Manipur. IV: Species of *Paramylonchulus* Jairajpuri and Khan, 1982. *Uttar Pradesh Journal of Zoology* 17, 113-118.

ANDRÁSSY, I. (1956). Eine interessante Nematodenfauna der Gerste. Nematologische Notizen. *Acta Zoologica Academiae Scientiarum Hungaricae* 2, 307-317.

ANDRÁSSY, I. (1958). Über das System der Mononchiden (Mononchidae Chitwood, 1937; Nematoda). *Annales Histoirico-Naturalis Musei Nationalis Hungarici* 50, 151-171.

ANDRÁSSY, I. (1959). Die Mundhöhlentypen der Mononchiden und der Schlussel der *Mylonchulus*-Arten (Nematoda). *Opuscula Zoologica Budapest* 3, 3-12.

ANDRÁSSY, I. (1964). Süsswasser-Nematoden aus den Grossen Gebirgagegenden Ostafrikas. *Acta Zoologica Academiae Scientiarum Hungaricae* 10, 1-59.

ANDRÁSSY, I. (1967). Ergebnisse der Zoologischen Forschungen von Dr. Z. Kaszab in der Mongolei. 92. Weitere Bodennematoden aus den Jahren 1964 und 1965. *Opuscula Zoologica Budapest* 6, 203-233.

ANDRÁSSY, I. (1968). The scientific results of the Hungarian soil zoological expedition to the Brazzaville-Congo. 31. Nematoden aus grundwasser. *Annales Universitatis Scientiarum Budapestensis de Rolando Eötvös Nominatae Sectio Biologica* 9-10, 1-27.

ANDRÁSSY, I. (1970). Nematodens aus einigen Fluss-Systemen Sudafrikas. *Opuscula Zoologica Budapest* 10, 179-219.

ANDRÁSSY, I. (1972). Zwei neue Gattungen von Bodennematoden. *Annales Universitatis Scientiarum Budapestensis de Rolando Eötvös nominatae Sectio Biologica* 14, 87-92.

ANDRÁSSY, I. (1973). 100 neue Nematodenarten in der ungarischen Fauna. *Opuscula Zoologica Budapest* 11, 7-48.

ANDRÁSSY, I. (1976). *Evolution as a basis for the systematization of nematodes*. London, UK, Pitman Publishing, 288 pp.

ANDRÁSSY, I. (1983). The free-living nematode fauna of Hortobagy National Park. In: *The fauna of the Hortobagy National Park*. Budapest, Hungary, Akademinai Kiado, pp. 31-46.

ANDRÁSSY, I. (1985). Three new species of Mononchoidea (Nematoda) from the southern hemisphere. *Opuscula Zoologica Budapest* 21, 23-30.

ANDRÁSSY, I. (1986). Fifteen new nematode species from the southern hemisphere. *Acta Zoologica Hungarica* 32, 1-33.

ANDRÁSSY, I. (1992). A taxonomic survey of the family Mylonchulidae (Nematoda). *Opuscula Zoologica Budapest* 25, 11-35.

ANDRÁSSY, I. (1993a). A taxonomic survey of the family Mononchidae (Nematoda). *Acta Zoologica Hungarica* 39, 13-60.

ANDRÁSSY, I. (1993b). A taxonomic survey of the family Anatonchidae (Nematoda). *Opuscula Zoologica Budapest* 26, 9-52.

ANDRÁSSY, I. (1996). *Tricaenonchus caucasius* gen. *et* sp. n., a remarkable nematode species from the Caucasus (Nematoda: Mononchidae). *Zoosystematica Rossica* 4, 19-21.

ANDRÁSSY, I. (2003). New and rare nematodes from Alaska II. Four species of the order Mononchida. *Journal of Nematode Morphology and Systematics* 5, 61-72.

ANDRÁSSY, I. (2008). Four new species of Mononchida (Nematoda) from tropical regions. *Opuscula Zoologica Budapest* 39, 3-13.

ANDRÁSSY, I. (2009). *Free-living nematodes of Hungary (Nematoda errantia)*. Vol. III. Pedozoologica Hungarica No. 5 (Eds: Csuzdi, C. & Mahunka, S.). Budapest, Hungary, Hungarian Natural History Museum and Systematic Zoology Research Group of the Hungarian Academy of Sciences, 608 pp.

ARPIN, P. & KILBERTUS, G. (1981). Ultrastructure of intestinal contents and epithelium of some predatory (Mononchida) and bacteriophagous nematodes. *Revue de Nématologie* 4, 131-143.

AZMI, M.I. (1991). *Mylonchulus sarmini* n. sp. (Mononchida: Nematoda) from India. *Current Nematology* 2, 193-194.

AZMI, M.I. & JAIRAJPURI, M.S. (1978). Studies on nematode behaviour IV. Comparative activity of *Tylenchorhynchus mashhoodi* and *Rhabditis* sp. *Indian Journal of Nematology* 6, 96-98.

AZMI, M.I. & JAIRAJPURI, M.S. (1979). Effect of mechanical stimulation and crowding on *Mylonchulus dentatus*. *Nematologia Mediterranea* 7, 203-207.

BANAGE, W.B. (1961). The ecological importance of free living soil nematodes with special reference to those of moorland soils. *Journal of Animal Ecology* 32, 133-140.

BAQRI, Q.H. & BAQRI, S.Z. (1983). On the locations of the oesophageal gland nuclei in the order Mononchida (Nematoda). *Records of the Zoological Survey of India* 80, 341-354.

BAQRI, Q.H., BAQRI, S.Z. & JAIRAJPURI, M.S. (1978). Studies on Mononchida. XI. Two new species of *Iotonchus*, *Cobbonchus indicus* n. sp. and *Anatonchus ginglymodontus* Mulvey, 1961. *Nematologica* 24, 436-444.

BAQRI, Q.H., DAS, P.K. & AHMAD, N. (1981). Nematodes from West Bengal (India). X. On a new species of *Anatonchus* (Cobb, 1916) De Coninck, 1939 (Mononchida: Nematoda). *Bulletin of the Zoological Survey of India* 4, 209-211.

BAQRI, S.Z. & JAIRAJPURI, M.S. (1974). Studies on Mononchida V. The mononchs of El-Salvador with descriptions of two new genera, *Actus* and *Paracrassibucca*. *Nematologica* 19, 326-333.

BASTIAN, H.C. (1865). Monograph on the Anguillulidae, free nematoids, marine, land, and freshwater; with descriptions of 100 new species. *Transactions of the Linnean Society of London* 25, 73-184.

BILGRAMI, A.L. & JAIRAJPURI, M.S. (1985). Cannibalism in *Mononchus aquaticus*. *Indian Journal of Nematology* 14, 202-203.

BILGRAMI, A.L., AHMAD, I. & JAIRAJPURI, M.S. (1983). Some factors influencing predation of *Mononchus aquaticus*. *Revue de Nématologie* 6, 325-326.

BILGRAMI, A.L., AHMAD, I. & JAIRAJPURI, M.S. (1984). Observations on the predatory behaviour of *Mononchus aquaticus*. *Nematologia Mediterranea* 12, 41-45.

BILGRAMI, A.L., AHMAD, I. & JAIRAJPURI, M.S. (1988). Interaction between mononchs in non-sterile agar plates. *Indian Journal of Nematology* 17, 339-340.

BLAXTER, M.L., DE LEY, P., GAREY, J.R., LIU, L.X., SCHELDEMAN, P., VIERSTRAETE, A., VANFLETEREN, J.R., MACKEY, L.Y., DORRIS, M., FRISSE, L.M., VIDA, J.T. & THOMAS, W.K. (1998). A molecular evolutionary framework for the phylum Nematoda. *Nature* 392, 71-75.

BORGONIE, G., VAN DRIESSCEHE, R. & COOMANS, A. (1995). Scanning electron microscopy of the outer and inner surface of the buccal cavity of some Mononchida. *Fundamental and Applied Nematology* 18, 1-10.

DE BRUIN, S. & HEYNS, J. (1992a). Mononchida (Nematoda) from southern Africa: genus *Iotonchus* (Cobb, 1916) Altherr, 1950. *Phytophylactica* 24, 39-60.

DE BRUIN, S. & HEYNS, J. (1992b). Mononchida (Nematoda) from southern Africa: genera *Mononchus* Bastian, 1865, *Clarkus* Jairajpuri, 1970 and *Coomansus* Jairajpuri & Khan, 1977. *Phytophylactica* 24, 61-73.

BRZESKI, M.W. (1960). Drei neue freilbende Nematoden aus Polen. *Bulletin of the Polish Academy of Sciences* 8, 261-264.

BUANGSUWON, D.K. & JENSEN, H.J. (1966). A taxonomic study of Mononchidae (Enoplida: Nemata) inhabiting cultivated areas of Thailand. *Nematologica* 12, 259-274.

BÜTSCHLI, O. (1873). Beitrage zur Kenntnis der freilebenden Nematoden. *Nova Acta Academiae Caesareae Leipold-Carolinae Nat Curios* 36, 1-24.

DE CARVALHO, J.C. (1951). Uma nova espécie de *Mononchus* (Nematoda, Mononchidae). *Bragantia* 11, 51-54.

DE CARVALHO, J.C. (1953). *Mononchus* – um predator voraz. *Revista do Instituto Adolfo Lutz* 13, 75-82.

DE CARVALHO, J.C. (1955). *Mononchus risoceiae*, nova espécie (Nematoda, Mononchidae). *Revista do Instituto Adolfo Lutz* 15, 129-134.

DE CARVALHO, J.C. (1956a). "*Mononchus soutoi*" sp. n. (Nematoda, Mononchidae). *Revista Brasileira de Biologia* 16, 133-134.

DE CARVALHO, J.C. (1956b). *Mononchus coronatus* n. sp. (Nematoda, Mononchidae). *Revista do Instituto Adolfo Lutz* 16, 151-153.

DE CARVALHO, J.C. (1960a). *Mononchus* (*Cobbonchus*) *incultus*. *Revista do Instituto Adolfo Lutz* 20, 177-180.

DE CARVALHO, J.C. (1960b). *Mononchus* (*Iotonchus*) *piracicaboides* n. sp. *Revista do Instituto Adolfo Lutz* 20, 181-183.

CASSIDY, G.H. (1931). Some mononchs of Hawaii. *Hawaii Planters' Record* 35, 305-339.

CASSIDY, G.H. (1933). A bivulvular specimen of the nematode *Mononchus muscorum* (Dujardin) Bastian. *Journal of the Washington Academy of Sciences* 23, 141-144.

CHAVES, E. (1990). Mononchida (Nematoda) from Argentina. *Nematologica* 36, 181-193.

CHITWOOD, B.G. (1937). A revised classification of Nematoda, In: *Papers in Helminthology, 30 years Jubileum of K.I. Skrjabin*. Moscow, USSR, Nauka, pp. 67-79.

CHITWOOD, B.G. & CHITWOOD, M.B. (1937). *An introduction to nematology*. Baltimore, MD, USA, Monumental Printing, 240 pp.

CHOI, Y.E. & KHAN, Z. (2000). Descriptions of two new species of *Iotonchus* and first record of *Coomansus zschokkei* (Mononchida) from Korea. *Annales Zoologici* (*Special issue: M.W. Brzeski (1937-1999) in memoriam*) 50, 183-191.

CHOI, Y.E., KHAN, Z. & LEE, S.M. (1999). Descriptions of new species of predatory nematodes (Mononchida) from Korea. *Korean Journal of Soil Zoology* 4, 89-100.

CHOI, Y.S. & CHOI, Y.E. (1987). A taxonomical and morphological study of predatory nematodes (mononchs) in Korea. *Korean Journal of Plant Protection* 26, 209-219.

CHOUDHURY, B.N. & SIVAKUMAR, C.V. (1998). Three new species of Mononchoidea (Nematoda: Mononchida) from Tamil Nadu. *Current Nematology* 9, 69-73.

CHRISTIE, J.R. (1960). Biological control – predaceous nematodes. In: Sasser, J.N. & Jenkins, W.R. (Eds). *Nematology.* Chapel Hill, NC, USA, University of North Carolina Press, pp. 466-468.

CLARK, W.C. (1960a). The oesophago-intestinal junction in the Mononchidae (Enoplida, Nematoda). *Nematologica* 5, 178-183.

CLARK, W.C. (1960b). Redescription of *Mononchus truncatus* Bastian, *M. papillatus* Bastian and *Prionchulus muscorum* (Dujardin) (Enoplida, Nematoda). *Nematologica* 5, 184-198.

CLARK, W.C. (1960c). New Zealand Mononchidae (Enoplida, Nematoda) I. The genera *Mononchus* Bastian and *Prionchulus* Cobb. *Nematologica* 5, 199-214.

CLARK, W.C. (1961a). The Mononchidae (Enoplida, Nematoda) of New Zealand II. The genus *Iotonchus* (Cobb, 1916) Altherr, 1950. *Nematologica* 5, 260-274.

CLARK, W.C. (1961b). The Mononchidae (Enoplida, Nematoda) of New Zealand III. A review of genus *Cobbonchus* Andrássy, 1958 with descriptions of new species. *Nematologica* 5, 275-284.

CLARK, W.C. (1961c). The Mononchidae (Enoplida, Nematoda) of New Zealand IV. The genus *Mylonchulus* (Cobb, 1916) Pennak, 1953. *Nematologica* 6, 1-6.

CLARK, W.C. (1961d). A revised classification of the order Enoplida (Nematoda). *New Zealand Journal of Science* 4, 123-150.

CLARK, W.C. (1962). Evolution within the family Mononchidae (Enoplida, Nematoda). *Nematologica* 7, 252-255.

CLARK, W.C. (1963). Notes on Mononchidae (Nematoda) of the New Zealand region with description of new species. *New Zealand Journal of Science* 6, 612-632.

COBB, M.V. (1915). Some freshwater nematodes of the Douglas Lake Region of Michigan, USA. *Transactions of the American Microscopical Society* 34, 21-47.

COBB, N.A. (1893a). Nematode worms found attacking sugarcane. In: *Plant diseases and their remedies. Agricultural Gazette of New South Wales* 4, 808-833.

Cobb, N.A. (1893b). *Nematodes mostly Australian and Fijian. McLeay memorial volume.* Linnean Society of New South Wales, pp. 252-308.
Cobb, N.A. (1904). Free-living and fresh water New Zealand nematodes. *Proceedings of the Cambridge Philosophical Society* 12, 363-374.
Cobb, N.A. (1906). *Fungus maladies of the sugar cane. IX. Free-living nematodes inhabiting the soil about the roots of cane, and their relation to root diseases.* Hawaiian Sugar Planters Association, Division of Pathology and Physiology, Bulletin 5, 163-195.
Cobb, N.A. (1913). New nematode genera found inhabiting fresh water and non-brackish soils. *Journal of the Washington Academy of Sciences* 3, 432-444.
Cobb, N.A. (1916). Notes on new genera and species of nematodes. Four subdivisions of *Mononchus. Journal of Parasitology* 2, 195-196.
Cobb, N.A. (1917). The mononchs (*Mononchus* Bastian, 1865). A genus of free-living predatory nematodes. *Soil Science* 3, 431-486.
Cobb, N.A. (1918). Filter-bed nemas: nematodes of the slow sand filter-beds of American cities. *Contributions to a Science of Nematology* 7, 189-212.
Cobb, N.A. (1920). Transference of nematodes (mononchs) from place to place for economic purposes. *Science* 51, 640-641.
Coetzee, V. (1965). South African species of the genus *Cobbonchus* Andrássy, 1958 (Nematoda: Mononchidae). *Nematologica* 11, 281-290.
Coetzee, V. (1966). Species of the genera *Granonchulus* and *Cobbonchus* (Mononchidae) occurring in Southern Africa. *Nematologica* 12, 302-312.
Coetzee, V. (1967a). Species of the genus *Mylonchulus* (Nematoda: Mononchidae) occurring in Southern Africa. *Nematologica* 12, 557-567.
Coetzee, V. (1967b). Species of the genus *Iotonchus* (Nematoda: Mononchidae) occurring in Southern Africa. *Nematologica* 13, 367-377.
Coetzee, V. (1968). Southern African species of the genera *Mononchus* and *Prionchulus* (Mononchidae). *Nematologica* 14, 63-76.
Cohn, E. & Mordechai, M. (1974). Experiments in suppressing citrus populations by use of marigold and a predaceous nematode. *Nematologia Mediterranea* 2, 43-53.
Coomans, A. & van der Heiden, A. (1978). The systematic position of the family Ironidae and its relation to the Dorylaimida. *Annales de la Société Royale Zoologique de Belgique* 108, 5-11.
Coomans, A. & Khan, S.H. (1981). Mononchida from Mount Kenya. *Biologisch Jaarboek Dodonaea* 49, 64-75.
Coomans, A. & Lima, M.B. (1965). Description of *Anatonchus amiciae* n. sp. (Nematoda: Mononchidae) with observations on its juvenile stages and anatomy. *Nematologica* 11, 413-431.
Coomans, A. & Loof, P.A.A. (1970). Morphology and taxonomy of Bathyodontina (Dorylaimida). *Nematologica* 16, 180-196.

von Daday, E. (1894). Újabb adatok a Balaton mikrofaunájának ismeretéhez (Nematoda, Rotatoria, Entomostraca). *Mathematikai és Természettudomanyi Értesitö Budapest* 12 (4-5), 122-145.

von Daday, E. (1899). Új-guineai szabadon élő nematodok. *Mathematikai és Természettudomanyi Értesitö Budapest* 17, 557-572.

von Daday, E. (1901). Mikroskopische Süsswasserthiere aus Deutsch-Neu-Guinea. *Természetrajzi Füzetek* 24, 1-56.

von Daday, E. (1908). Adatok német-kelet Afrika Mikrofauna. *Mathematikai és Természettudomanyi Értesitö Budapest* 26, 1-57.

De Coninck, L.A.P. (1939). Les nématodes libres de la grotte de Han (Han-sur-lesse, Belgique). Note de biospéléologie. *Bulletin du Musée Royal d'Histoire Naturelle de Belgique* 15, 1-40.

De Coninck, L.A.P. (1965). Systematique de nématodes. In: Grassé, P.P. (Ed.). *Traité de zoologie, anatomie, systematique, biologie.* Paris, France, Masson Cie 4, pp. 586-681.

De Ley, P. & Blaxter, M.L. (2002). Systematic position and phylogeny. In: Lee, D.L. (Ed.). *The biology of nematodes.* London, UK, Taylor & Francis, pp. 1-30.

De Ley, P. & Blaxter, M.L. (2004). A new system for Nematoda: combining morphological characters with molecular trees, and translating clades in ranks and taxa. In: Cook, R. & Hunt, D.J. (Eds). *Proceedings of the Fourth International Congress of Nematology, Tenerife, Spain. 8-13 June 2002. Nematology monographs and perspectives 2.* Leiden, The Netherlands, Brill, pp. 633-653.

Dhanachand, Ch., Renubala, K. & Mohilal, N. (1991). The new genus *Morenchus denticulus* n. gen., n. sp. and a new species of *Parahadronchus* from Moreh, Manipur, India. *Current Nematology* 2, 135-138.

Dhanachand, Ch., Rombati, N. & Anandi, Y. (1995). Two new species of *Paramylonchulus* and description of *Mononchus truncatus* Bastian, 1865 from Manipur. *Uttar Pradesh Journal of Zoology* 15, 182-188.

Dhanachand, Ch., Romoni, H. & Pramodini, M. (2006). Two new species of mononch (Nematoda: Mononchida) from soil around the rhizosphere of bamboo from Manipur. *Indian Journal of Nematology* 36, 32-36.

Dhanam, M. & Jairajpuri, M.S. (1998). Five new species of mononchs from Malnad tracts of Karnataka, India. *Nematologica* 44, 21-35.

Dhanam, M. & Jairajpuri, M.S. (1999). Three new species of predatory nematodes (Mononchida: Nematoda) from Malnad tracts of Karnataka, India. *Journal of Coffee Research* 27, 93-102.

Dhanam, M. & Jairajpuri, M.S. (2002). Two new species of the genus *Iotonchus* Cobb, 1916 with a report on males of *Iotonchus silvallus* Ahmad & Jairajpuri, 1983 (Nematoda: Mononchida) from Malnad tracts of Karnataka, India. *Journal of Coffee Research* 28, 49-56.

DITLEVSEN, H. (1911). Danish free-living nematodes. *Videnskabelige Meddelelser fra Dansk Naturhistorisk Forening Kobenhavn* 63, 213-256.

DOUCET, M.E. (1980). Description d'une nouvelle espèce du genre *Mylonchulus* (Nematoda: Dorylaimida). *Nematologia Mediterranea* 8, 37-42.

DUJARDIN, F. (1845). *Histoire naturelle des helminthes ou vers intestinaux.* Paris, France, Librairie Encyclopédique de Roret, 654 pp.

DZHURAEVA, L.M. & KANKINA, V.K. (1997). [Predatory nematodes from the family Anatonchidae Jairajpuri, 1969.] *Izvestiya Akademii Nauk Respubliki Tadzhikistan Otdelenie Biologicheskikh Nauk* 2-4134, aprel'-dekabr', 31-34.

EISENDLE, U. (2008). Description of *Mononchus sandur* n. sp. (Nematoda: Mononchidae) and remarks on *M. truncatus* Bastian, 1865 and *M. aquaticus* Coetzee, 1968 from a glacial flood plain reach (Großglockner region, Hohe Tauern, Austria). *Nematology* 10, 809-818.

ELBADRI, G.A.A., KHAN, Z., MOON, I.S., LEE, D.W. & CHOO, Y.H. (2008). Descriptions of three new species of soil nematodes from Sudan. *International Journal of Nematology* 18, 4-12.

ELIAVA, I.YA. (1978). [The systematics of nematodes of the order Dorylaimida.] In: *Fitogel'mintologicheskie issledovaniya.* Moscow, USSR, Nauka, pp. 157-167.

EROSHENKO, A.S. (1972). [New species of carnivorous nematodes (Mononchida) in the Primorsk Territory.] *Zoologichesky Zhurnal* 51, 13-20.

EROSHENKO, A.S. (1975). [Ten new nematode species of the order Mononchida Jairajpuri, 1969 from coniferous forests of Primorye.] *Institute of Biology and Pedology, Far-East Science Centre, USSR Academy of Sciences Proceedings, New Series* 26, 152-169.

ESSER, R.P. (1963). Nematode interactions in plates of non-sterile water agar. *Proceedings of the Soil Crop Sciences Society of Florida* 23, 121-138.

ESSER, R.P. & SOBERS, E.K. (1964). Natural enemies of nematodes. *Proceedings of the Soil Crop Sciences Society of Florida* 24, 326-352.

FIELDING, M.J. (1950). Three new predacious nematodes. *The Great Basin Naturalist* 10, 45-50.

FILIPJEV, I.N. (1934). The classification of free-living nematodes and their relation to parasitic nematodes. *Smithsonian Miscellaneous Collections* 89, 1-63.

FUCHS, A.G. (1930). Neue an Borken-und Rüsselkäfer gebundene Nematoden, aus Moos und Walderde in Borken-und Rüsselkäfergängen. *Zoologische Jahrbücher Systematik* 59, 505-646.

GAGARIN, V.G. (1984). [New species of free-living nematodes from the Angara River Basin.] *Zoologichesky Zhurnal* 63, 607-609.

GAGARIN, V.G. (1991). [Seven new species of freshwater nematodes.] *Zoologichesky Zhurnal* 70, 20-27.

GAGARIN, V.G. (1993). [*Free living nematodes from fresh waters of Russia and adjacent areas.*] St Petersburg, Russia, Gidrometeoizdat, 352 pp.

GAGARIN, V.G. (2000). Novye vidy nematode iz vodoemov russkoj Arktiki. *Biologia Vnutrinnykh vod* 3, 9-15.

GAGARIN, V.G. (2001). [New species of free-living freshwater nematodes from Eurasia.] *Zoosystematica Rossica* 9, 11-17.

GAGARIN, V.G. & MATAPHONOV, P.V. (2004). [Free-living nematodes from water bodies of the Ivano-Arakhleiskaya lake system in the Transbaikal region.] *Biologii Wnutrennih Wod* 4, 29-38.

GAMBHIR, R.K. & DHANACHAND, CH. (1990). Nematodes of fruit plants in Manipur VII: descriptions of males of *Coomansus icarus* and *Iotonchus longicaudatus* and a new species *Iotonchus manipuriensis. Current Nematology* 1, 53-58.

GOODEY, J.B. (1963). *Soil and freshwater nematodes*, 2nd edition. London, UK, Methuen, 544 pp.

GOODEY, T. (1951). *Soil and freshwater nematodes.* London, UK, Methuen, 390 pp.

GROOTAERT, P. & MAERTENS, D. (1976). Cultivation and life cycle of *Mononchus aquaticus. Nematologica* 22, 173-181.

GROOTAERT, P. & WYSS, U. (1979). Ultrastructure and function of the anterior feeding apparatus in *Mononchus aquaticus. Nematologica* 25, 163-173.

HEYNS, J. & LAGERWEY, G. (1965). South African species of the genus *Iotonchus* Cobb, 1916 (Nematoda: Mononchidae). *South African Journal of Agricultural Science* 8, 775-784.

HOLLINGS, C.S. (1966). The functional response of invertebrate predators to prey density. *Memoirs of the Entomological Society of Canada* 48, 1-86.

HOLTERMAN, M., RYBARCZYK, K., VAN DER WURFF, A., VAN DEN ELSEN, S., VAN MEGEN, H., MOOYMAN, P., BONGERS, T., HOLOVACHOV, O., BAKKER, J. & HELDER, J. (2006). Phylum-wide analysis of SSU rDNA reveals deep phylogenetic relationships among nematodes and accelerated evolution towards crown clades. *Molecular Biology and Evolution* 23, 1792-1800.

HOLTERMAN, M., RYBARCZYK, K., VAN DEN ELSEN, S., VAN MEGEN, H., MOOYMAN, P., PEÑA-SANTIAGO, R., BONGERS, T., BAKKER, J. & HELDER, J. (2007). A ribosomal DNA-based framework for the detection and quantification of stress-sensitive nematode families in terrestrial habitats. *Molecular Ecology Notes* 7, 1-12.

HOPPER, B.E. & CAIRNS, E.J. (1959). *Taxonomic keys to plant, soil and aquatic nematodes.* Auburn, AL, USA, Alabama Polytechnic Institute, 176 pp.

HUNT, D.J. (1978). List of Mononchida from St Lucia and Dominica. *Revue de Nématologie* 1, 265-266.

IVANOVA, T.S. & DZHURAEVA, L.M. (1971). [A new genus of a new family Tigronchidae n. fam. (Nematoda: Mononchida) from Tadzhikistan.] *Izvestiya Akademii Nauk Tadzhikskoĭ SSR* 4, 89-93.

S'JACOB, J.J. & LOOF, P.A.A. (1962). The taxonomic status of the genera *Bathyodontus* Fielding, 1950 and *Mirolaimus* Andrássy, 1959. *Nematologica* 8, 66-74.

JAIN, S.K., SAXENA, R. & SHARMA, R.K. (1993). *Mylonchulus esculentus* sp. n. (Nematoda: Mononchida) and occurrence of *M. lacustris* in Rohilkhand Division, Uttar Pradesh, India. *Annals of Plant Protection Sciences* 1, 85-88.

JAIRAJPURI, D.S. (1982). *Mylonchulus sessus* n. sp. (Nematoda: Mononchida) from Brunei. *Systematic Parasitology* 4, 185-187.

JAIRAJPURI, M.S. (1969). Studies on Mononchida of India. I. The genera *Hadronchus*, *Iotonchus* and *Miconchus* and a revised classification of Mononchida, new order. *Nematologica* 15, 557-581.

JAIRAJPURI, M.S. (1970a). Studies on Mononchida of India. II. The genera *Mononchus*, *Clarkus* n. gen. and *Prionchulus* (Family Mononchidae Chitwood, 1937). *Nematologica* 16, 213-221.

JAIRAJPURI, M.S. (1970b). Studies on Mononchida of India. III. The genus *Mylonchulus* (Family Mylonchulidae Jairajpuri, 1969). *Nematologica* 16, 434-456.

JAIRAJPURI, M.S. (1971). Studies on Mononchida of India. IV. The genera *Sporonchulus*, *Bathyodontus* and *Oionchus* (Nematoda). *Nematologica* 17, 407-412.

JAIRAJPURI, M.S. & AZMI, M.I. (1978). Some studies on the predatory behaviour of *Mylonchulus dentatus*. *Nematologia Mediterranea* 6, 205-212.

JAIRAJPURI, M.S. & KHAN, W.U. (1975). Studies on Mononchida of India. VII. Excretory system of *Prionchulus muscorum*. *Nematologica* 21, 409-410.

JAIRAJPURI, M.S. & KHAN, W.U. (1977). Studies on Mononchida of India. IX. Further division of the genus *Clarkus* Jairajpuri, 1970 with the proposal of *Coomansus* n. gen. (Family Mononchidae Chitwood, 1937) and descriptions of two new species. *Nematologica* 23, 89-96.

JAIRAJPURI, M.S. & KHAN, W.U. (1982). *Predatory nematodes (Mononchida)*. New Delhi, India, Associated Publishing, 131 pp.

JAIRAJPURI, M.S. & LOOF, P.A.A. (1966). The systematic position of *Mononchulus* Cobb, 1918. *Nematologica* 11, 496-500.

JAIRAJPURI, M.S. & RAHMAN, M.F. (1984). *Hadronchoides microdenticulatus* n. gen., n. sp. (Nematoda: Iotonchidae) from India. *Nematologica* 29, 266-268.

JAIRAJPURI, M.S., AHMAD, W. & STURHAN, D. (1998). Descriptions of three new species of Mononchida. *Journal of Nematode Morphology and Systematics* 1, 99-106.

JAIRAJPURI, M.S., TAHSEEN, Q. & CHOI, Y.E. (2000). Description of *Iotonchus onchus* n. sp. and detailed observations on *I. litoralis* Coetzee, 1967 (Anatonchoidea: Mononchida) from Korea. *Journal of Asia-Pacific Entomology* 3, 19-24.

JAIRAJPURI, M.S., TAHSEEN, Q. & CHOI, Y.E. (2001a). *Micatonchus reversus* gen. n., sp. n. (Mononchida: Anatonchidae), a unique mononch from Korea. *International Journal of Nematology* 11, 77-80.

JAIRAJPURI, M.S., TAHSEEN, Q. & CHOI, Y.E. (2001b). *Parkellus parkus* gen. n., sp. n. and *Miconchus koreanus* sp. n. (Mononchida), two new predaceous nematodes from Korea. *International Journal of Nematology* 11, 98-103.

JANA, T., CHATTERJEE, A. & MANNA, B. (2006). *Nullonchus rafiqi* sp. n., a new mononchid species from West Bengal, India. *Nematologia Mediterranea* 34, 183-186.

JANA, T., CHATTERJEE, A. & MANNA, B. (2008). Two new and a known species of the genus *Miconchus* Andrássy, 1958 (Nematoda: Anatonchidae) from West Bengal, India. *Opuscula Zoologica Budapest* 39, 77-90.

JENSEN, H.J. & MULVEY, R.H. (1968). *Predaceous nematodes (Mononchidae) of Oregon.* Oregon State Monographs, Studies in Zoology No. 12, 57 pp.

JIMÉNEZ-GUIRADO, D. (1994). Mononchid nematodes from Spain. *Iotonchus parageminus* sp. n. and its relationship with *I. geminus* Heyns & Lagerwey, 1965 and *I. rinae* Coetzee, 1967. *Fundamental and Applied Nematology* 17, 347-356.

JIMÉNEZ-GUIRADO, D. & MURILLO-NAVARRO, R. (2008). Mononchid nematodes from Spain. Two species of *Mylonchulus* Cobb, 1916 inhabiting the coastal dunes of the Cádiz Gulf. *Journal of Nematode Morphology and Systematics* 11, 137-146.

JIMÉNEZ-GUIRADO, D., PEÑA-SANTIAGO, R. & CASTILLO-CASTILLO, P. (1993). Mononchid nematodes from Spain. One known and another new species of the genus *Miconchus baeticus* Andrássy, 1958. *Fundamental and Applied Nematology* 16, 63-72.

JIMÉNEZ-GUIRADO, D., ALHAMA, J.C. & GUTIERREZ, M.D.G. (1997). Mononchid nematodes from Spain. Six known species and *Miconchus baeticus* sp. n. occurring in southern fir forests. *Fundamental and Applied Nematology* 20, 371-383.

JIMÉNEZ-GUIRADO, D., WOUTS, W.M. & BELLO, A. (1998). Morphometric studies on *Coomansus* Jairajpuri & Khan, 1977 (Nematoda: Mononchida) and descriptions of two new species from the subantarctic region. *Fundamental and Applied Nematology* 21, 527-546.

JONES, F.G.W. (1974). Control of nematode pests, background and outlook for biological control. In: Price-Jones, D. & Soloman, M.E. (Eds). *Biology in pest and disease control*. Oxford, UK, Blackwell Publishing, pp. 249-268.

KANNAN, S. (1961a). Further additions to the list of *Mononchus* from Madras City. *Journal of Madras University B* 31, 59-62.

KANNAN, S. (1961b). Soil nematodes of the genera *Mononchus* and *Mononchulus*. *Journal of Madras University B* 31, 63-68.

KANNAN, S. (1961c). Nematodes of the genus *Mononchus* from Madras City. *Journal of Madras University B* 31, 69-74.

KHAN, H.A. (1975). A new species of the genus *Mylonchulus* (Cobb, 1916) Altherr, 1953 (Enoplida: Mononchidae) from Karachi. *Pakistan Journal of Zoology* 7, 143-144.

KHAN, H.A. & SAEED, M. (1987). *Pakmylonchulus*, new genus (Mononchida: Mylonchulidae) with notes on *Mylonchulus nainitalensis* Jairajpuri, 1970 and *M. sigmaturus* (Cobb, 1917) Altherr, 1953 from Pakistan. *Pakistan Journal of Zoology* 19, 313-320.

KHAN, M.L. (1992). Species of predaceous nematodes (Mononchida: Nematoda) and the report of males of *Iotonchus risocieae* (Carvalho, 1955) Andrássy, 1958 from India. *Current Nematology* 2, 185-188.

KHAN, S.H. (1972). On the mononchid nematode genera *Bathyodontus* Fielding, 1950 and *Mirolaimus* Andrássy, 1956 with a description of *M. indicus* n. sp. *Annales de Zoologie-Ecologie Animale* 4, 507-512.

KHAN, W.U. & JAIRAJPURI, M.S. (1979). Studies on Mononchida of India XII. The genus *Mylonchulus* (Cobb, 1916) Altherr, 1953 with descriptions of three new species. *Nematologica* 25, 406-418.

KHAN, W.U. & JAIRAJPURI, M.S. (1980). Studies on Mononchida of India XIII. The genus *Iotonchus* (Cobb, 1916) Altherr, 1950 with a key to the species. *Nematologica* 26, 1-9.

KHAN, W.U., AHMAD, S. & JAIRAJPURI, M.S. (1978). Studies on Mononchida of India X. Two new species of the genus *Miconchus* Andrássy, 1958. *Nematologica* 24, 321-327.

KHAN, Z. & ARAKI, M. (2002). A new and two first recorded species of mononchids (Nematoda: Mononchida) from Japan. *Nematologia Mediterranea* 30, 167-173.

KHAN, Z., ARAKI, M. & BILGRAMI, A.L. (2000). *Iotonchus sagaensis* sp. n., *I. arcuatus* sp. n. and *Miconchus japonicus* sp. n. (Nematoda: Mononchida) from Japan. *International Journal of Nematology* 10, 143-152.

KHAN, Z., CHOI, Y.E., LEE, S.M. & CHOI, J.S. (2000). Descriptions of four new and an unknown species of predatory nematodes (Mononchida) from Korea. *Korean Journal of Soil Zoology* 5, 71-83.

KHAN, Z., PARK, S.D., BAE, S.G. & SHIN, Y.S. (2002). Two new and two first recorded species of predatory soil nematodes (Nematoda: Mononchida) from Korea. *Korean Journal of Soil Zoology* 7, 51-57.

KHAN, Z., LEE, S.M., CHOI, J.S. & PARK, S.D. (2003). Descriptions of two new and a known species of predatory nematodes (Nematoda: Mononchida) from Korea. *Nematologia Mediterranea* 31, 207-213.

KHAN, Z., YOSHIGA, T., OKUMURA, E. & TANAKA, R. (2008). Descriptions of four new species of predatory soil nematodes (Nematoda: Mononchida) from Japan. *International Journal of Nematology* 18, 144-150.

KIRJANOVA, E.S. & KRALL, E.L. (1969). [*Plant parasitic nematodes and their control.*] Vol. 1, Leningrad, USSR, Nauka, 447 pp.

KREIS, H.A. (1924). Contribution à la connaissance de nématodes libres du Surinam (Guyane hollandaise). *Annales de Biologie Lacustre* 13, 123-136.

KREIS, H.A. (1932). Beitrage zur Kenntnis Pflanzen parasitischer Nematoden. *Zeitschrift für Parasitenkunde* 5, 184-194.

LAL, A. & KHAN, E. (1988). Taxonomic studies on nematodes associated with the rhizosphere of forest trees in India. I. Two new nematode species of the Order Mononchida. *Nematologia Mediterranea* 16, 1-3.

LOOF, P.A.A. (1964). Free living and plant parasitic nematodes from Venezuela. *Nematologica* 10, 201-300.

LOOF, P.A.A. (1993). On the status of *Mylonchulus incurvus* Cobb, 1917 and *M. subtenuis* Cobb, 1917 with redescription of *M. sexcristatus* (Merzheevskaya, 1951), description of *M. inflatus* n. sp. and comment on the genera *Paramylonchulus* Jairajpuri & Khan, 1982 and *Pakmylonchulus* Khan & Sayeed, 1987 (Nematoda: Mononchina). *Nematologica* 39, 153-176.

LOOF, P.A.A. (2006). Mononchina (Dorylaimida) from Western Malaysia. *Nematology* 8, 287-310.

LOOF, P.A.A. & COOMANS, A. (1970). On the development and location of the oesophageal gland nuclei in Dorylaimina. *Proceedings of the IX International Nematology Symposium, Warsaw, Poland, 1967*, pp. 79-161.

LOOF, P.A.A. & WINISZEWSKA-SLIPIŃSKA, G. (1993). Systematic observations on some species of *Coomansus* Jairajpuri & Khan, 1977 and *Iotonchus* Cobb, 1916 (Nematoda: Mononchida). *Nematologica* 39, 183-217.

LOOF, P.A.A., BAROOTI, S. & KHEYRI, A. (1990). Predatory nematodes (Mononchina) from Iran. *Applied Entomology and Phytopathology* 57, 27-36.

LORDELLO, L.G.E. (1953). *Contribuicao ao conhecimento dos nematodeos do solo de algumas regioes do Estado de São Paulo.* Ph.D. Thesis, Escola Superior de Agricultura 'Luiz de Queiroz', Piracicaba, 75 pp.

LORDELLO, L.G.E. (1959). Nota sobre ao genero *Mononchus* de nematodes predadores. *Anais Escola Superior de Agricultura 'Luiz de Queiroz'* (1957-1958) 14/15, 119-124.

LORDELLO, L.G.E. (1970). Research on nematodes of the family Mononchidae encountered in Brazil. "Pesquisas sobre nematoides da famalia Mononchidae encontrados no Brasil". *Anais Escola Superior de Agricultura 'Luiz de Queiroz'* 17, 15-48.

LORENZEN, S. (1981). Entwurf eines phylogenetischen Systems der freilebenden Nematoden. *Veröffentlichungen des Instituts fur Meeresforschung Bremerhaven, Supplement* 7, 1-472.

MAERTENS, D. (1976). Observations on the life cycle of *Prionchulus punctatus* (Cobb, 1917) cultural conditions. *Biologisch Jaarboek Dodonaea* 43, 197-218.

MAGGENTI, A.R. (1983). Nematode higher classification as influenced by species and family concepts. In: Stone, A.R., Platt, H.M. & Khalil, L.F. (Eds). *Concepts in nematode systematics.* The Systematics Association Special Volume 22. London, UK, Academic Press, pp. 25-40.

MAGGENTI, A.R. (1991). Nemata: higher classification. In: Nickle, W.R. (Ed.). *Manual of agricultural nematology*. New York, NY, USA, Marcel Dekker, pp. 147-187.

MALCEVSCHI, S. (1981). Mononchidae of Ivory Coast savannahs: genera *Iotonchus* (Cobb, 1916) and *Miconchus* Andrássy, 1958. *Nematologia Mediterranea* 9, 63-74.

DE MAN, J.G. (1876). Onderzoekingen over vrij in de aarde levende Nematoden. *Tijdschrift Nederlandsche dierkundige Vereeniging* 2, 78-196.

DE MAN, J.G. (1880). Die einheimischen, frei in der reinen Erde und im süssen Wasser lebenden Nematoden. *Tijdschrift Nederlandsche dierkundige Vereeniging* 5, 1-104.

DE MAN, J.G. (1904). Nématodes libres. In: *Résultats du voyage du S.Y. Belgica en 1897-1898-1899 sous le commandement de A. de Gerlach de Gomery, etc. Zoologie*, 55 pp., pls. 1-111.

VAN MEGEN, H., VAN DEN ELSEN, S., HOLTERMAN, M., KARSSEN, G., MOOYMAN, P., BONGERS, T., HOLOVACHOV, O., BAKKER, J. & HELDER, J. (2009). A phylogenetic tree of nematodes based on about 1200 full length small subunit ribosomal DNA sequences. *Nematology* 11, 927-950.

MELDAL, B.H.M., DEBENHAM, N.J., DE LEY, P., TANDINGAN DE LEY, I., VANFLETEREN, J.R., VIERSTRAETE, A.R., BERT, W., BORGONIE, G., MOENS, T., TYLER, P.A., AUSTEN, M.C., BLAXTER, M.L., ROGERS, A.D. & LAMBSHEAD, P.J.D. (2007). An improved molecular phylogeny of the Nematoda with special emphasis on marine taxa. *Molecular Phylogenetics and Evolution* 42, 622-636.

MENZEL, R. (1913). *Mononchus zschokkei* n. sp. und einige weing beKannte, für die Schweiz neue freilebende Nematoden. *Zoologischer Anzeiger* 42, 408-413.

MERZHEEVSKAYA, O.I. (1951). [Nematodes of peaty and mineral soils and their importance to agriculture.] *Sbornik Nauchnykh Trudov Akademiya Nauk Belorusskoĭ SSR* 2, 112-120.

MEYL, A.H. (1954). Die bisher in Italien gefundenen freilebenden Erd und SüsswasserNematoden. *Archivio Zoologico Italiano* 39, 161-264.

MEYL, A.H. (1955a). Über einige an den deutschen Kusten vorkommende Arten der Nematodengattung *Mononchus* Bastian, 1865. *Kieler Meeresforsch* 11, 80-85.

MEYL, A.H. (1955b). Nematoden aus dem Psammon des Tanganyika-Sees. *Zoologischer Anzeiger* 155, 135-138.

MEYL, A.H. (1957a). Free living nematodes. In: *Résultats scientifiques de l'exploration hydrobiologique du lac Tanganika (1946-1947). Institut Royal des Sciences Naturelles de Belgique Mémoires* 3, 27-51.

MEYL, A.H. (1957b). Beitrage zur freilebenden Nematoden-fauna Brasiliens. II. Weitare neue oder wenig bekannte Nematodenarten. *Kieler Meeresforschungen* 13, 125-133.

MEYL, A.H. (1961). Die freilebenden Erd-und Süsswassernematoden (Fadenwürmer). In: *Die Tierwelt Mitteleuropas*. Leipzig, Germany, Quelle & Meyer, 164 pp.

MICOLETZKY, H. (1922). Die freilebende Erd Nematoden. *Archiv für Naturgreschichte, Abteilung A* 87, 1-650.

MICOLETZKY, H. (1925). Zur Kenntnis tropischer, freilebender Nematoden RUS Surinam, Trinidad, und Ostafrika. *Zoologischer Anzeiger Leipzig* 64, 1-28.

MOHANDAS, C. (1979). *Iotonchus prabhooi* sp. n. (Mononchida: Nematoda) a new predatory nematode from the soils of Kerala. *Current Science* 48, 460-461.

MOHANDAS, C. & PRABHOO, N.R. (1979). New predatory nematodes of the genus *Iotonchus* (Iotonchidae: Mononchida) from the soils of Kerala (India). *Proceedings of the Indian Academy of Sciences* 88, 433-440.

MOHANDAS, C. & PRABHOO, N.R. (1982). On a new genus, *Sporonchuloides*, with notes on *Sporonchulus* (Cobb, 1917) Pennak, 1953 (Mononchida: Nematoda). *Records of the Zoological Survey of India* 79, 359-362.

MOHILAL, N. & DHANACHAND, CH. (1997). Three new species of mononchs (Nematoda: Mononchida). *Indian Journal of Nematology* 27, 179-186.

MOHILAL, N., ANANDI, Y. & DHANACHAND, CH. (2000). Studies on soil nematodes of Manipur-IX. Two new species and a male record of Iotonchinae. *Uttar Pradesh Journal of Zoology* 20, 175-181.

MOHILAL, N., PRAMODINI, M. & ROMONI, H. (2006). Studies on soil nematodes of Manipur-XII. Species of *Mylonchulus* Cobb, 1916. *Uttar Pradesh Journal of Zoology* 26, 349-357.

MULLIN, P.G., HARRIS, T.S. & POWERS, T.O. (2005). Phylogenetic relationships of Nygolaimina and Dorylaimina (Nematoda: Dorylaimida) inferred from small subunit ribosomal DNA sequences. *Nematology* 7, 59-79.

MULVEY, R.H. (1961a). The Mononchidae: a family of predaceous nematodes I. Genus *Mylonchulus* (Enoplida: Mononchidae). *Canadian Journal of Zoology* 39, 665-696.

MULVEY, R.H. (1961b). The Mononchidae: a family of predaceous nematodes II. Genus *Anatonchus* (Enoplida: Mononchidae). *Canadian Journal of Zoology* 39, 807-826.

MULVEY, R.H. (1962). The Mononchidae: a family of predaceous nematodes III. Genus *Miconchus* (Enoplida: Mononchidae). *Canadian Journal of Zoology* 40, 65-81.

MULVEY, R.H. (1963a). The Mononchidae: a family of predaceous nematodes IV. Genus *Iotonchus* Enoplida: Mononchidae). *Canadian Journal of Zoology* 41, 79-98.

MULVEY, R.H. (1963b). The Mononchidae: a family of predaceous nematodes V. Genera *Sporonchulus*, *Granonchulus* and *Prionchuloides* n. gen. (Enoplida: Mononchidae). *Canadian Journal of Zoology* 41, 763-774.

MULVEY, R.H. (1967a). The Mononchidae: a family of predaceous nematodes VI. Genus *Mononchus* (Nematoda: Mononchidae). *Canadian Journal of Zoology* 45, 915-940.

MULVEY, R.H. (1967b). The Mononchidae: a family of predaceous nematodes VII. Genus *Prionchulus* (Nematoda: Mononchidae). *Canadian Journal of Zoology* 45, 941-953.

MULVEY, R.H. (1978). Predaceous nematodes of the family Mononchidae from the Mackenzie and Porcupine river systems and Somerset Island, N.W.T., Canada. *Canadian Journal of Zoology* 56, 1847-1868.

MULVEY, R.H. & DICKERSON, O.J. (1970). *Miconchus kansasensis* n. sp. (Mononchidae: Nematoda) from Kansas, United States. *Canadian Journal of Zoology* 48, 231-234.

MULVEY, R.H. & JENSEN, H.J. (1967). The Mononchidae of Nigeria. *Canadian Journal of Zoology* 45, 667-727.

NELMES, A.J. (1974). Evaluation of the feeding behaviour of *Prionchulus punctatus* (Cobb): a nematode predator. *Journal of Animal Ecology* 43, 553-565.

NGUYEN VU THANH (2006). Two new predatory nematode species *Iotonchus helicus* sp. nov. and *Iotonchus thui* sp. nov. (Nematoda: Mononchida) from Vietnam. *Tap Chi Sinh Hoc* 28, 10-16.

OLIA, M., CHOUDHARY, M., AHMAD, W. & JAIRAJPURI, M.S. (2004). Nematodes of the Order Mononchida from Iran with description of *Anatonchus kafii* sp. n. *International Journal of Nematology* 14, 135-138.

OLIA, M., AHMAD, W., ARAKI, M., MINAKA, N., OBA, H. & OKADA, H. (2008). *Actus salvadoricus* Baqri & Jairajpuri (Mononchida: Mylonchulidae) from Japan with comment on the phylogenetic position of the genus *Actus* based on 18S rDNA sequences. *Japanese Journal of Nematology* 38, 57-69.

OLIVEIRA, J. (1940). Plant-parasitic and free-living nematodes in Hawaii. *Occasional Paper, Bernice Pauahi Bishop Museum* 15, 361-373.

ORSELLI, L. & VINCIGUERRA, M.T. (2007). Three new and a rare species of Mononchida (Nematoda) from Ecuador. *Journal of Nematode Morphology and Systematics* 9 (2006), 137-146.

PATIL, K.J. & KHAN, E. (1982). Taxonomic studies on nematodes of Vidarbha region of Maharashtra, India IV. *Sporonchulus grandis* sp. n. and *Iotonchus shamimi* sp. n. (Nematoda: Mononchida). *Indian Journal of Nematology* 12, 161-166.

PEÑA-SANTIAGO, R. & JIMÉNEZ-GUIRADO, D. (1991). Mononchid nematodes from Spain. Description of *Iotonchus rotundicaudatus* sp. n. and observations on *I. zschokkei* (Menzel, 1913) Altherr, 1955. *Revue de Nématologie* 14, 353-360.

PENEVA, V., NEILSON, R. & NEDELCHEV, S. (1999). Mononchid nematodes from oak forests in Bulgaria. 1. The subfamily Anatonchinae Jairajpuri, 1969 with descriptions of *Anatonchus genovi* sp. n. and *Tigronchoides quercus* sp. n. *Nematology* 1, 37-53.

PENNAK, R.W.P. (1953). *Fresh-water invertebrates of the United States.* New York, NY, USA, Ronald Press, 769 pp.

PHUKAN, P.N. & SANWAL, K.C. (1981). Two new species of *Hadronchus* Mulvey & Jensen, 1967 from Assam (India). *Journal of Research, Assam Agriculture University* 2, 80-85.

POPOVICI, I. (1990). Studies on Mononchida (Nemata) from Romania. *Nematologica* 36, 161-180.

RAHM, G.F. (1938). Freilebende und saprophytische Nematoden der Insel Hainan (Mit besonderer Berucksichtigung der bisher bekannt gewordenen Nematoden Nordchinas und Japans). *Annotationes Zoologicae Japonenses* 17, 646-667.

RAHMAN, M.F. & JAIRAJPURI, M.S. (1983). Two new species of Mononchida from India. *Nematologica* 29, 126-131.

RAWAT, V.S. & AHMAD, M. (2000). Mononchida of Garhwal Himalayas (U.P.) India II. Two known and two new species of the genus *Mononchus* Bastian, 1865 (Nematoda: Mononchida). *Indian Journal of Nematology* 30, 62-66.

RAY, S. & DAS, S.N. (1983). *Hadronchulus shamimi* n. gen., n. sp. and *Cobbonchus artemisiae* (Mononchida: Nematoda) from Orissa, India. *Nematologica* 28 (1982), 247-252.

RAZZHIVIN, A.A. (1971). New species of nematodes of the family Mononchidae from soils of apple forests in Dzhungarian Alatau. *Zoologichesky Zhurnal* 50, 1574-1578.

RENUBALA, K. & DHANACHAND, CH. (1992). Two new species of the genus *Parahadronchus* Mulvey, 1978 from Manipur. *Current Nematology* 3, 79-82.

RITTER, M. & LAUMOND, C. (1975). [Review of the use of nematodes in biological control programmes against parasites and pests of cultivated plants.] *Bulletin des Recherches Agronomiques de Gembloux, Semaine d'Etude Agriculture et Hygiène des Plantes*, pp. 331-344.

RUSSIN, L.Y., ALESHIN, V.V., TCHESUNOV, A.V. & ATRASHKEVICH, G.I. (2003). The 18s ribosomal RNA gene of *Soboliphyme baturini* Petrow, 1930 (Nematoda: Dioctophymida) and its implication for phylogenetic relationships within Dorylaimida. *Nematology* 5, 615-628.

SAHA, M., LAL, M. & SINGH, M. (2004). Description of two new species of predatory nematodes belonging to Mylonchulidae and Ironidae from Delhi, India. *Annals of Plant Protection Sciences* 12, 171-176.

SAHA, M., LAL, M. & SINGH, M. (2006). Three new species of nematodes belonging to Mononchida Jairajpuri 1969, from rhizosphere of mango in Western Uttar Pradesh, India. *Annals of Plant Protection Sciences* 14, 177-183.

SCHNEIDER, W. (1925). Freilebende Susswassernematoden aus ostholsteinischcn Seen. Nebst Bemerkungan uber die Nematodenfauna des Madu-und Schaalsees. *Archiv für Hydrobiologie* 15, 536-584.

SCHNEIDER, W. (1937). Freilebende Nematoden der Deutschen Limnologischen Sundaexpedition nach Sumatra, Java und Bali. *Archiv für Hydrobiologie* 15 (Supplement), 30-108.

SCHNEIDER, W. (1939). Wurmer oder Vermes. II. Fadenwunner oder Nematoden. I. Freilebende und Pflanzenparasitiche Nematoden. *Die Tierwelt Deutschlands* 36, 1-260.

SCHNEIDER, W. (1940). Neue freilebende Nematoden aus Hohlen und Brunnen. 1. Nematoden ausjugoslawischen Hohlen. *Zoologischer Anzeiger, Leipzig* 132, 84-94.

SCHUURMANS STEKHOVEN, J.H. (1951). Nématodes saprozoaires et libres du Congo Belge. *Mémoires de l'Institut Royal des Sciences Naturelles de Belgique* 2, 1-79.

SCHUURMANS STEKHOVEN, J.H. & TEUNISSEN, R.J.H. (1938). Nématodes libres terrestres. *Exploration du Parc National Albert, Mission de Witte* (1933-35), 229 pp.

SHARMA, R.K. & SAXENA, V. (1980). Two new species of the genus *Mylonchulus* (Cobb, 1916) Altherr, 1953 (Mononchida: Nematoda) from north India. *Istituto Lombardo Accademia di Scienze e Lettere, Rendiconti Classe di Scienze (B)* 114, 18-26.

SHARMA, R.K. & SAXENA, V. (1981). *Granonchulus helicus* sp. n. (Nematoda: Mononchida) from north India. *Nematologia Mediterranea* 9, 159-162.
SIDDIQI, M.R. (1983). Phylogenetic relationships of the soil nematode orders Dorylaimida, Mononchida, Triplonchida and Alaimida, with a revised classification of the subclass Enoplia. *Pakistan Journal of Nematology* 1, 79-110.
SIDDIQI, M.R. (1984a). Four new genera and four new species of mononchs (Nematoda). *Pakistan Journal of Nematology* 2, 1-13.
SIDDIQI, M.R. (1984b). Two new species of *Iotonchus* Cobb (Nematoda: Mononchida) from Fiji Islands. *Nematologia Mediterranea* 12, 81-85.
SIDDIQI, M.R. (1984c). *Anatonchus ferox* n. sp., with miscellaneous notes on Anatonchidae (Mononchida). *Pakistan Journal of Nematology* 2, 43-47.
SIDDIQI, M.R. (1984d). *Nullonchus*, a new genus of toothless mononchs, with comments on the phylogenetic relationship of Nullonchinae n. subfam. (Nematoda: Mononchida). *Nematologica* 30, 11-21.
SIDDIQI, M.R. (2001). Study of *Iotonchus* species (Mononchida) from West Africa with descriptions of eleven new species. *International Journal of Nematology* 11, 104-123.
SIDDIQI, M.R. & JAIRAJPURI, M.S. (2002). *Crestonchulus crestus* gen. n., sp. n. (Nematoda: Mononchida) from Cameroon. *International Journal of Nematology* 12, 85-88.
SINHA, B., CHOUDHURY, A. & BAQRI, Q.H. (1989). A new nematode of the genus *Mylonchulus* (Cobb, 1916) Altherr, 1953 (Mylonchulidae: Mononchida) from littoral sands of Sunderbans, West Bengal, India. *Indian Journal of Helminthology* (new series) 6, 5-6.
SKWARRA, E. (1921). Diagnosen neuer freilebender Nematoden Ostpreussens. *Zoologischer Anzeiger, Leipzig* 53, 66-74.
SMALL, R.W. (1979). The effects of predatory nematodes on populations of plant parasitic nematodes in pots. *Nematologica* 25, 94-103.
SMALL, R.W. & GROOTAERT, P. (1983). Observations on the predation abilities of some soil dwelling predatory nematodes. *Nematologica* 29, 109-118.
SOLOMON, M.F. (1949). The natural control of animal population. *Journal of Animal Ecology* 18, 1-35.
SONI, G.R. & NAMA, H.S. (1980). On a new species of *Mylonchulus* (Cobb, 1916) Altherr, 1956 (Nematoda: Mylonchulidae). *Current Science* 49, 750-751.
SONI, G.R. & NAMA, H.S. (1983). *Mononchus sinensis* n. sp. (Nematoda: Mononchida) from India. *Current Science* 52, 75-76.
STEFANSKI, W. (1915). Nouvelles espèces de nématodes provenant de Pologne. *Zoologischer Anzeiger, Leipzig* 45, 346-349.

STEINER, G. (1914). Freilebende Nematoden aus der Schweiz. 1. Teil einer vorlaufigen Mitteilung. *Archiv für Hydrobiologie und Planktonkunde* 9, 259-276.

STEINER, G. & HEINLY, H. (1922). The possibility of control of *Heterodera radicicola* and other plant infesting nemas by means of predatory nemas, especially by *Mononchus papillatus* Bastian. *Journal of Washington Academy of Sciences* 12, 367-386.

SUKUL, N.C. (1969). *Cryptonchulus*, a new genus of soil nematode allied to *Cryptonchus* from Andamans, India. *Indian Journal of Helminthology* 21, 23-26.

SUKUL, N.C. (1971). New free-living nematodes from the thermal spring Tantloya, Bihar. *Indian Journal of Helminthology* 23, 135-145.

SULTANALIEVA, G.B. (1983). [New species of soil nematodes in the fauna of Kirghizia.] *Zoologichesky Zhurnal* 62, 1897-1902.

SUSULOVSKY, A. (1988). [A new species of predatory nematodes from genus *Iotonchus* (Mononchida).] *Trudy Zoologicheskogo Instituta, Akademiya Nauk SSR* 180, 80-81.

SUSULOVSKY, A. (1994). A new species of the genus *Clarkus* from Ukraine (Nematoda: Mononchida). *Zoosystematica Rossica* 2, 225-228.

SUSULOVSKY, A. (2000). *Mylonchulus politus* sp. nov. (Nematoda: Mylonchulidae) from Ukraine. *Annales Zoologici* 50, 277-280.

SUSULOVSKY, A. (2001). A new species of the genus *Comiconchus* Jairajpuri & Khan, from Ukraine with redescription of *C. trionchus* (Thorne) (Nematoda, Mononchida: Mononchidae). *Zoosystematica Rossica* 9, 261-266.

SUSULOVSKY, A. & WINISZEWSKA, G. (2002). *Prionchulus fistulosus* sp. nov. (Nematoda: Mononchidae) from Ukraine. *Annales Zoologici* 52, 483-488.

SUSULOVSKY, A. & WINISZEWSKA, G. (2006). Two new species of the genus *Prionchulus* Cobb, 1916 (Nematoda: Mononchina). *Annales Zoologici* 56, 241-248.

SUSULOVSKY, A., WINISZEWSKA, G. & GAGARIN, V.G. (2003). One new and another known species of the genus *Prionchulus* Cobb, 1916 from Kamchatka Peninsula, Russia. *Annales Zoologici* 53, 749-754.

SUSULOVSKY, A.S. (1998). *Iotonchus aequabilis* sp. n. from Ukraine. In: *Proceedings of the State Natural History Museum, vol. 14.* Lviv, Ukraine, National Academy of Sciences of Ukraine, pp. 175-177.

THONG, C.H.S. (1971). Five species of Mononchidae (Nematoda) from Singapore Island. *Nematologica* 16, 470-476.

THORNE, G. (1924). Utah nemas of the genus *Mononchus. Transactions of the American Microscopical Society* 43, 157-171.

THORNE, G. (1927). The life history, habits and economic importance of some mononchs. IV. *Journal of Agriculture Research* 34, 265-286.

THORNE, G. (1929). Nematodes from the summit of Long's Peak, Colorado. *Transactions of the American Microscopical Society* 48, 181-195.

THORNE, G. (1935). Notes on free-living and plant-parasitic nematodes. II. Higher classification of Dorylaimoidea. *Proceedings of the Helminthological Society of Washington* 2, 96-98.

THORNE, G. (1939). A monograph of the nematodes of the superfamily Dorylaimoidea. *Capita Zoologica* 8, 1-261.

TIMM, R.W. (1957). Some soil nematodes of East Pakistan. *Pakistan Journal of Biological and Agricultural Science* 1, 6-10.

TSALOLIKHIN, S.YA. (1988). [Tropical nematodes – new and rare species.] *Trudy Zoologicheskogo Instituta, Akademiya Nauk SSR* 180, 59-67.

TYSOWSKI, A. (1915). Wolnożyjace nicienie (nematodes) zebrine w powiecie sokalskim. *Rozprawy i Wiadomości z Muzeum im. Dzieduszyckich* 1, 65-92.

VINCIGUERRA, M.T. (1986). Description of two new species and remarks on some known species of nematodes from Sardinia. *Animalia (Catania)* 11, 127-134.

VINCIGUERRA, M.T. & LA ROSA, G. (1990). A new species of *Clarkus* Jairajpuri, 1970 (Nematoda: Mononchida) from subantarctic America. *Animalia* 17, 99-104.

VINCIGUERRA, M.T. & ORSELLI, L. (2000). Nematodes from Italian sand dunes. 5. Two new species of Dorylaimida and three known species of Enoplida. *Nematologia Mediterranea* 28, 301-312.

VINCIGUERRA, M.T. & ORSELLI, L. (2006). *Iotonchus aequatorialis* sp. n. (Nematoda: Mononchida) from Ecuador with a key to the species of *Iotonchus* Cobb, 1916 and remarks on the identity of the genus. *Nematology* 8, 837-846.

VOLCY, C. (1988). Study of the mononchid nematodes in Antioquia, Colombia. II. Superfamily Mononchoidea (Chitwood, 1937) Clark, 1961. *Fitopatologia Colombiana* 12, 29-33.

WEBSTER, J.M. (1972). Nematodes and biological control. In: Webster, J.M. (Ed.). *Economic nematology*. London, UK, Academic Press, pp. 469-496.

WILLIAMS, J.R. (1958). Studies on the nematode soil fauna of sugar cane fields in Mauritius. I. The genus *Mononchus* (Trilobidae, Enoplida). *Mauritius Sugar Industry Research Institute, Occasional Paper* 1, 1-13.

WINISZEWSKA, G. (1998). *Mononchus loofi* sp. nov. (Nematoda: Mononchidae) from Poland. *Annales Zoologici* 48, 99-101.

WINISZEWSKA, G. (2002). *Prionchulus brevicaudatus* sp. nov. (Nematoda: Mononchidae) from Poland. *Annales Zoologici* 52, 343-346.

WINISZEWSKA, G. & SUSULOVSKY, A. (2003). Revision of the genus *Prionchulus* Cobb, 1916 (Nematoda: Mononchina). I. *Prionchulus muscorum* (Dujardin, 1845) Cobb, 1916 and related species. *Annales Zoologici* 53, 559-577.

WINISZEWSKA, G. & SUSULOVSKY, A. (2004). Revision of the genus *Prionchulus* Cobb, 1916 (Nematoda: Mononchina). II. *Prionchulus spectabilis* (Ditlevsen, 1911) Cobb, 1916 and related species. *Annales Zoologici* 54, 491-509.

WINISZEWSKA-SLIPIŃSKA, G. (1989). Two new species of *Anatonchus* Cobb, 1916 (Nematoda: Mononchida). *Annales Zoologici* 42, 197-202.

WU, H.W. & HOEPPLI, R.J.C. (1929). Free-living nematodes from Fookien and Chekiang. *Archiv für Schiffs- und Tropen-Hygiene* 33, 35-43.

YEATES, G.W. (1967). Studies on nematodes from dune sands. B. Oncholaimidae, Ironidae. Alaimidae and Mononchidae. *New Zealand Journal of Science* 10, 299-321.

YEATES, G.W. (1987). Distribution of Mononchoidea (Nematoda: Enoplea) in pasture soils, with description of *Iotonchus stockdilli* n. sp. *New Zealand Journal of Zoology* 14, 351-358.

YEATES, G.W. (1992). Nematodes from New Caledonia. 1. Introduction and Mononchoidea. *Fundamental and Applied Nematology* 15, 101-126.

ZELL, H. (1985). Nematodes of a beechwood soil. 3. *Prionchulus muscorum* (Nematoda: Mononchida). *Carolinea* 42, 57-74.

ZULLINI, A. & PENEVA, V. (2006). Order Mononchida. In: Eyualem-Abebe, Andrássy, I. & Traunspurger, W. (Eds). *Freshwater nematodes: ecology and taxonomy.* Wallingford, UK, CABI Publishing, pp. 468-496.

ZULLINI, A., LOOF, P.A.A. & BONGERS, T. (2002). Free-living nematodes from nature reserves in Costa Rica. 2. Mononchina. *Nematology* 4, 1-23.

Index to taxa

www.ingramcontent.com/pod-product-compliance
Lightning Source LLC
Chambersburg PA
CBHW071307180926
51613CB00001B/149
9789004174641